Workbook in ORGANIC CHEMISTRY

A series of books in chemistry

Linus Pauling and Harden M. McConnell, editors

Workbook in ORGANIC CHEMISTRY

EXERCISES IN THE PROPERTIES, BEHAVIOR, AND SYNTHESIS OF ORGANIC COMPOUNDS

T. A. Geissman
UNIVERSITY OF CALIFORNIA
Los Angeles

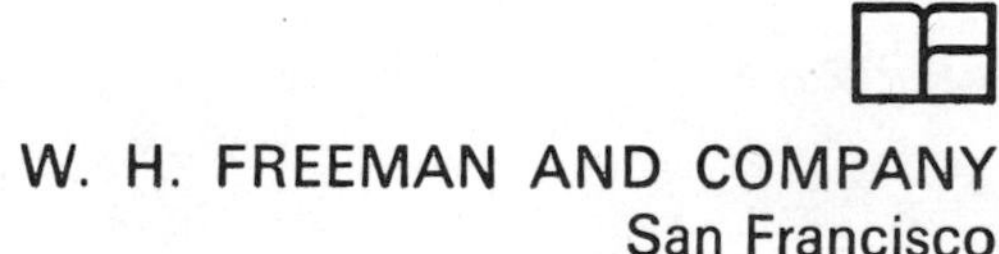

W. H. FREEMAN AND COMPANY
San Francisco

Printed in the United States of America

International Standard Book Number: 0-7167-0167-7

9 8 7 6 5 4 3 2 1

Contents

Preface

The purpose of this workbook is to provide the student in the introductory course in organic chemistry with the opportunity to review and reinforce his understanding of many of the fundamental modes of behavior of organic compounds, and to help him to learn to approach the solution of problems of synthesis and structure analysis in a logical and rational way. It is important for the student to consolidate his learning into a habit pattern of ready recognition of the principal features of the behavior of organic compounds. He should develop the ability to perceive the reactive capabilities of functional groups and structural assemblies with ease and without conscious and repeated reference to formal schemes of classification. The development of this fluency and perceptiveness can best be achieved by the working out of problems that are designed to provide exercises that are at once repetitive and varied. What is important to the student in developing his competence in organic chemistry is not so much to learn reactions by names or even by types, but to develop the ability to recognize the ways in which a compound will behave under a variety of experimental conditions.

The accumulation of factual information—the building of an active memory bank—is not alone sufficient for a creative attack upon old and new problems in the sciences; but it is not to be despised. Just as a good vocabulary helps in both the understanding and expression of ideas, so does a store of knowledge about the reactive potentialities of functional parts of organic molecules enable its possessor to devise means of attack upon chemical problems and to read and study with quick comprehension.

The plan of this book is to provide succinct reviews of the salient aspects of various fundamental features of molecular constitution and behavior. Included in each Part are a brief exposition of the subject, specific examples of actual cases, and problems chosen to illustrate further the topic being dealt with. An answer to an exercise is given occasionally in the text, and an answer section at the back of the workbook provides answers to the other exercises. Most of the synthetic procedures and sequences illustrated in the exercises represent actual examples that have been carried out and are described in the chemical literature. Many of these have been altered and often improved upon in later studies, and it is possible (moreover, to be hoped) that the student who works out his solutions may sometimes arrive at answers to the questions that are simpler or otherwise superior to those given in the answer section.

The sections on the reactions of organic compounds are often categorized by the names by which the reaction types are commonly known. Although students often question the necessity for memorizing "name reactions," they will find that such names carry within them valuable summaries of behavior. Most practicing organic chemists use the convenience of "name reactions" in their thinking and conversation. To speak of a "Friedel-Crafts reaction," a "Markownikov addition," or a "Hofmann degradation" is not necessarily to substitute a name for understanding; it is to compress a concept into a readily communicable form. It is not the author's intention to urge the student to memorize the names of reactions; but it is almost certain that in time the student will come to know them and to profit by the knowledge.

Because this workbook is planned to assist instruction and study in the usual one-year course in organic chemistry, the author has limited its content to matters appropriate to instruction at that level. Many of the specialized topics found in more advanced courses have been omitted; and it is recognized that the course of instruction the student is engaged in will include many topics omitted from this workbook. The author's experience in teaching introductory organic chemistry assures him that it it unlikely the average student will find the time to work *all* of the problems during the course. The instructor may therefore find it desirable to assign parts of each exercise, perhaps giving extra credit for additional accomplishment.

March 1972 *T. A. Geissman*

Introduction

The exercises and problems to be found in the following parts of this workbook are designed to provide instruction and practice in a number of fundamental concepts and reactions that are central to large areas of organic chemistry. Many of the organic syntheses and reaction sequences that are illustrated require additional manipulations—most of them relatively simple in concept—along with the principal procedure under discussion. Among these are saturation of unsaturated linkages by hydrogenation; saponification and hydrolysis of acid derivatives (ester, amides, nitriles); reduction of carbonyl and nitro groups by metal hydride and metal reducing agents; dehydration of alcohols and dehydrohalogenation of halides; and addition of halogen and hydrogen halides to carbon–carbon double bonds. These reactions are adequately dealt with in the usual course in organic chemistry and in the popular textbooks, and it will be assumed in most of the exercises that additional steps such as these must be added in a synthetic sequence as they are required. For example, if an ester is the desired product, the synthesis of the corresponding acid will often accomplish the purpose of the question, esterification being an unexceptional final step.

It is to be stressed also that the student recognize the necessity for adjustment of acidity or basicity of a reaction mixture in order that the product can be isolated. For example, when a haloform reaction is carried out, the carboxylic acid that is formed is present in the reaction medium as its sodium salt, and a final step of acidification will be necessary to permit its isolation by crystallization of solvent extraction; yet the usual formulation of the reaction shows the carboxylic acid as the direct product. Similarly, Grignard reactions lead to the formation of halomagnesium salts (of alcohols, for example), and a final step of treatment (usually referred to as "decomposition" of the reaction mixture) with aqueous mineral acid is required; yet in the usual formulation of a Grignard reaction the alcohol is shown as the product. The student should always keep clearly in mind the simple concepts of acidity and basicity and recognize the state of ionization in which a compound exists in solution, and the manipulations that are necessary to convert it into a form in which it can be isolated.

Workbook in ORGANIC CHEMISTRY

PART 1

Establishing the constitution of an organic compound

A. INTRODUCTION

The study of an organic compound of unknown identity usually begins with the application of a variety of analytical procedures with the aim of establishing

(a) the elementary composition, expressed as a molecular formula;
(b) the presence of characteristic groups or structural features.

B. ELEMENTAL ANALYSIS

(a) A fundamental analytical procedure of the organic chemist is the determination of the composition of a compound, expressed as the percentage of each of the elements it contains. The results of such determinations are expressed in the following way:

Compound *A*

Analysis. Calculated for $C_5H_8O_2$: C, 60.00; H, 8.01.

Found: C, 59.89; H, 8.15.

Compound *B*

Analysis. Calculated for $C_6H_{11}O_2Br$: C, 36.90; H, 5.64; Br, 41.00.

Found: C, 36.82; H, 5.71; Br, 41.21.

The symbol "%" or "percent" may or may not be used; the figure "C, 59.89" is understood to mean "59.89% carbon."

(b) Since all organic compounds contain carbon, and the majority of them contain hydrogen, the most generally applicable analytical determination is the *carbon-hydrogen analysis.* Separate analyses for nitrogen, halogens, sulfur, and so forth, may be carried out when these elements are present.

The carbon-hydrogen determination is in principle very simple. The compound to be analyzed is burned in a current of pure oxygen, and the carbon dioxide and water that are formed are collected in suitable absorption tubes, which are accurately weighed before and after the combustion. In modern practice a sample of 2–5 mg is used, and highly sensitive, accurate microbalances are used. The advent of new (and often automated) apparatus for combustion analysis has made this determination speedy and reliable.

(c) An example of the actual combustion of an organic compound and of the yields of CO_2 and H_2O that are formed is the following:

$$C_{20}H_{28}O_2 + \text{excess } O_2 \longrightarrow 20\,CO_2 + 14\,H_2O$$

			20×44	14×18
mol wts:	300		(880)	(252)
sample used:	3.00 mg	$\longrightarrow$	8.80 mg CO_2	+ 2.52 mg H_2O

From these results* the analyst calculates the composition of the compound:

$$8.80 \text{ mg } CO_2 \times \frac{12}{44} = 2.40 \text{ mg C}$$

$$\text{percent C} = \frac{2.40 \text{ mg C}}{3.00 \text{ mg sample}} \times 100 = 80.00$$

$$2.52 \text{ mg } H_2O \times \frac{2}{18} = 0.280 \text{ mg H}$$

$$\text{percent H} = \frac{0.280}{3.00} \times 100 = 9.33$$

* Since the analysis is an experimental operation that depends upon instrumental accuracy, the "found" figures for an analysis may vary between certain tolerable limits. Accomplished analysts can ordinarily be relied upon to provide results that are within less than ±0.20% of the expected values. Thus, the "found" value for a compound with 80.00% C may lie between 79.80 and 80.20%.

In a compound containing only C, H, and O, percent of O is usually found by difference. In this example, percent of C = 80.00; percent of H = 9.33; percent of O (difference) = 10.67.

That this experimental result agrees with what is expected can be seen by calculation of the composition of the compound $C_{20}H_{28}O_2$:†

$$C_{20}: 20 \times 12 = 240$$

$$H_{28}: 28 \times 1 = 28$$

$$O_2: 2 \times 16 = \underline{32}$$
$$300$$

$$\text{percent C} = \frac{240}{300} \times 100 = 80.00$$

and so forth.

(d) Elements other than carbon and hydrogen can be determined in separate analyses. Nitrogen, for example, is determined by heating the N-containing sample with copper oxide in a current of pure CO_2, and passing the products of combustion over heated copper (a part of the packing material in the combustion tube). Oxides of nitrogen are reduced to nitrogen. The mixture of nitrogen and carbon dioxide (the "sweeping" gas) is passed through KOH solution, which absorbs the CO_2, and then into an accurately calibrated buret, where the nitrogen volume is measured directly.

Exercise 1.1

1. Calculate the percent composition (% C, % H, and so on) for:
   (*a*) $C_{20}H_{28}O$
   (*b*) $C_{12}H_{12}$
   (*c*) $C_6H_6O_4$
   (*d*) $C_6H_{12}O_6$
   (*e*) C_5H_5N
   (*f*) CCl_4
   (*g*) CH_2Cl_2
   (*h*) C_5H_4NBr
   (*i*) C_2H_5NO
   (*j*) $C_{10}H_7N_2O_2Br$
   (*k*) $C_4H_{10}SO$
2. How many mg of CO_2 will be formed by the combustion of:
   (*a*) 3.60 mg of a compound, $C_6H_{12}O_6$
   (*b*) 4.60 mg of a compound, C_8H_8NOBr
   (*c*) 3.45 mg of a compound, $C_6H_{12}O$
   (*d*) 4.84 mg of a compound, $C_{12}H_{22}O_{11}$
   (*e*) 5.11 mg of a compound, $C_{10}H_8$
3. Calculate the weight of CO_2 and H_2O that will be formed in the complete combustion of:
   (*a*) 1 U.S. gallon of gasoline (assume C_7H_{16})
   (*b*) 1 U.S. ounce of ethanol
   (*c*) 3.34 mg of pure filter paper (cellulose)
   (*d*) 100 g of sucrose ($C_{12}H_{22}O_{11}$)
   (*e*) 2.15 mg of diamond

---

† Although professional analysts use the accurate atomic weights for the elements (that is, C = 12.01), the examples here will be calculated with the use of C = 12, H = 1, N = 14, and so forth, to ordinary slide rule accuracy.

C. THE USE OF ANALYTICAL DATA IN DETERMINING THE EMPIRICAL AND MOLECULAR FORMULAS OF A COMPOUND

(a) The results of an elemental analysis must be translated into an expression for the molecular formula of the compound.

In the first example of Section B(c), the analytical values found in the combustion analysis were C = 80.0%, H = 9.33%, O (difference) = 10.67%.

Consider 100 g of the compound. This will contain

$$\begin{aligned} 80.00 \text{ g C} &= 80/12 \text{ g-atoms C} = 6.67 \text{ g-atoms C} \\ 9.33 \text{ g H} &= 9.33/1 \text{ g-atoms H} = 9.33 \text{ g-atoms H} \\ 10.67 \text{ g O} &= 10.67/16 \text{ g-atoms O} = 0.667 \text{ g-atoms O} \end{aligned}$$

Thus, the *atomic ratio* C/H/O = 6.67/9.33/0.667. This must, of course, be reduced to a ratio of whole numbers. This is most simply done by the following procedure:

$$\begin{aligned} \text{atoms C/atom O} &= 6.67/0.667 = 10 \\ \text{atoms H/atom O} &= 9.33/0.667 = 14 \\ (\text{atoms O/atom O} &= 0.667/0.667 = 1) \end{aligned}$$

Thus, the formula for the compound can be written as $C_{10}H_{14}O$, which is an *empirical formula*. It is obvious that the same relative proportions of C, H, and O are present in $C_{10}H_{14}O$, $C_{20}H_{28}O_2$, $C_{30}H_{42}O_3$, etc. The selection one of these as the correct *molecular formula* depends upon a determination of the molecular weight. This is not inherent in the elemental analysis and must be accomplished by special techniques.* In the following discussions we shall arbitrarily ignore molecular weights and assume the molecular formulas are those calculated from the analytical data.

* Because the molecular weights for the three formulas given ($C_{10}H_{14}O$ and its multiples) are 150, 300, and 450, clearly the selection of the correct molecular formula requires only an approximate value for the molecular weight. For certain compounds more accurate values are needed. Adequate approximations can often be made very simply; for example, if an unknown liquid hydrocarbon had the empirical composition C_3H_7, the choice between C_6H_{14} and $C_{12}H_{28}$ could be made at once from its boiling point. (*Query:* why would C_9H_{21} be excluded from consideration?)

(b) Illustrative Examples.

(1) A halogen-free, N-containing compound gives the following analytical values:

% C, 65.12; H, 8.38; N, 16.85; O (difference), 9.65

atoms	atoms/atom O
C 65.12/12 = 5.43	5.43/0.60 = 9
H 8.38/1 = 8.38	8.38/0.60 = 14
N 16.85/14 = 1.20	1.20/0.60 = 2
(O 9.65/16 = 0.60)	(0.60/0.60 = 1)

Thus, the composition is expressed by $C_9H_{14}N_2O$. This is checked as follows:

C_9 108, % C = (108/166) 100 = 65.05; found: 65.12% C

H_{14} 14, % H = (14/166) 100 = 8.43; found: 8.38% H

N_2 28, % N = (28/166) 100 = 16.83; found: 16.85% N

(2) A compound containing only C, H, and O gives the following analytical results:

percent C, 82.34; H, 10.62; O (difference), 7.04

	atoms	atoms/atom O
C	82.34/12 = 6.85	6.85/0.44 = 15.5
H	10.62/ 1 = 10.62	10.62/0.44 = 24.2
(O	7.04/16 = 0.44)	(0.44/0.44 = 1)

These results are in accord with $C_{31}H_{48}O_2$. A calculation of the percentage composition from this formula shows that the analysis agrees with the "expected" values.

C_{31}	31 × 12 = 372	(372/452)100 = 82.35 % C
H_{48}	48 × 1 = 48	(48/452)100 = 10.60 % H
O_2	2 × 16 = 32	(7.07 % O)
	452	

However, it will be seen from the following calculations that the formulas $C_{30}H_{46}O_2$ and $C_{32}H_{50}O_2$ also fit the "found" values:

C_{30}	360	82.20% C	C_{32}	384	82.40% C
H_{46}	46	10.50% H	H_{50}	50	10.70% H
O_2	32	(7.30% O)	O_2	32	(6.90% O)
	438			466	

It is clear that only an accurate molecular weight determination can resolve this dilemma; carbon-hydrogen analyses alone cannot always establish the correct formula.

Exercise 1.2*

1. Calculate a satisfactory empirical formula for each of the following analytical results. The compounds contain only C, H, O (and, when a value for N is given, nitrogen):
   (*a*) C, 40.12; H, 6.75
   (*b*) C, 65.58; H, 9.52
   (*c*) C, 64.88; H, 8.18
   (*d*) C, 77.61; H, 7.45; N, 15.21
   (*e*) C, 63.33; H, 5.18; N, 15.01
   (*f*) C, 71.15; H, 6.80; N, 10.25
2.† Calculate a satisfactory empirical formula for each of the following analytical results:
   (*a*) C, 75.12; H, 9.52
   (*b*) C, 51.68; H, 7.02
   (*c*) C, 41.55; H, 2.95; Br, 46.33
   (*d*) C, 65.82; H, 4.71; S, 29.44

Partial answers to exercise 1.2.1

(*e*)	percent by weight	g-atoms/100 g	atoms/atom O
C	63.33	63.33/12 = 5.28	5.1
H	5.18	5.18/1 = 5.18	5.0
N	15.01	15.01/14 = 1.07	1.0
(O	16.47	16.47/16 = 1.03	1.0)

Thus, empirical formula is C_5H_5NO.

(*f*)	percent by weight	g-atoms/100 g	atoms/atom O
C	71.15	5.93	8.0
H	6.80	6.80	9.2
N	10.25	0.73	1.0
(O	11.80	0.74	1.0)

Thus, empirical formula is C_8H_9NO.

* Slide rule calculations, with the use of C = 12 and H = 1, are acceptable.
† Check answers by calculation from formula arrived at from data.

D. THE PRELIMINARY INTERPRETATION OF ANALYTICAL DATA

(a) The molecular formula of an organic compound provides little specific information about the chemical nature of the substance—its structure, the kinds of functional groups it contains, the chemical behavior that it may be expected to show. Nevertheless, an examination of the molecular formula provides a starting point for further inquiry.

Consider the paraffin hydrocarbon with seven carbon atoms:

$$CH_3CH_2CH_2CH_2CH_2CH_2CH_3 \quad (C_7H_{16})$$

It is seen that this composition can be expressed by the general formula C_nH_{2n+2}. All of the homologs and isomers of heptane have the same C/H ratio:

methane	CH_4	(C_1H_{2+2})
propane	C_3H_8	$(C_3H_{(2\times 3)+2})$
2-methylhexane	$CH_3\underset{\displaystyle CH_3}{\underset{\vert}{C}}HCH_2CH_2CH_2CH_3$	(C_7H_{16})
neopentane	$(CH_3)_4C$	(C_5H_{12})

and so on.

Saturated alcohols (hydroxyalkanes) have the same C/H ratio as alkanes:

$$CH_3CH_2OH \qquad (C_2H_6O)$$

$$CH_3\underset{\displaystyle OH}{\underset{\vert}{C}}HCH_2\underset{\displaystyle CH_3}{\underset{\vert}{C}}HCH_3 \quad (C_6H_{14}O)$$

as do ethers containing only saturated alkyl groups:

$$CH_3CH_2OCH_3 \qquad (C_3H_8O)$$

$$CH_3\underset{\displaystyle CH_3}{\underset{\vert}{C}}HOCH_2\underset{\displaystyle CH_3}{\underset{\vert}{C}}HCH_3 \quad (C_7H_{16}O)$$

It will be noted that these compounds *contain only single bonds and no rings.* Consequently, if we found an unknown compound had the composition $C_8H_{18}O$ we could conclude that it was a saturated acyclic alcohol (an alkanol) or ether. Moreover, this rule is also obeyed for multifunctional compounds in which no rings or double bonds are present:

$$HOCH_2\underset{\displaystyle OH}{\underset{\vert}{C}}HCH_2OH \qquad (C_3H_8O_3)$$

$$CH_3OCH_2CH_2OCH_2CH_2OCH_3 \quad (C_6H_{14}O_3)$$

and so on.

Exercise 1.3*

1. Write the molecular formula for each of the following:
   (*a*) 2-propanol
   (*b*) di-*n*-butyl ether
   (*c*) nonane
   (*d*) $(CH_3)_2CHCH_2OCH_2CH_2OCH_2CH_3$
   (*e*) $(HOCH_2)_4C$
   (*f*) cholesterol
   (*g*) β-carotene
   (*h*) adamantane

   Do these conform to the C/H ratio C_nH_{2n+n}? If not, explain.
2. Write two possible structural formulas for each of the following:
   (*a*) C_2H_6O
   (*b*) $C_4H_{10}O_2$
   (*c*) $C_6H_{14}O_6$
   (*d*) C_5H_{12}
   (*e*) $C_8H_{18}O_3$
   (*f*) $C_5H_{12}O_4$
   (*g*) C_3H_8O
   (*h*) $C_3H_8O_2$
   (*i*) $C_2H_6O_2$
   (*j*) $C_{10}H_{22}O$
3. Ethylamine is $CH_3CH_2NH_2$; di-*n*-propylamine is $(CH_3CH_2CH_2)_2NH$; trimethylamine is $(CH_3)_3N$.

   Calculate a general expression (C_xH_yN) for saturated, acyclic amines, and write two possible structural formulas for each of the following:
   (*a*) C_2H_7NO
   (*b*) $C_3H_{10}N_2O$
   (*c*) $C_4H_{11}NO$
   (*d*) $C_5H_{11}N$
   (*e*) C_5H_9NO
4. Which of the following molecular formulas are those of saturated, acyclic compounds:
   (*a*) C_6H_{12}
   (*b*) $C_8H_{18}O_3$
   (*c*) $C_5H_{10}O$
   (*d*) C_4H_6O
   (*e*) $C_5H_{12}O_2$
   (*f*) CH_4O
   (*g*) $C_{10}H_{20}O_3$
   (*h*) $C_4H_{11}NO$
   (*i*) C_5H_5N
   (*j*) $C_5H_{11}N$
5. Draw *possible* structural formulas for 4(*a, b, h, i, j*).

Partial answers to exercises 1.3.3

(*a*) Since a saturated amine, such as $C_2H_5NH_2$, $C_3H_7NHCH_3$, has the general atomic ratio expressed by $C_nH_{2n+3}N$, the compound C_2H_7NO ($n = 2$; $2n + 3 = 7$) is fully saturated. A possible formula is $HOCH_2CH_2NH_2$.

(*e*) The fully saturated C_5 compound would be $C_5H_{13}NO$. Thus, C_5H_9NO lacks four hydrogens of the fully saturated compound and has two double bonds, two rings, one double bond and one ring, or one triple bond. One possible formula is $(CH_3)_2NCH{=}CHCH{=}O$.

* Compounds for which names are given should be looked up in a textbook or reference book.

Exercise 1.4

1. Draw five acceptable (that is, obeying the valence requirements for C, H, O) structures for $C_7H_{10}O$.
2. Write an acceptable structure for each of the following and show how they conform to a general C/H ratio:
   (*a*) C_6H_6 (*Answer:* the fully saturated, open-chain C_6 compound would be C_6H_{14}. Since C_6H_6 contains eight fewer H atoms, four rings or double bonds are present.

Thus, C_6H_6 could be, for example, benzene or $CH_2{=}CHCH{=}C{=}C{=}CH_2$.*

(*b*) C_6H_7N
(*c*) C_7H_5N
(*d*) $C_{14}H_{10}$
(*e*) $C_6H_4O_2$
(*f*) $C_{14}H_{10}O_2$
(*g*) $C_2H_2O_2$
(*h*) $C_2H_2O_4$
(*i*) $C_4H_6O_2$
(*j*) $C_{24}H_{12}$ (*Note:* although numerous structures obeying the valence rules could be written for $C_{24}H_{12}$, so great a deficiency in hydrogen atoms is most commonly associated with polycyclic aromatic structures. The formula $C_{24}H_{12}$ is actually that of *coronene*, which the student should look up in an appropriate reference source.)

3. Look up the structures of the following compounds in a textbook or handbook. Write their molecular formulas and show how they conform to the rule of the C/H ratio that is under discussion:

(*a*) Phenanthrene
(*b*) Fluorenone
(*c*) Benzilic acid
(*d*) Azobenzene
(*e*) Quinoline
(*f*) Styrene
(*g*) Stilbene
(*h*) Vitamin A
(*i*) β-Carotene
(*j*) Cholesterol

Partial answer to exercise 1.4.3

(*a*) Phenanthrene, $C_{14}H_{10}$: $C_{14}H_{30} - C_{14}H_{10} = H_{20}$; thus, ten rings or double bonds, or both.

(*d*) Azobenzene, $C_{12}H_{10}N_2$: saturated, acyclic compound would be $C_{12}H_{28}N_2$; thus, $28 - 10 = 18$, or nine rings or double bonds, or both. (*Note:* although, as the student will learn, benzene rings do not possess discrete double bonds, they may be regarded as having three double bonds for the purposes of these calculations.)

---

* It has not been explicitly stated above, but the triple bond is equivalent, in these terms, to two double bonds. Thus: $HC{\equiv}CCH_2CH_3$ is C_4H_6 (compare with C_4H_{10}); and C_6H_6 could also be $HC{\equiv}C{-}C{\equiv}C{-}CH_2CH_3$ or $HC{\equiv}C{-}CH{=}CH{-}CH{=}CH_2$ (and others).

PART 2

Bonding and the shapes of organic molecules

A. BOND ANGLES AND HYBRIDIZATION OF ORBITALS

Organic compounds are composed of atoms that occupy definite positions in space with respect to one another; they are joined by bonds of specific lengths, which are disposed at angles that are characteristic of the type of compound.

Molecules (and ions) that consist of four atoms or atom groups surrounding a central atom are represented by these typical examples:

methane, CH_4
ammonium ion, NH_4^+
fluoborate ion, BF_4^-
silicon tetrachloride, $SiCl_4$

In all of these compounds the four bonds are disposed in a tetrahedral arrangement. The bonds are all identical and form angles of 109.5° with each other:

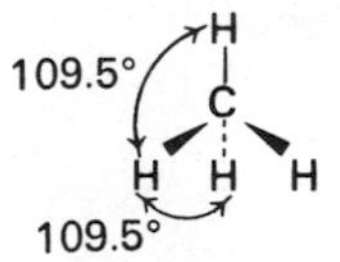

109.5° 109.5°
methane

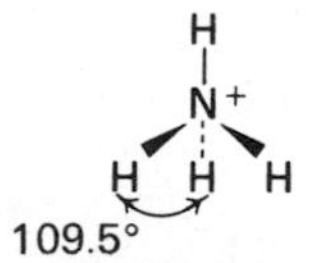

109.5°
ammonium ion

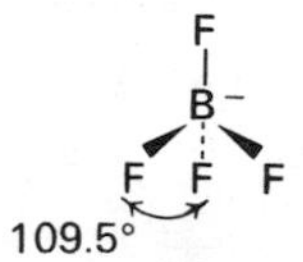

109.5°
fluoborate ion

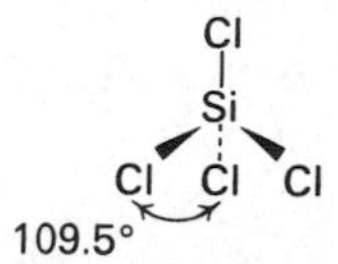

109.5°
silicon tetrachloride

Three orbitals are most favorably disposed about the central atom when they are in the same plane and are equidistant: that is, at angles of 120°. Thus, the shapes of the following molecules are as shown:

Molecule	Structure	Shape
BF_3 boron trifluoride	B—F (three F at 120°, 120°, 120°)	planar, and trigonal: B with three F
$H_2C{=}O$ formaldehyde	C=O with two H (120°, 120°, 120°)	planar
$CH_2{=}CH_2$ ethylene	$H_2C{=}CH_2$	planar: H—C—H angles, near 120°
SO_3 sulfur trioxide	^-O—$\overset{+}{S}$=O, S=O	planar: O—S—O angle, 120°
NO_3^- nitrate ion	^-O—$\overset{+}{N}$=O, N—O^-	planar: O—N—O angle, 120°
CO_3^{--} carbonate ion	^-O—C=O, C—O^-	planar: O—C—O angle, 120°

Two orbitals will tend to move to diametrically opposite sides of the central atom to give a linear molecule.

Molecule	Structure	Shape
CO_2 carbon dioxide	O=C=O	linear: O—C—O angle, 180°
HCN hydrocyanic acid	H—C≡N	linear: H—C—N angle, 180°
HC≡CH acetylene	H—C≡C—H	linear: H—C—C angle, 180°

In the same way, from the simple conclusions drawn after a consideration of coulombic repulsions and the solid geometry of a system, the following shapes are to be expected for the molecules discussed above and for some further examples:

Bond arrangement	Bond description	Example	Shape
linear	*sp*	CO_2	
trigonal	sp^2	BF_3	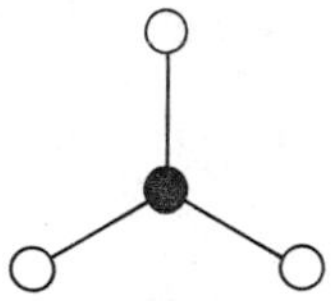
tetrahedral	sp^3	CH_4	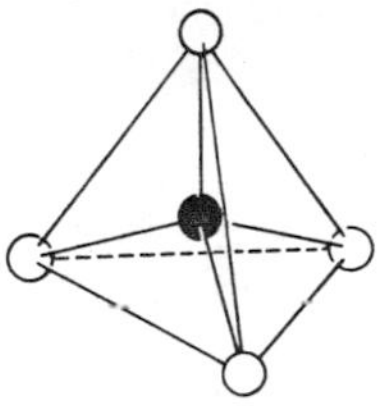
trigonal bipyramidal	dsp^3	PCl_5	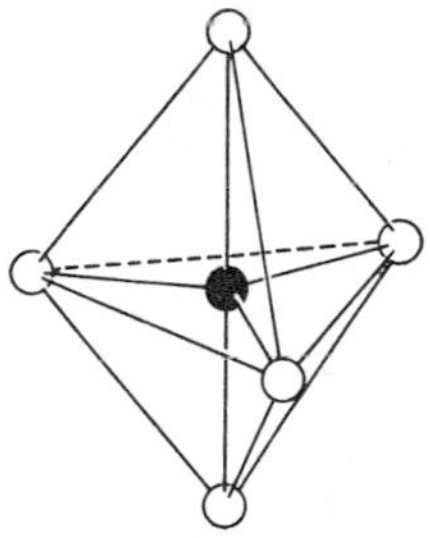
octahedral	d^2sp^3	$Fe(CN)_6{}^{3-}$	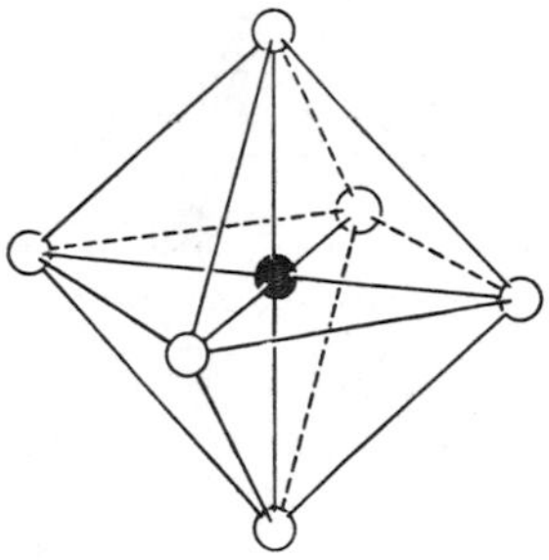

Exercise 2.1

What would be the expected angles in the following:

(*a*) O—S—O angle in sulfate ion, $SO_4{}^{--}$
(*b*) C—B—C angle in trimethylboron, $B(CH_3)_3$
(*c*) C—Be—C angle in dimethylberyllium, $Be(CH_3)_2$
(*d*) C—N—C angle in tetramethylammonium ion, $N(CH_3)_4{}^+$
(*e*) O—C—O angle in carbonate ion, $CO_3{}^{--}$
(*f*) Br—C—Br angle and H—C—H angle in the transition state complex in the reaction $Br^- + CH_3Br \longrightarrow BrCH_3 + Br^-$. (*Note:* The Br atoms may be distinguished, if necessary to aid the student's comprehension, by assuming one of them to be a radioactive isotope.)

Electron pairs in excess of those involved in covalent bond formation occupy orbitals that are disposed in definite relationship to the covalent bonds. For example, in ammonia the nitrogen nucleus is surrounded by eight electrons, six of them in the three N—H bonds, two of them in a fourth orbital (an "unshared pair"). The H—N—H bond angle is found by experimental measurement to be about 107°. This suggests that the four orbitals are essentially tetrahedral in their disposition, with the unshared pair in an orbital as shown in the following diagram:

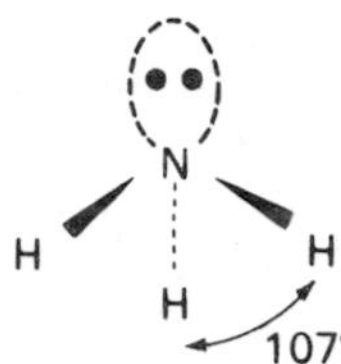

In water, the H—O—H bond angle is about 104°, with the two unshared pairs having a spatial disposition as in a slightly distorted tetrahedron:

It is only in symmetrical molecules with occupied molecular orbitals that accurate bond dispositions can be predicted *a priori*; for example, CH_4, CCl_4, BCl_3, CO_2, acetylene, BF_4^-, etc. When the molecules are not perfectly symmetrical, small deviations from the ideal angles may be expected. In CH_2Cl_2, for example, the bonds are tetrahedral but the bond angles are not all identical.

Although a knowledge of the precise values of bond angles is not always essential for an understanding of the principles of molecular geometry, these data do emphasize that such factors as steric interactions between adjacent atoms and unequal sharing of electrons in covalent bonds introduce small distortions from the ideal angles observed in symmetrical molecules.

Exercise 2.2

Estimate the angular disposition of the bonds around the central atoms in the following compounds. Draw perspective sketches of the molecules:

(*a*) SiF_4
(*b*) carbon disulfide
(*c*) hydronium ion, H_3O^+
(*d*) nitrogen trichloride
(*e*) acetone (around C of C=O)
(*f*) $Ni(CO_4)$
(*g*) $Fe(CO)_5$
(*h*) PCl_6^-
(*i*) $ZnCl_4^{--}$
(*j*) monomeric $AlCl_3$

(*k*) $Si(CH_3)_4$
(*l*) SiO_2
(*m*) $POCl_3$ (around P)
(*n*) $Hg(CH_3)_2$
(*o*) $Ag(NH_3)_2^+$
(*p*) $N(CH_3)_4^+$
(*q*) $CH_3MgCl \cdot 2Et_2O$ (around Mg)
(*r*) $(CH_3)_3N^+—BF_3^-$ (around N and B)
(*s*) $(CH_3)_3NO$

B. THE SHAPES OF MOLECULES WITH TETRAHEDRAL BONDS

The structure of ethane, CH_3CH_3, appears at first sight to be simple and obvious:

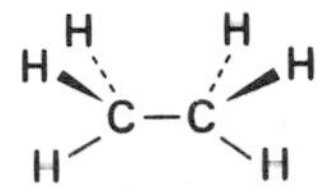

It is apparent, however, that two principal arrangements of the hydrogen atoms are possible; looking at the molecule from one end, down the C—C axis, we see

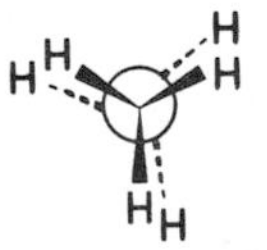

"eclipsed" position

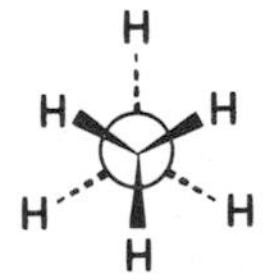

"staggered" position

It is apparent that when the hydrogens are "eclipsed" there is internuclear and interorbital interaction between them; this is relieved when one CH_3 group is rotated slightly to the "staggered" position. The energy barrier that must be surmounted when one CH_3 group is rotated *through* the eclipsed arrangement is small but real (approximately 3 kcal) and so the population of molecules in a specimen of ethane would consist largely of "staggered" molecules in which the hydrogen atoms and the C—H bond orbitals are arranged at positions of minimum interaction.

Exercise 2.3

Draw three-dimensional sketches (thicker lines can indicate nearness, dotted lines can indicate distance) showing the most probable conformation of the following molecules:

(*a*) butane
(*b*) 1,2-dibromopropane
(*c*) 2,2-dimethylbutane
(*d*) cyclopropane
(*e*) cyclopentane
(*f*) cyclohexane

In cyclic compounds such as cyclohexane all of the carbon atoms carry four bonds in regular tetrahedral arrangement. This requires that the six carbon atoms be in a nonplanar array (for if the molecule were planar, the C—C—C bond angles would necessarily be 120°). Two principal arrangements are the following:

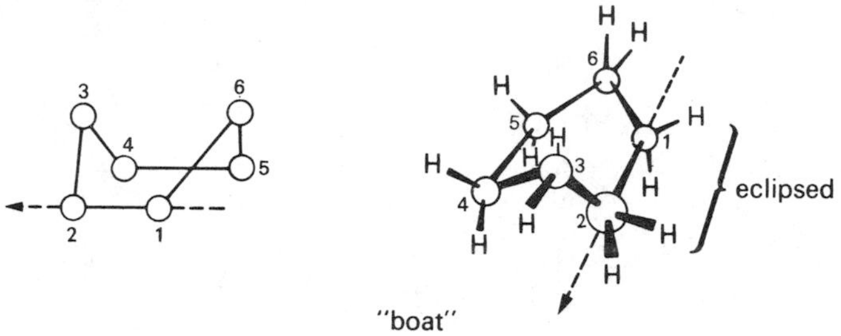

"boat"

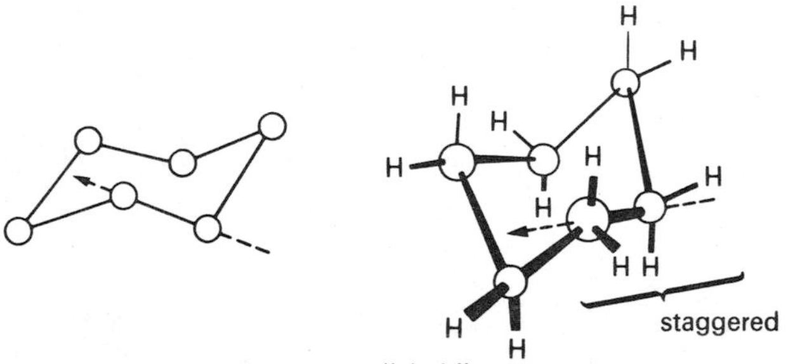

eclipsed

"chair"

staggered

It can be seen that the most favorable conformation is the "chair" form, for it is in this arrangement that the twelve H atoms are disposed with minimal interaction. The "eclipsing" strain and the interaction of the two "inner" H atoms on the third and sixth carbon atoms of the boat form introduce repulsive forces which raise the energy of this conformation.

In cyclopentane a perfectly flat molecule would require but very little departure from the tetrahedral angle (108° instead of 109.5°), but in this arrangement all of the H atoms are fully "eclipsed." By cocking one point in the ring slightly out of plane these interactions are reduced. Cyclopentane actually has a slightly nonplanar conformation. In rings of seven or more C atoms the flexibility of the rings permits tetrahedral angles to be maintained.

Exercise 2.4

What would you predict about the shapes (conformations) and interatomic interactions in

(*a*) cyclooctane (*b*) cyclodecane (*c*) cyclooctadecane.

Exercise 2.5

1. With the aid of molecular models, predict which of the following compounds could be prepared:
   (*a*) cyclohexyne
   (*b*) cyclonona-1,2-diene
   (*c*) cyclooctene
   (*d*) *trans*-cyclohexene
   (*e*) *trans*-cyclodecene
   (*f*) *trans*-cyclohexadecene.
2. Which, if any, of the compounds in Part 1 of this Exercise could exist in two stable, nonidentical forms? (*Note:* It is not correct to say that *cis*- and *trans*-cyclohexene, for example, are nonidentical *forms of the same compound*; they are quite different structures.)

PART 3

Stereochemistry

A. INTRODUCTION

The topics discussed in Part 2 are stereochemical in nature, for the term stereochemistry refers to the three-dimensional features of molecular structure. Stereochemistry, however, includes more than merely the shapes of individual molecules; it deals to a large extent with isomerism due to the spatial distribution of the groups within molecules.

Three principal divisions of this area are:

cis-trans isomerism
optical isomerism
configurational (conformational) isomerism.

B. *CIS-TRANS* ISOMERISM

Cis-trans isomerism exists when isomers differing only in the spatial disposition of constituent groups have independent existence because there are large energy barriers to their interconversion.

(1) H, Br / C=C / Br, H — *trans*-1,2-dibromoethylene

H, H / C=C / Br, Br — *cis*-1,2-dibromoethylene (2)

These are *cis-trans* (sometimes called "geometrical") isomers. They cannot be interconverted easily, for the rotation of one of the =CHBr groups can be accomplished only by the expenditure of the energy needed to break a π-bond.*

When a series of double bonds is present in a molecule, numerous geometrical isomers are possible. The compound deca-2,4,6,8-tetraene, (3), which can be

(3) 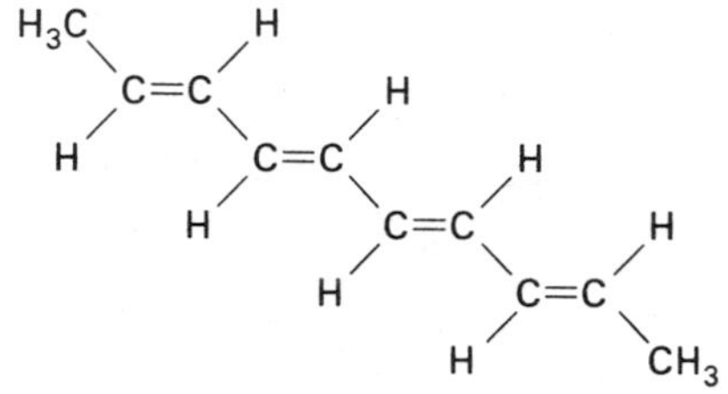 (4)

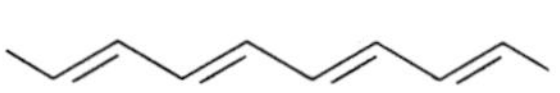

written in the condensed manner in (4), is represented in (3) and (4) as an all-*trans* isomer. Other isomers are possible; for example, the 4-*cis*-2,6,8-*trans* form at (5):

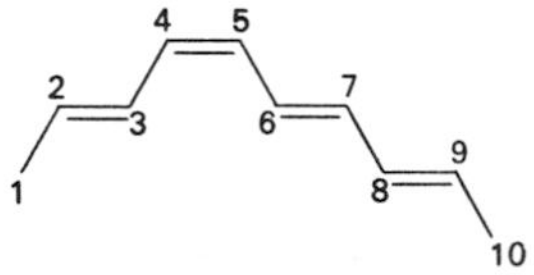

(5)

* Since this energy would be recovered upon reformation of the bond, the energy needed for interconversion is an *activation energy* required to overcome the energy barrier to rotation.

Exercise 3.1

Draw the structures of the following (look up named compounds in an appropriate textbook or reference book):
(*a*) 2-butene (*cis*- and *trans*-)
(*b*) all *trans*- vitamin A
(*c*) all possible *cis-trans* isomers of 2,4-hexadiene
(*d*) two additional *cis-trans* isomers of the decatetraenes shown above as (3), (4), and (5)
(*e*) *cis*- and *trans*-2-buten-4-yne
(*f*) 3-pentene (*cis*- and *trans*-)
(*g*) 1-pentene
(*h*) 2,3-pentadiene (an allene).

Answer to exercise 3.1(*c*)

The 2,4-hexadienes that can be written are

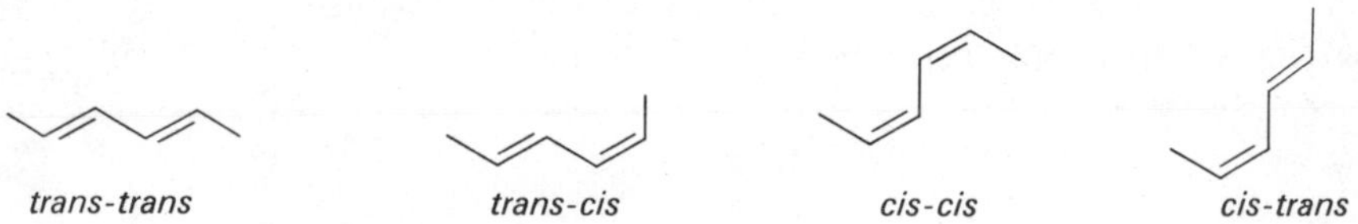

trans-trans *trans-cis* *cis-cis* *cis-trans*

But *trans-cis* and *cis-trans* are identical; thus, three 2,4-hexadienes are possible.

Cis-trans-isomerism is also encountered in cyclic systems, in which the ring defines the relative positions of the groups as, for example, in (6), (7), (8), and (9):

(6) *trans*-1,2-dimethylcyclohexane

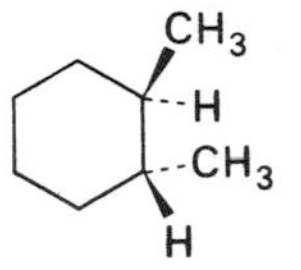

(7) *cis*-1,4-dimethylcyclohexane

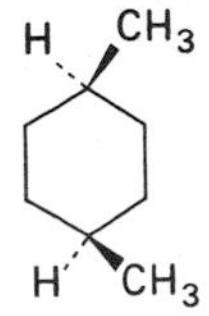

(8) *cis*-2-bromocyclo-propanecarboxylic acid

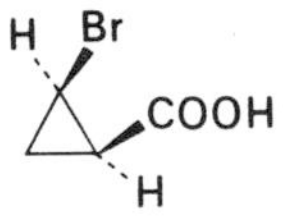

(9) *trans*-cyclopentene-3,5-diol

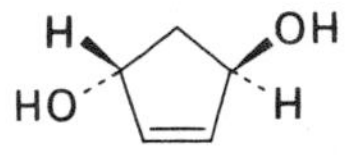

Note that in *trans*-cyclopentene-3,5-diol the term *trans* necessarily refers to the relative position of the hydroxyl groups, for a *trans*- double bond cannot exist in so small a ring.

Exercise 3.2

Write structures such as those above for the following compounds:
(*a*) *cis*-1,2-cyclohexanediol
(*b*) *cis*-1,2-*trans*-2,3-cyclohexanetriol
(*c*) *cis*-cyclobutanc-1,3-dicarboxylic acid
(*d*) *trans*-hexahydrophthalic acid (phthalic acid will be found in a textbook of organic chemistry)
(*e*) *cis*-4-methylcyclohexanol
(*f*) *trans*-2,3-dimethylbutanedioic anhydride.

It should be noted that a compound such as *trans*-cyclohexane-1,2-dicarboxylic acid can exist in two forms, (10) and (11), which are enantiomeric (see Section D).

(10) and 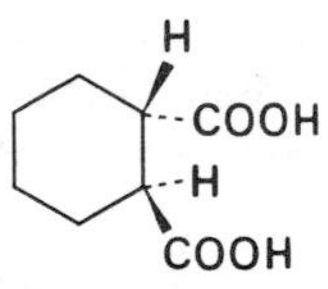 (11)

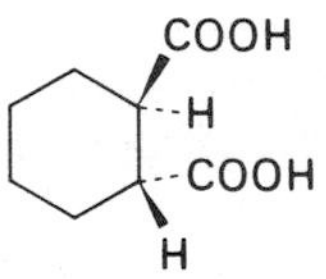

Consequently, the complete name for one of these must include a term that indicates whether it is the (+) or the (−) form. The method of doing this is described in Section D.6.

C. REPRESENTATION OF CONFORMATION

Since the cyclohexane ring is not planar (neither are larger carbocyclic rings), a perspective representation shows the actual angular dispositions of the ring and its substituents. For example, *trans*-cyclohexane-1,2-dicarboxylic acid could exist as (12) or (13):

(12) 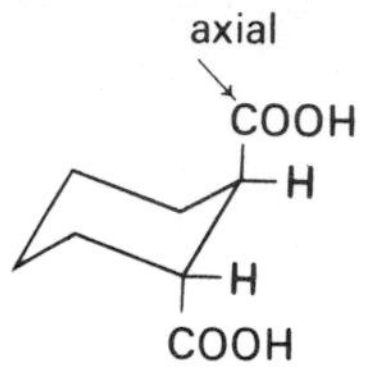 axial COOH, H, H, COOH — *trans*-diaxial

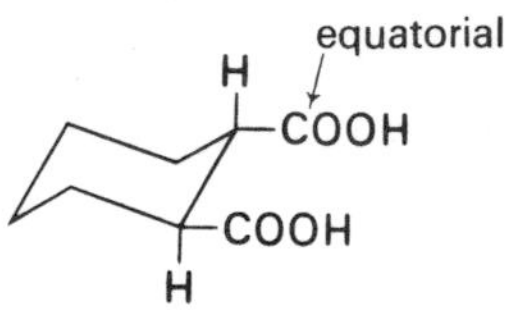

(13) equatorial H, COOH, COOH, H — *trans*-diequatorial

Note that these rings are made interconvertible simply by altering the shape of the ring. This is best seen by making a model of (12) and "flipping" it into (13) without opening any C—C bonds.

In most cases the preferred (most stable) conformation is that in which *the bulky groups are equatorial*, for in the diaxial conformation there exist transannular interactions between axial substituents and axial hydrogen atoms.

Exercise 3.3

1. Draw the two chair conformations for each of the following, and state which you believe would be the more stable of each pair:
   (*a*) *cis*-1,4-dimethylcyclohexane
   (*b*) *trans*-1,4-dimethylcyclohexane
   (*c*) *cis*-1,3-dibromocyclohexane
   (*d*) all *trans*-1,2,3,4,5,6-hexachlorocyclohexane
   (*e*) dimethyl *cis*-1,2-cyclohexanedicarboxylate.
2. It is possible to form the lactone of *cis*-3-hydroxycyclohexanecarboxylic acid. Draw the preferred conformational structures of the hydroxyacid and its lactone and comment on the formation of the latter.
3. Draw the structures for the most stable conformations of the following:
   (*a*) *cis*-decahydronaphthalene
   (*b*) *trans*-decahydronaphthalene
   (*c*) perhydrophenanthrene
   (*d*) perhydroanthracene. (*Note:* "Perhydro" aromatic hydrocarbons are the fully reduced (hydrogenated) compounds. "Decahydronaphthalene" is perhydronaphthalene.

D. OPTICAL ISOMERISM

1. Molecular asymmetry

Molecules that lack elements of symmetry can exist as nonidentical mirror images. These are called "enantiomorphs" or "enantiomers" and are distinguished only in

(a) their effect upon polarized light and (b) their chemical behavior toward an optically active reagent.

The following compounds possess an element of symmetry and exist in one form only—they possess a *plane of symmetry*: CH_2Cl_2, CH_3CH_2COOH, $(CH_3)_2CHCOOH$, *cis*- or *trans*-1,4-dimethylcyclohexane (see (14)), *cis*-1,2-dimethylcyclohexane, cyclohexene, *cis*-2-butene, *trans*-2-butene.

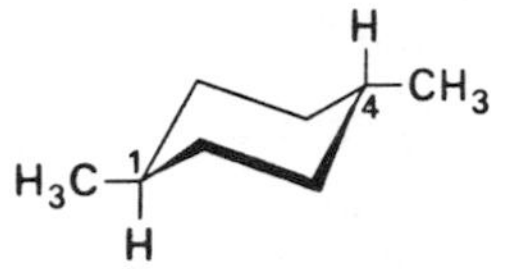

(14)

trans-1,4-dimethylcyclohexane; a symmetry plane passes through C-1 and C-4 in the plane of the page

In a molecule *Cabde*, enantiomorphic forms are possible when *a*, *b*, *d*, and *e* are different; if any two of the substituents are the same, a plane of symmetry is present.

Examples of enantiomeric compounds are the following:

(15) 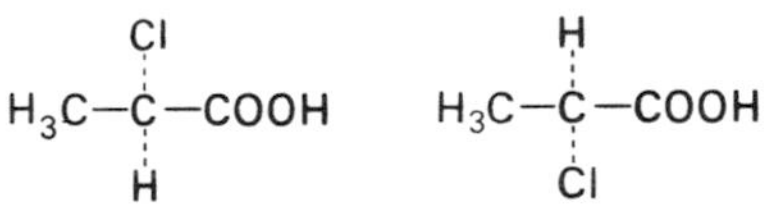 $H_3C-C(Cl)(H)-COOH$ $H_3C-C(H)(Cl)-COOH$

(±) pair

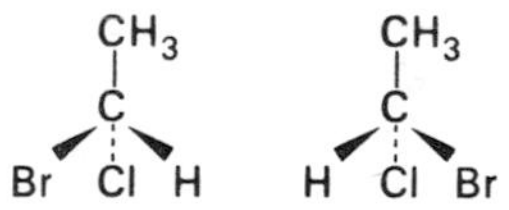

(16)

(±) pair

(17)

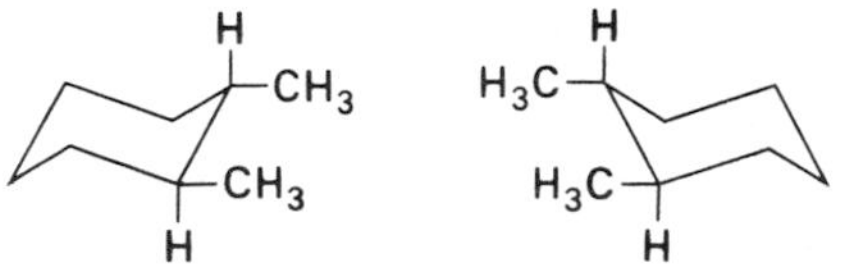

(±) pair

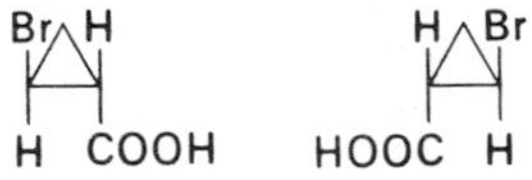

(18)

(±) pair

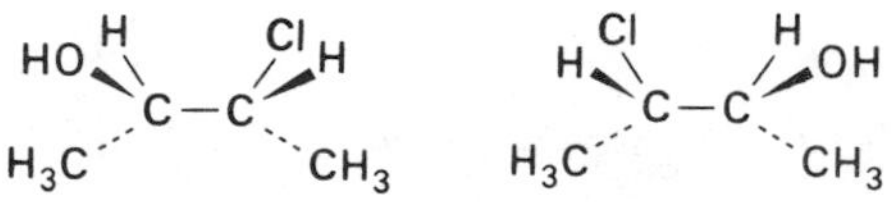

(19)

(±) pair

$CH_3CH{=}C{=}CHCOOH$ (20)

(Draw the two forms, using a conventional 3-dimensional projection.)

Study of examples (15–20) will disclose that none possesses a plane of symmetry, and that the (+) and (−) forms of each pair are not superimposable.*

* It is not possible to state which of the members of such pairs is the (+) form and which is the (−) form by inspection of their formulas. The direction of rotation is a property that is determined experimentally.

Exercise 3.4

Which of the following compounds can exist in enantiomeric forms:

(*a*) 2-bromopropanoic acid
(*b*) 2-methylbutanoic acid
(*c*) crotonic acid
(*d*) toluene
(*e*) *o*-xylene
(*f*) *cis*-1,3-dibromocyclohexane
(*g*) *trans*-1,2-dibromocyclohexane
(*h*) 1-chloro-2-bromopropane
(*i*) tartaric acid
(*j*) vitamin A
(*k*) 4-chloro-2-pentenoic acid
(*l*) *cis*-2-butene
(*m*) 3-methylhexanoic acid
(*n*) 1,2-dimethylsuccinic acid
(*o*) α-methylglutaric acid
(*p*) *β*-methylglutaric acid
(*q*) *p*-bromotoluene
(*r*) cyclohexene
(*s*) 3-bromocyclohexanol
(*t*) 4-bromocyclohexanol.

---

2. The Fischer transformation. Recognition of enantiomers

A useful device in inspecting the formulas of optical isomers is that known as the Fischer transformation. This is based upon an experimental interconversion, first described by Emil Fischer,* in which a compound

(21) $a-\overset{d}{\underset{e}{C}}-b$, was converted into its enantiomer, $a-\overset{d}{\underset{b}{C}}-e$

(21) (22)

Note that in going from (21) to (22), *b* and *e* were interchanged. This simple observation can be utilized advantageously in situations such as the following.

Suppose it is asked:

Is $HOOC-\overset{CH_3}{\underset{H}{C}}-Br$ identical, or enantiomorphic, with $H_3C-\overset{H}{\underset{Br}{C}}-COOH$?

Changing $a\overset{b}{\underset{e}{+}}d$ to $a\overset{d}{\underset{e}{+}}b$ *inverts* the configuration (from (+) to (−) or

(23) (24)

vice versa).

---

* This manner of presenting configuration depends upon the convention that the *horizontal bonds always project toward the viewer*; that is,

$$a\overset{b}{\underset{e}{(C)}}d = a\overset{b}{\underset{e}{C}}d = a-\overset{b}{\underset{e}{C}}-d = a\overset{b}{\underset{e}{+}}d$$

Changing $\begin{array}{ccc} & d & \\ a & + & b \\ & e & \end{array}$ (25) to $\begin{array}{ccc} & d & \\ b & + & a \\ & e & \end{array}$ (26) inverts it again.

Thus, (26) and (23) are identical, and so, applying these inversions to the compound

$$\begin{array}{ccc} & CH_3 & \\ HOOC & + & Br \\ & H & \end{array} \xrightarrow[\text{HOOC and Br}]{\text{interchange}} \begin{array}{ccc} & CH_3 & \\ Br & + & COOH \\ & H & \end{array} \xrightarrow[\text{and H}]{\text{interchange } CH_3}$$

$$\begin{array}{ccc} & H & \\ Br & + & COOH \\ & CH_3 & \end{array} \xrightarrow[\text{and } CH_3]{\text{interchange Br}} \begin{array}{ccc} & H & \\ H_3C & + & COOH \\ & Br & \end{array}$$

Thus, $\begin{array}{c} CH_3 \\ | \\ HOOC-C-Br \\ | \\ H \end{array}$ (27) and $\begin{array}{c} H \\ | \\ H_3C-C-COOH \\ | \\ Br \end{array}$ (30) are enantiomers.

In the same way we see that $\begin{array}{c} CH_3 \\ | \\ HOOC-C-Br \\ | \\ H \end{array}$ (27) and $\begin{array}{c} COOH \\ | \\ H_3C-C-H \\ | \\ Br \end{array}$ (31) are identical.

Exercise 3.5

Select the identical and enantiomorphic structures from the following sets:

(*a*) $\begin{array}{ccc} & C_2H_5 & \\ H_3C & + & NH_2 \\ & H & \end{array}$ (*i*) $\begin{array}{ccc} & NH_2 & \\ H & + & CH_3 \\ & C_2H_5 & \end{array}$ (*ii*) $\begin{array}{ccc} & H & \\ H_2N & + & CH_3 \\ & C_2H_5 & \end{array}$ (*iii*) $\begin{array}{ccc} & CH_3 & \\ H_2N & + & H \\ & C_2H_5 & \end{array}$ (*iv*) $\begin{array}{ccc} & H & \\ H_3C & + & C_2H_5 \\ & NH_2 & \end{array}$ (*v*)

(*b*) $\begin{array}{ccc} & CH_3 & \\ H & + & Cl \\ & COOH & \end{array}$ (*i*) $\begin{array}{ccc} & COOH & \\ Cl & + & H \\ & CH_3 & \end{array}$ (*ii*) $\begin{array}{ccc} & H & \\ H_3C & + & Cl \\ & COOH & \end{array}$ (*iii*) $\begin{array}{ccc} & CH_3 & \\ Cl & + & H \\ & COOH & \end{array}$ (*iv*)

(*c*) $\begin{array}{ccc} & H & \\ H_3C & + & OH \\ H & + & OH \\ & CH_3 & \end{array}$ (*i*) $\begin{array}{ccc} & CH_3 & \\ HO & + & H \\ HO & + & CH_3 \\ & H & \end{array}$ (*ii*) $\begin{array}{ccc} & H & \\ H_3C & + & OH \\ HO & + & CH_3 \\ & H & \end{array}$ (*iii*) $\begin{array}{ccc} & CH_3 & \\ H & + & OH \\ H_3C & + & OH \\ & H & \end{array}$ (*iv*)

Answer to exercise 3.5(*c*)

The Fischer transformation can be performed at one center of asymmetry without affecting others that may be present. Consider the structures

$$\begin{array}{ccc} & H & \\ H_3C & \overset{1}{+} & OH \\ H & \underset{2}{+} & OH \\ & CH_3 & \end{array} \qquad \text{and} \qquad \begin{array}{ccc} & H & \\ H_3C & \overset{1}{+} & OH \\ HO & \underset{2}{+} & CH_3 \\ & H & \end{array}$$

(*i*) (*ii*)

Form *ii* is obtained from form *i* by the following operation: At carbon 2, exchange CH_3 for H, then exchange CH_3 for OH. Since the two exchanges at carbon 2 return the configuration to its original condition, and since no change was made at carbon 1, compound *ii* is identical with compound *i*.

Consider the structures

$$\begin{array}{ccc} & H & \\ H_3C & + & OH \\ H & + & OH \\ & CH_3 & \end{array} \qquad \text{and} \qquad \begin{array}{ccc} & CH_3 & \\ H & + & OH \\ H_3C & + & OH \\ & H & \end{array}$$

(*i*) (*iv*)

Form *iv* is obtained from form *i* by exchanging H and CH_3 at carbon 1, and exchanging H and CH_3 at carbon 2. Since these operations invert the configurations at *both* carbon atoms, form *iv* is the enantiomer of (*i*).

By similar manipulations, it can be shown that (*ii*) and (*iii*) are identical.

---

3. Compounds with more than one asymmetric center

In the foregoing sections some of the examples have been those of compounds with more than one asymmetric center. For example, the formula

$$CH_3\overset{OH}{\overset{|}{C}}H-\overset{Cl}{\overset{|}{C}}HCH_3$$

can represent four different compounds; that is, two enantiomeric pairs. These are

$$\underbrace{\begin{array}{ccc} & CH_3 & \\ H & + & Cl \\ H & + & OH \\ & CH_3 & \\ & (32) & \end{array} \quad \begin{array}{ccc} & CH_3 & \\ Cl & + & H \\ HO & + & H \\ & CH_3 & \\ & (33) & \end{array}}_{(\pm)_1} \qquad \text{and} \qquad \underbrace{\begin{array}{ccc} & CH_3 & \\ Cl & + & H \\ H & + & OH \\ & CH_3 & \\ & (34) & \end{array} \quad \begin{array}{ccc} & CH_3 & \\ H & + & Cl \\ HO & + & H \\ & CH_3 & \\ & (35) & \end{array}}_{(\pm)_2}$$

It can be seen that in (32) and (34) the configurations at the CH_3CHOH- centers are identical, but that the CH_3CHCl- centers are of opposite configurations. Thus (32) and (34) are not mirror images of each other. This can be seen by observing that (32) and (33) *are* mirror images, and (33) is clearly not the same as (34). When three asymmetric centers are present the four forms can be written this way:

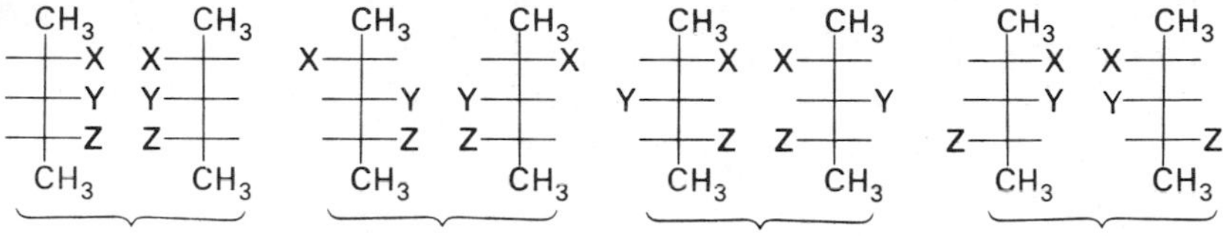

$(\pm)_1$ $(\pm)_2$ $(\pm)_3$ $(\pm)_4$

In summary, when one asymmetric carbon atom is present, there are two optically active isomers; when two asymmetric carbon atoms are present, there are four; and when three asymmetric carbon atoms are present, there are eight. When n asymmetric carbon atoms are present, 2^n optical isomers are possible. This holds true whether the asymmetric centers are adjacent (as in the above examples) or at nonadjacent position. Thus, there are 2^4, or sixteen possible forms—that is, eight $(\pm)$ pairs—of the compound in (36).

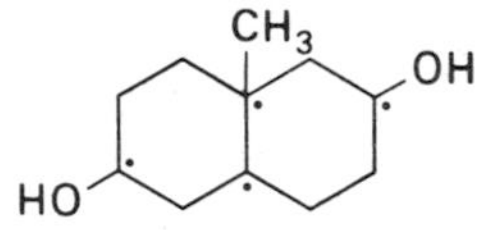

(36)

dots mark asymmetric centers

Exercise 3.6

How many optically active forms are possible for each of the following:

(*a*) 3-methylcyclohexanol
(*b*) 2,3-dibromopentanoic acid
(*c*) 2-chloro-1-butanol
(*d*) 1,2,3-tribromopropane
(*e*) *cis*-4-methylcyclohexanol
(*f*) 3,4-dimethylcyclohexanol
(*g*) glycerol 1-monoacetate
(*h*) glycerol 2-monoacetate
(*i*) ethylene glycol monoacetate
(*j*) *m*-chlorotoluene
(*k*) *sec*-butylbenzene
(*l*) borneol.

Partial answer to exercise 3.6

In answering questions of this kind, it is necessary to examine the structure for the presence of carbon atoms that carry four different substituents, or to look for elements of molecular symmetry or asymmetry. For example, part *g*: $HOCH_2CH(OH)CH_2OAc$: the central carbon atom carries the four substituents, $HOCH_2—$, H—, OH—, and $—CH_2OAc$. Thus, two enantiomeric forms are possible. Part *e*: 4-methylcyclohexanol (37) possesses what may at first sight appear to be two asymmetric carbon atoms, the $>CHCH_3$ group and the $>CHOH$ group. But it will be seen that there is a plane of symmetry at right angles to the ring, cutting through the 1 and 4 carbon atoms. Therefore, *cis*- (or *trans*-) 4-methylcyclohexanol cannot exist in two enantiomeric forms. Note the conformational isomers in (37) and (38) are not enantiomeric and each possesses a plane of symmetry. They are, moreover, interconvertible by a ring flip that requires overcoming only a low energy barrier, and so would not be experimentally separable as distinct substances.

(37) 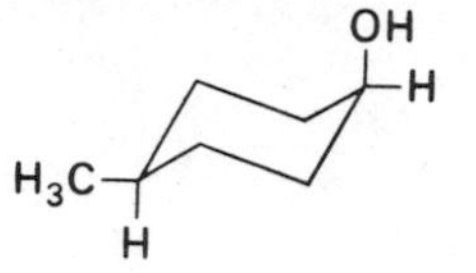 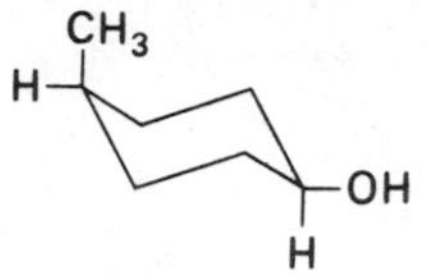 (38)

Different *conformations* of 4-methylcyclohexanol.
In both, the CH_3— and —OH groups are *cis* to each other.

Part *f*: In 3,4-dimethylcyclohexanol there is no plane of symmetry and optical isomers can exist. Indeed, because of the presence of the 3-methyl group all *three* of the carbon atoms bearing substituents become asymmetric, and there are eight forms (four enantiomeric pairs) possible.

---

4. Diastereomers

The compounds (39) and (40)

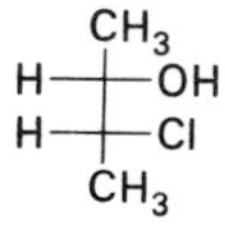

are not enantiomers, although the difference between them lies only in the configurations of the two asymmetric carbon atoms. They are therefore stereoisomers, and are called "diastereomers." Diastereomers are optical isomers that are not enantiomers.

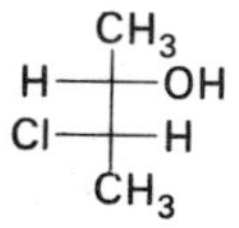

Diastereomers often have similar chemical properties, but are never identical in their chemical behavior. For example, both (39) and (40) react with alkali to give epoxides (41):

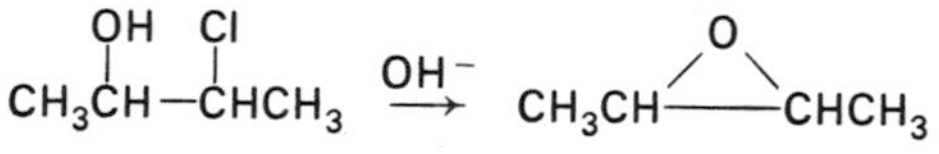

$$CH_3CH(OH)-CH(Cl)CH_3 \xrightarrow{OH^-} CH_3CH\overset{O}{—}CHCH_3 \quad (41)$$

but their rates of reaction are quite different, and the oxides are not identical:

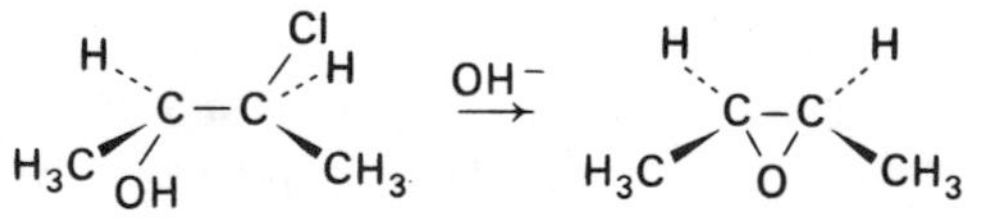

(42) $\xrightarrow{OH^-}$ *cis*-2,3-epoxybutane

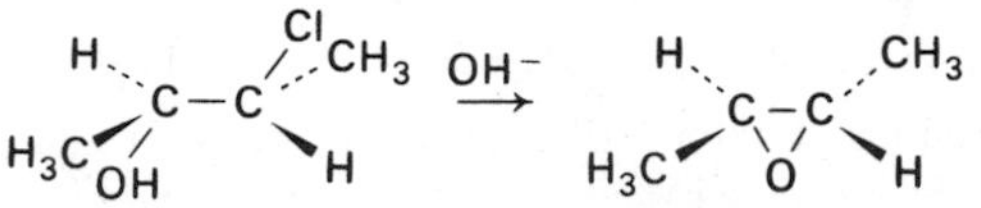

(43) $\xrightarrow{OH^-}$ *trans*-2,3-epoxybutane

Exercise 3.7

Which of the 3-chloro-2-butanols drawn as (42) and (43) is (39)?

Answer to exercise 3.7

Consider the chlorohydrin drawn as (43). This can be redrawn in the Fischer convention as follows:

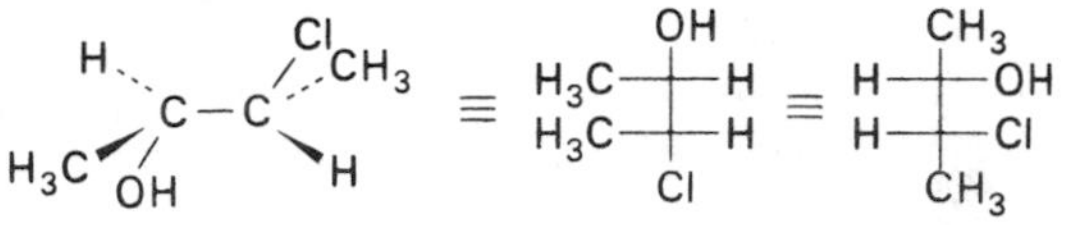

H, Cl, CH_3, C–C, H_3C, OH, H ≡ Fischer projection (OH top; H_3C–C–H; H_3C–C–H; Cl bottom) ≡ Fischer projection (CH_3 top; H–C–OH; H–C–Cl; CH_3 bottom)

(43) (39)

Do the same for the other diastereomer (42).

Exercise 3.8

Draw two diastereomers for each of the following; use three-dimensional notation as in (42) and (43):

(*a*) 2-chlorocyclohexanol
(*b*) 3-chloro-4-methylhexane
(*c*) cyclohexane-1,3-dicarboxylic acid monomethyl ester
(*d*) 2,3,4-trihydroxyheptane
(*e*) the product formed by addition of HOCl to 3-hexene.

Exercise 3.9

One of the diastereomeric 2-chlorocyclohexanols reacts with alkali to give 1,2-epoxycyclohexane (also called cyclohexene oxide); the other does not give the oxide when treated with alkali. Describe this reaction and show why the diastereomers react differently.

---

5. Optical isomerism due to molecular asymmetry

The most common source of optical activity is molecular asymmetry due to the presence of an asymmetric carbon atom, that is, the general conditions symbolized by C *a d b e*. Molecular asymmetry due to other structural features is another condition for optical activity. *The important condition in both cases is the dissymmetry of the molecule as a whole.*

Allenes possess the general structure (44)

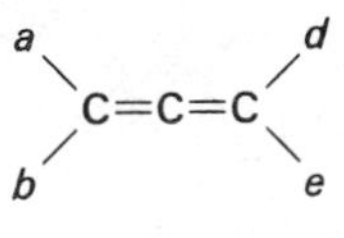

$$\begin{matrix} a & & & & d \\ & \diagdown & & \diagup & \\ & & C{=}C{=}C & & \\ & \diagup & & \diagdown & \\ b & & & & e \end{matrix} \qquad (44)$$

A model of an allene will show clearly that the plane of $\underset{b}{\overset{a}{>}}C$ is at a right angle to that of $C\underset{e}{\overset{d}{<}}$

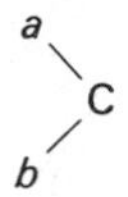

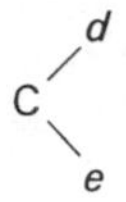

Thus, if $a = b$ or $d = e$, the molecule has a plane of symmetry. But if $a \neq b$ and $d \neq e$ two enantiomeric forms are possible. For example, the following drawing shows such forms:

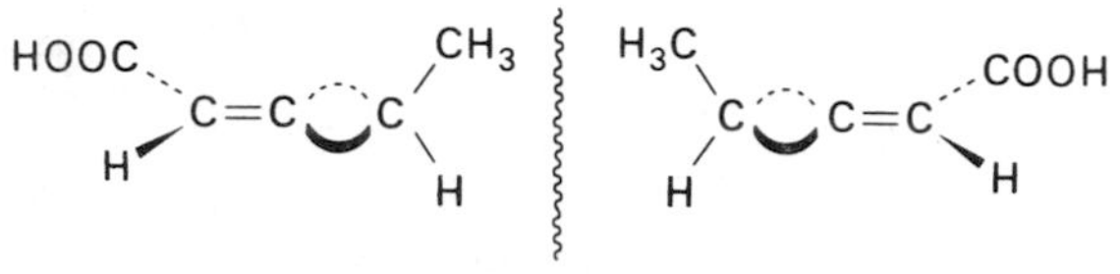

HOOC, H, C=C=C, CH₃, H | H₃C, H, C=C=C, COOH, H

enantiomers: each is optically active

Ring systems can create similar conditions. In the class of compounds known as "spirans" the role of the double bonds is filled by the rings. For example,

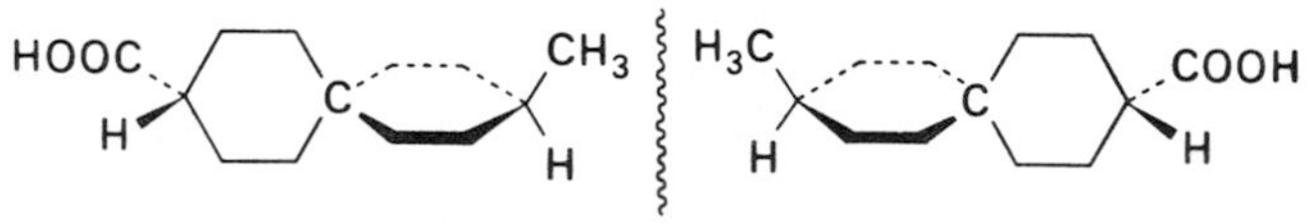

HOOC, H, C, CH₃, H | H₃C, H, C, COOH, H

enantiomers

Exercise 3.10

Draw perspective formulas for allenes of the types $R_2C{=}C{=}CH_2$, $RCH{=}C{=}CH_2$, $R_2C{=}C{=}CHR$, and $R_2C{=}C{=}CR_2$ (where R's are the same). Show whether enantiomers are possible, and, if not, what element of symmetry is present.

Another kind of molecular asymmetry exists if free rotation about a single bond is hindered. The most widely studied examples of this kind are found in derivatives of biphenyl. When sufficiently large (bulky) substituents are present in the four *ortho* positions, rotations of the rings is inhibited, and enantiomeric forms can be separated. For example, the biphenyl in (45) prepared (by any means) will be found to consist of two enantiomers that can be separated (the process is called "resolution") into individual (+) and (−) forms:

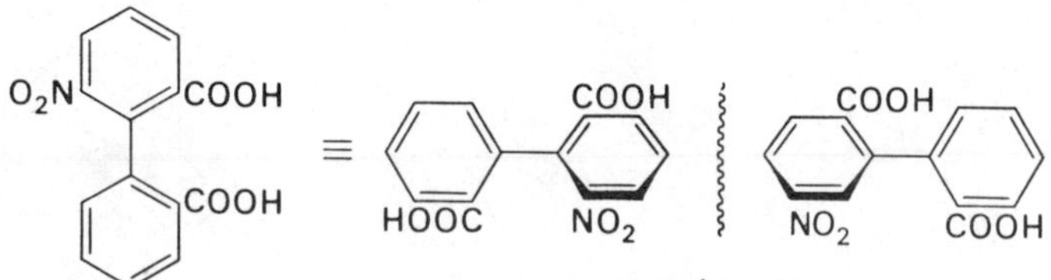

O_2N, COOH, COOH ≡ HOOC, COOH, NO_2 | NO_2, COOH, COOH

(45)

enantiomers: nonsuperimposable mirror images

It will be noted that if the two aromatic rings could become coplanar, the plane of the rings would be a plane of symmetry. The optical activity depends upon the fact that the three bulky groups (COOH, COOH, and NO_2) prevent coplanarity and permit the existence of stable enantiomers.

If only two *ortho* substituents are present, hindrance to coplanarity is usually absent and resolution (separation into (+) and (−) forms) is not possible. The following compounds cannot be resolved:

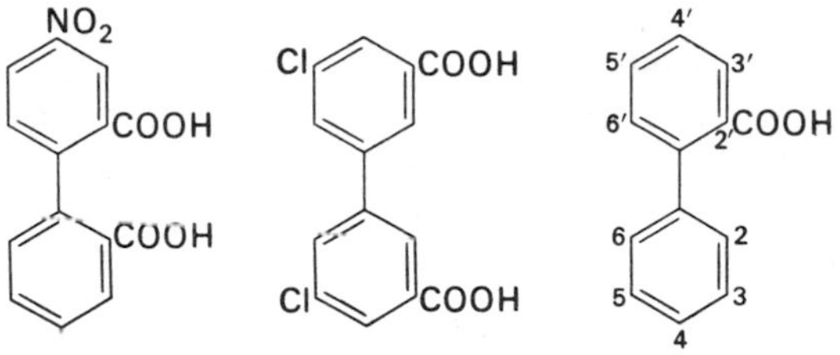

nonresolvable biphenyl derivatives

Exercise 3.11

Which of the following biphenyls would you expect to find are resolvable?

(*a*) 2,6,2′,6′-tetra COOH
(*b*) 2-Cl-6-COOH-2′-Cl-6′-NO_2
(*c*) 2-NO_2-6-COOH-2′,6′-diOCH_3
(*d*) 2-NO_2-6-COOH-2′-COOH-4′-NO_2
(*e*) 2-F-2′-F-6′-COOH.

Note: It will be seen from some of these examples that even if noncoplanarity is maintained, symmetry can still be realized. Notice that the following compound (46) has a plane of symmetry *in* the plane of ring *A* and *vertical* to the plane of ring *B* (an angular relationship of the rings of 90° to each other can be assumed).

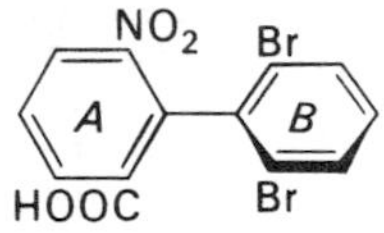

(46)

plane of symmetry *in the plane* of ring *A* passes through 1 and 4 positions of ring *B*

---

6. Optical notation—nomenclature

Optical rotation and configurational series. Optically active compounds were in the past referred to by the letters *d* (for dextro) and *l* (for levo). Thus, *d*-lactic acid; *l*-asparagine; and *d*-2-octanol. This practice, while still used occasionally, is no longer recommended; in this book the signs (+) and (−) will be used instead of the letters *d*- and *l*-. Thus, (+)-lactic acid, (−)-asparagine, and (+)-2-octanol are preferred.

It should be stressed that the signs of rotation ((+) or (−)) are those *observed* by measurement of optical rotation. They cannot be assigned *a priori* (except in certain restricted groups of closely related compounds) by examination of the structure.

Configuration (as distinct from optical rotatory power) can be indicated in naming an optically active compound. The nomenclature of configuration depends upon arbitrary conventions and comprises two general systems.*

Relative configuration. In one system, configuration is assigned by reference to a standard substance. This standard is, for may classes of compounds, the compound glyceraldehyde, of which there are, of course, two enantiomeric forms. These are named

CHO (top); H, OH (wedges); CH_2OH (bottom) or H—C—OH with CHO above and CH_2OH below

D-glyceraldehyde

(47)

CHO (top); HO, H (wedges); CH_2OH (bottom) or HO—C—H with CHO above and CH_2OH below

L-glyceraldehyde

(48)

Notice that the small capital letters D and L are used. These do not necessarily mean *dextro* and *levo*. Although D-glyceraldehyde is indeed (+)-glyceraldehyde (and thus would properly be called D-(+)-glyceraldehyde), compounds that belong to the D-series may in fact be (−), or levorotatory, compounds; and compounds of the L-series may be dextrorotatory (+). The following are examples:

COOH / H—C—OH / HO—C—H / COOH

L-(+)-tartaric acid

COOH / H_2N—C—H / CH_2COOH

L-(−)-aspartic acid

COOH / HO—C—H / CH_3

L-(+)-lactic acid

The designation of the configurational series D- and L- is arrived at by relating the compound to glyceraldehyde; for instance, by the conversion of D-glyceraldehyde to lactic acid by the following operations (which are not described here):

CHO / H—C—OH / CH_2OH —(oxidize −CHO to −COOH)→ COOH / H—C—OH / CH_2OH —(reduce $-CH_2OH$ to $-CH_3$)→ COOH / H—C—OH / CH_3 (49)

D-glyceraldehyde D-lactic acid

Similar transformations have been carried out with many compounds, and configurations *relative* to D-glyceraldehyde can be assigned to many compounds. These

* The system of configurational nomenclature known as the R— and S— system is not dealt with here.

are also *absolute* configurations, for the *absolute* configuration of (+)-glyceraldehyde is now known to be that expressed in formula 47.

It should be recognized that the original assignment of the D-configuration to the substance known as (+)-glyceraldehyde by measurement of its rotation was arbitrary. The later discovery (by x-ray analysis of (+)-tartaric acid) that (+)-glyceraldehyde had indeed the configuration of D-glyceraldehyde was fortuitous—and convenient.

Exercise 3.12

Assign each of the following compounds to the D- or L-series (refer to the configurations given above for D-lactic acid, L-lactic acid, L-asparagine, etc.

$$\begin{array}{c} COOH \\ H-\!\!+\!\!-OH \\ CH_2COOH \end{array} \qquad \begin{array}{c} NH_2 \\ H-\!\!+\!\!-CH_3 \\ COOH \end{array} \qquad HO-C_6H_4-\begin{array}{c} H \\ CH_2-\!\!+\!\!-COOH \\ NH_2 \end{array}$$

malic acid — alanine — tyrosine

proline (pyrrolidine ring with N–H; ring carbon bearing H (dashed) and COOH (heavy))

$$\begin{array}{c} NH_3 \\ HOCH_2-\!\!+\!\!-COOH \\ H \end{array}$$

proline — serine

(*Note:* These are drawn in the Fischer convention, in which horizontal or heavy bonds project toward the viewer.)

Partial answer to exercise 3.12

Redraw alanine so that H and NH_2 are horizontally disposed:

$$\begin{array}{c} NH_2 \\ H-\!\!+\!\!-CH_3 \\ COOH \end{array} \xrightarrow[\text{thus no change in configuration)}]{\text{(two Fischer transformations,}} \begin{array}{c} COOH \\ H-\!\!+\!\!-NH_2 \\ CH_3 \end{array}$$

D-alanine

Proline can be redrawn in a partial structure showing only the asymmetric C atom:

$$\text{proline} = \begin{array}{c} CH_2\cdots \\ HOOC-\!\!+\!\!-H \\ NH\cdots \end{array}$$; one Fischer transformation gives $$\begin{array}{c} CH_2\cdots \\ NH-\!\!+\!\!-H \\ COOH \end{array}$$

and a second one gives $$\begin{array}{c} COOH \\ NH-\!\!+\!\!-H \\ CH_2\cdots \end{array}$$. This, the original configuration, is L-proline.

Exercise 3.13

Show by appropriate perspective formulas the stereochemical course of the following reactions. State whether the products are optically active (show (±) pairs) or inactive (*meso* forms). (*Note:* It will be recalled that Br_2 adds in the *trans* manner to a carbon-carbon double bond. For example, *trans*-2-butene + Br_2 ⟶ *meso*-2,3-dibromobutane.

(*a*) [3-bromocyclohexene: H, Br] + Br_2 ⟶

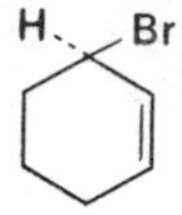

(*b*) *cis*- $CH_3CH{=}CHCH_3$ + Br_2 ⟶

(*c*) [3,6-dimethylcyclohexene: H, CH_3; H, CH_3] + Br_2 ⟶

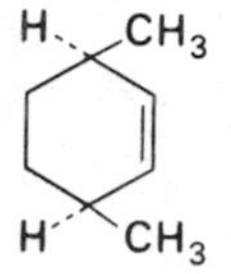

(*d*) *trans*- $CH_3CH{=}CHCH_2CH_3$ + Br_2 ⟶

(*e*) [vinylcyclohexane: H, $CH{=}CH_2$] + Br_2 ⟶

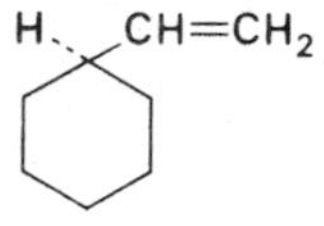

Answer to exercise 3.13(*a*)

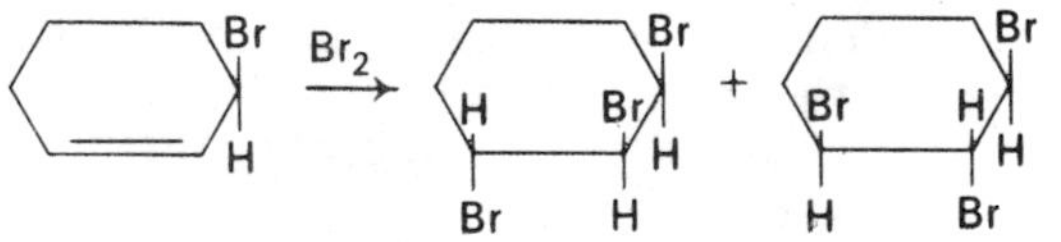

Br, H $\xrightarrow{Br_2}$ Br, H, Br, H, Br, H + Br, Br, H, H, H, Br

optically active   meso (inactive)

PART **4**

Synthetic methods I. Carbon-carbon bond formation by reactions of the aldol type

A. INTRODUCTION

The synthesis of an organic compound consists mainly of two parts: (1) the construction of the framework of atoms that comprises what can be called the "skeleton" of the molecule, and (2) the introduction of functional groups and other structural features that complete the structure of the compound.

One of the principal synthetic operations in organic chemistry is the formation of carbon-carbon bonds. In broad terms, this is accomplished in any of three ways: (1) by combination of free radicals:

$$-\overset{|}{\underset{|}{C}}\cdot + \cdot\overset{|}{\underset{|}{C}}- \longrightarrow -\overset{|}{\underset{|}{C}}-\overset{|}{\underset{|}{C}}-$$

(2) by attack of an anionic (nucleophilic) carbon atom upon an electron-deficient (electrophilic) carbon atom:

$$-\overset{|}{\underset{|}{C}}:^{-} \curvearrowright {}^{+}\overset{|}{\underset{|}{C}}- \longrightarrow -\overset{|}{\underset{|}{C}}-\overset{|}{\underset{|}{C}}-$$

or

$$-\overset{|}{\underset{|}{C}}:^- \longrightarrow \overset{|}{\underset{|}{C}}{=}X \longrightarrow -\overset{|}{\underset{|}{C}}-\overset{|}{\underset{|}{C}}-X^-$$

or

$$-\overset{|}{\underset{|}{C}}:^- \longrightarrow \overset{|}{\underset{|}{C}}-Y \longrightarrow -\overset{|}{\underset{|}{C}}-\overset{|}{\underset{|}{C}}- + :Y^-$$

(3) by attack of an electron-deficient carbon atom upon a nucleophilic (electron-donating) molecule:

$$-\overset{|}{\underset{|}{C}}^+ \quad \overset{|}{\underset{|}{C}}{=}\overset{|}{\underset{|}{C}} \longrightarrow -\overset{|}{\underset{|}{C}}-\overset{|}{\underset{|}{C}}-\overset{|}{\underset{|}{C}}^+$$

or

$$-\overset{|}{C}: \quad \overset{|}{\underset{|}{C}}{=}\overset{|}{\underset{|}{C}} \longrightarrow -\overset{|}{\underset{\cdot\cdot}{C}}-\overset{|}{\underset{|}{C}}-\overset{|}{\underset{|}{C}}^+$$

carbene

The first of these methods is less widely employed in synthesis than the other two, principally because of the relative dearth of general methods for producing carbon radicals with specifically located electron-deficient centers.

B. ALDOL CONDENSATIONS

Method 2 is found in many forms. One of the most widely used is the *aldol condensation* and its many variants. In its simplest form, the aldol condensation is the self-condensation of one carbonyl compound (a ketone, an aldehyde, an ester) with another like molecule (that is, with itself) under the influence of a strong base.

The role of the base is to abstract a proton from the methylene group *alpha* to the carbonyl group. For example, with acetone:*

$$CH_3COCH_3 + B:^- \rightleftharpoons CH_3COCH_2:^- + BH$$

The ability of acetone, a typical carbonyl compound, to be converted into its anionic conjugate base is due to the capacity of the carbonyl group to stabilize the anion by delocalization (resonance stabilization) of the negative charge. It will be seen that only the *alpha* carbon anion can be stabilized in this way. Therefore, β-, γ-, (and so forth) carbon atoms do not undergo this kind of ionization, for it is clear

* The symbol $B:^-$ for a base (and BH for its conjugate acid) will be used frequently, for in most cases any of several strong bases can be used. Common examples are sodium and potassium alkoxides, alkali amides, etc.

that a charge delocalization symbolized by the expression

$$\{CH_3\ddot{C}H{-}\underset{|}{C}{=}\ddot{O}: \longleftrightarrow CH_3CH{=}\underset{|}{C}{-}\ddot{O}:\}^-$$

is not possible with a β-anion

$$\ddot{C}H_2^-{-}CH_2{-}\underset{|}{C}{=}O$$

in which there is no alternate location for the electron pair.

The α-anion produced in this way is the conjugate base of an exceedingly weak acid (for example, acetone has pK_a about 20), and is therefore a strong base and an effective nucleophile. It can attack the electrophilic carbon atom of a carbonyl compound; for example, of a molecule of acetone itself:*

$$CH_3COCH_2:^- + (CH_3)_2C{=}O \rightleftharpoons CH_3COCH_2{-}C(CH_3)_2{-}O^-$$

This addition reaction is an equilibrium. The anionic product is also in proton exchange equilibrium with the final product:

$$CH_3COCH_2{-}C(CH_3)_2{-}O^- + BH \rightleftharpoons \underset{\text{diacetone alcohol}}{CH_3COCH_2{-}C(CH_3)_2{-}OH} + B:^+$$

Diacetone alcohol is the *product* of the base-catalyzed self-condensation of acetone. The equilibrium constant of this reaction is quite unfavorable, and the yield of the product is low unless the reaction is carried out with the use of special techniques, which are described in textbooks and will not be discussed here.

The equilibrium in the aldol condensation is usually unfavorable when the carbonyl compound *to which addition occurs* is a ketone, and this condition is exacerbated when the ketone is highly branched in the vicinity of the carbonyl group (that is, α-substituted).

When the acetone anion is produced in the presence of a carbonyl compound that is more susceptible to nucleophilic attack than is acetone itself, the reaction proceeds to yield the product of the addition of the anion of acetone to the second carbonyl compound; for example, in the presence of an aldehyde:

$$CH_3COCH_3 + B:^+ \rightleftharpoons CH_3COCH_2:^- + BH$$

$$CH_3COCH_2:^- + RCHO \rightleftharpoons CH_3COCH_2\overset{O^-}{\overset{|}{C}}HR$$

$$CH_3COCH_2\overset{O^-}{\overset{|}{C}}HR + B:H \rightleftharpoons CH_3COCH_2\overset{OH}{\overset{|}{C}}HR$$

* The student should always keep in mind the concept of the actual experimental conditions. The addition of a base to acetone converts only a portion of the ketone to the anion. The excess of the remaining, unionized, acetone can react with the carbon anions present in low concentration.

Now a final step can, and usually does, ensue. Loss of water from the β-hydroxy ketone gives rise to an unsaturated ketone, which, because of the added stability associated with the conjugated system, is usually the energetically preferred final product:

$$CH_3COCH_2\underset{|}{\overset{OH}{C}}HR \xrightarrow{-H_2O} CH_3COCH{=}CHR$$

Indeed, the aldol condensation tends to proceed to yield the α,β-unsaturated ketone; in most cases it is possible to isolate the intermediate hydroxyketone only by the use of special care in controlling the conditions of the reaction.

When the carbonyl compounds to be condensed in this way are closely similar in type and reactivity, "mixed" condensations can occur and synthetic utility is lost. Thus, while the condensation of acetaldehyde gives the single product

$$2\ CH_3CHO \longrightarrow CH_3\underset{OH}{\underset{|}{C}}HCH_2CHO \quad \text{or} \quad CH_3CH{=}CHCHO$$

the condensation of, say, propionaldehyde (propanal) with *n*-butyraldehyde (butanal) would yield four products.

Exercise 4.1

Formulate the base-catalyzed self-condensation of acetaldehyde showing each step in the series of equilibria that are involved.

Exercise 4.2

Formulate the following base-catalyzed condensations, showing all of the probable products:

(*a*) CH_3CHO and CH_3CH_2CHO
(*b*) CH_3COCH_3 and CH_3CHO
(*c*) $CH_3COC_6H_5$ and C_6H_5CHO
(*d*) $C_6H_5CH{=}CHCOCH_3$ and C_6H_5CHO
(*e*) cyclohexanone and C_6H_5CHO.

Partial answer to exercise 4.2

(*b*) The principal product will result from the addition of the anion from acetone ($CH_3CO\bar{C}H_2{:}^-$) to the carbonyl group of acetaldehyde, to yield as the final product (after loss of water) $CH_3COCH{=}CHCH_3$, but it would be expected that $CH_3CH{=}CHCHO$ would also be formed in appreciable amount.

(*d*) Only $C_6H_5CH{=}CHCOCH_3$ possesses an α-hydrogen atom. The hydrogen atom of benzaldehyde (C_6H_5CHO) is not α- to the carbonyl group; it is attached directly to the carbonyl carbon atom. Thus, the condensation can proceed only by attack of $C_6H_5CH{=}CHCOCH_2{:}^-$ upon the carbonyl group of the aldehyde, to give as the final product $C_6H_5CH{=}CHCOCH{=}CHC_6H_5$.

(*e*) The anion [cyclohexanone ring with C=O]$:^-$ formed from cyclohexanone attacks the carbonyl group benzaldehyde.

Two factors determine the way in which the condensation of two different aldehydes or ketones will occur: one is the relative acidity of the α-hydrogen atoms of the two compounds; the other is the relative reactivity of the two carbonyl groups to nucleophilic attack.

When an aromatic aldehyde (benzaldehyde, furfuraldehyde) is one of the reactants, it can act *only as the acceptor* of nucleophilic attack; since there is no α-hydrogen atom it cannot act as the nucleophilic component of the reacting pair. Formaldehyde, which possesses no α-hydrogen atoms, falls into the same category.

A special condition is met when formaldehyde is used as the acceptor of the attack of the carbon anion; for example, with acetaldehyde:

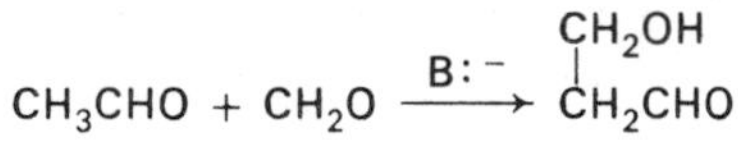

$$CH_3CHO + CH_2O \xrightarrow{B:^-} HOCH_2CH_2CHO$$

Not only can further addition occur (for the product of this first step still possesses two active (α-) hydrogen atoms),

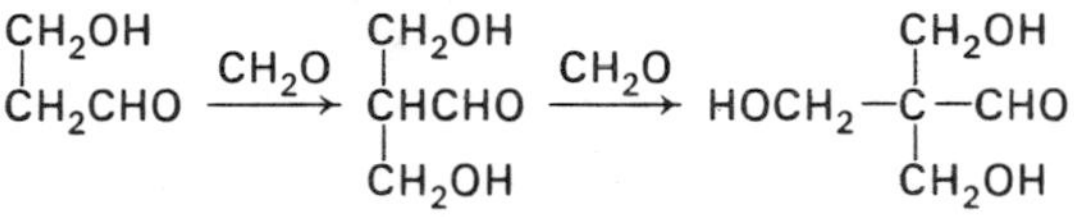

$$HOCH_2CH_2CHO \xrightarrow{CH_2O} (HOCH_2)_2CHCHO \xrightarrow{CH_2O} HOCH_2-C(CH_2OH)_2-CHO$$

but the formyl group of the final product is reduced by the excess formaldehyde that is present:

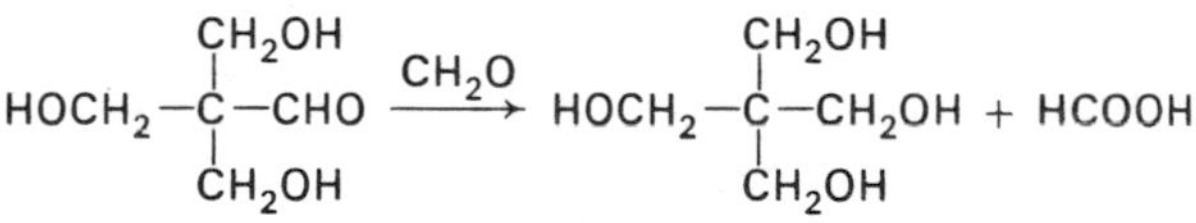

$$HOCH_2-C(CH_2OH)_2-CHO \xrightarrow{CH_2O} HOCH_2-C(CH_2OH)_2-CH_2OH + HCOOH$$

pentaerythritol

When only one α-hydrogen is present in the starting aldehyde, the final product is a 1,3-glycol:

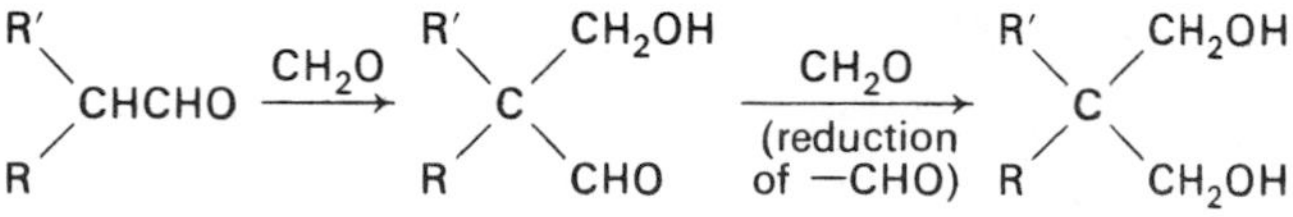

$$R'RCHCHO \xrightarrow{CH_2O} R'RC(CH_2OH)CHO \xrightarrow[\text{(reduction of }-CHO)]{CH_2O} R'RC(CH_2OH)_2$$

Exercise 4.3

Formulate the reactions for the synthesis of the following compounds (assume availability of starting materials prior to aldol condensation steps):

(*a*) $(CH_3)(CH_3CH_2)C(CH_2OH)_2$ (*b*) $CH_3-C(CH_2OH)_3$

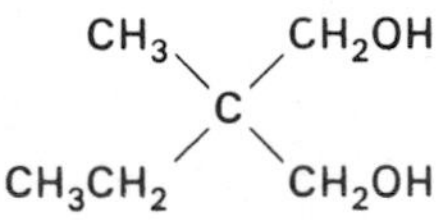

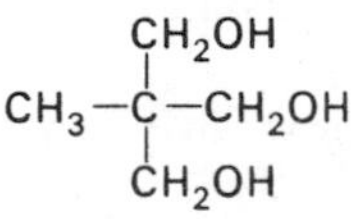

(c) $CH_3CH(OH)CH_2CH_2OH$

(d) [structure]

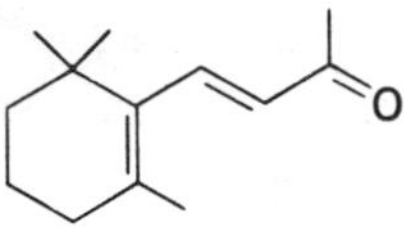

(e) 2-furyl–CH=CHCOCH=CH–2-furyl

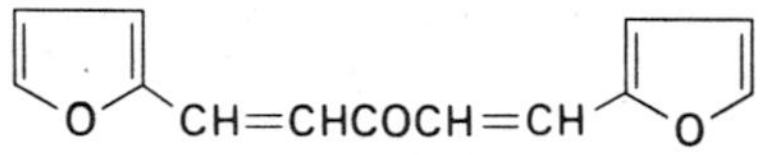

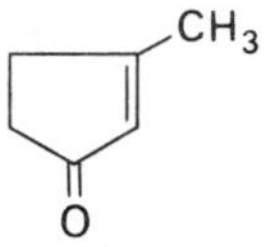

(f) $C_6H_5CH{=}C(CN)C_6H_5$

(g) [structure: CH_3, O]

(h) [structure: C_6H_5, C_6H_5, C_6H_5, C_6H_5, O]

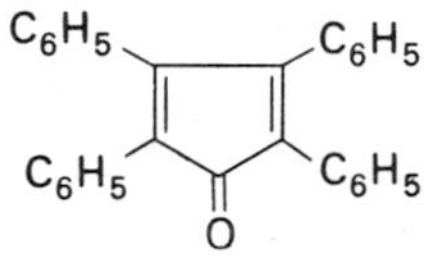

Exercise 4.4

Show how the aldol condensation (with benzaldehyde) can be employed as a diagnostic test for structure analysis, by describing how it can be used to distinguish between

(a) [structure: CH_3, O] and [structure: CH_3, O]

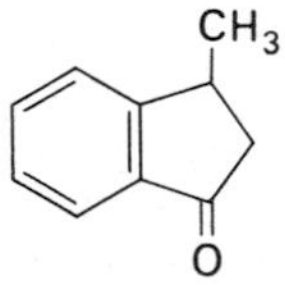

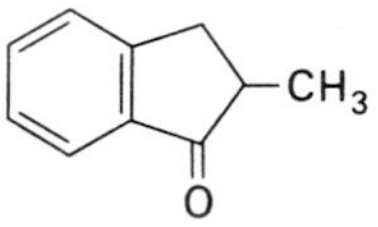

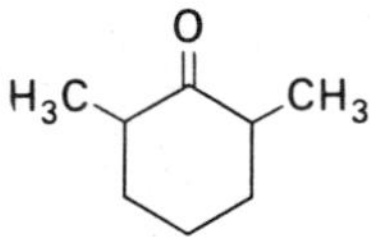

(b) [structure: H_3C, O, CH_3] and [structure: O, CH_3, CH_3]

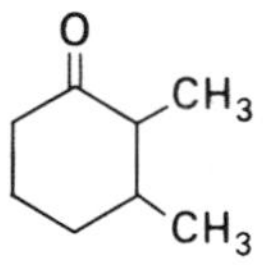

(c) $(CH_3)_2CHCOCH(CH_3)_2$ and $CH_3COCH(CH_3)CH_2CH_2CH_2CH_3$

(d) [structure: $COCH_3$, H_3C, CH_3] and [structure: $COCH(CH_3)_2$]

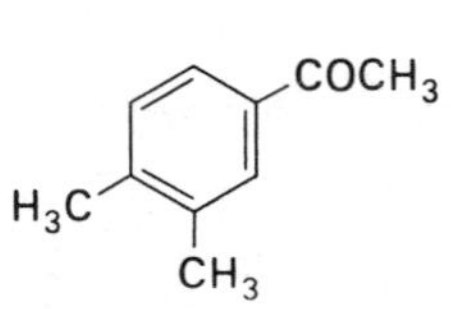

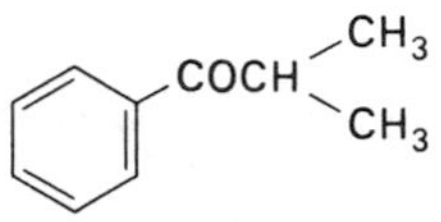

C. THE CLAISEN CONDENSATION

When the α-carbon anion attacks the carbonyl group of an ester, the reaction is known as the Claisen condensation; it proceeds as follows:*

* The ethyl ester is used in the example for simplicity in formulation. However, other esters may be used.

$$\mathrm{RCOCH_2:^- + \underset{OEt}{\overset{R'}{C}}{=}O \rightleftharpoons RCOCH_2{-}\underset{OEt}{\overset{R'}{C}}{-}O^-} \qquad \text{(C-1)}$$

$$\mathrm{RCOCH_2{-}\underset{OEt}{\overset{R'}{C}}{-}O^- \rightleftharpoons RCOCH_2\overset{R'}{C}{=}O + OEt^-} \qquad \text{(C-2)}$$

$$\mathrm{RCOCH_2COR' + OEt^- \rightleftharpoons \{RCO\ddot{C}HCOR'\}^- + EtOH} \qquad \text{(C-3)}$$

Final acidification of the reaction mixture gives the diketone:

$$\mathrm{\{RCOCHCOR'\}^- + H_3O^+ \longrightarrow RCOCH_2COR' + H_2O} \qquad \text{(C-4)}$$

When R = OEt and R′ = CH_3, Equations C-1–C-4 represent the self-condensation of ethyl acetate to yield ethyl acetoacetate. Although the Claisen condensation is in principle reversible, the final deprotonation shown in Equation C-3 leads to a product (which is now a highly stabilized anion of a β-keto ester or β-diketone) that is not susceptible to nucleophilic attack, and thus the reaction proceeds essentionally to completion to the right. Acidification of the final reaction mixture (Equation C-4) reforms the β-keto ester or ketone from its salt and permits its isolation by conventional means (e.g., extraction with ether).

When one of the participating reagents in the Claisen condensation is an ester that *lacks* α-hydrogen atoms, it is clear that this can act only as the *acceptor* of anionic attack. Examples of such esters are

(a) esters of aromatic acids, ArCOOR;
(b) esters of oxalic acid, ROOC-COOR;
(c) esters of formic acid, HCOOR;
(d) carbonic acid esters, $(RO)_2CO$.

Examples of the aldol-like Claisen condensation involving some of the above compounds are the following carbon-acylation reactions. The stepwise course of these reactions is the course described in Equations C-1–C-4:

$$\mathrm{C_6H_5COCH_3 + EtO^- \rightleftharpoons C_6H_5COCH_2:^- + EtOH}$$

$$\mathrm{C_6H_5COCH_2:^- + HCOOEt \longrightarrow C_6H_5COCH_2CHO + EtO^-}$$

$$\mathrm{CH_3CH_2COOEt + EtO^- \rightleftharpoons \{CH_3\ddot{C}HCOOEt\}^- + EtOH}$$

$$\mathrm{\{CH_3\ddot{C}HCOOEt\}^- + \underset{COOEt}{COOEt} \longrightarrow CH_3\overset{COOEt}{CH}{-}COCOOEt + EtO^-}$$

$$\text{cyclohexanone} + \mathrm{(COOEt)_2} \xrightarrow{\mathrm{NaOEt}} \text{2-(COCOOEt)cyclohexanone}$$

$$C_6H_5CH_2CN + (COOEt)_2 \xrightarrow{NaOEt} C_6H_5\underset{\displaystyle COCOOEt}{\underset{|}{C}}HCN$$

$$2\ CH_3COOEt \xrightarrow{NaOEt} CH_3COCH_2COOEt$$

Exercise 4.5

Formulate the following Claisen condensations. In each case show (1) the formation of the α-carbon anion, and (2) the attack of this anion upon the carbonyl group of the ester:*

(*a*) $CH_3COCH_3 + CH_3COOEt(NaOEt) \longrightarrow$
(*b*) $C_6H_5COCH_2CH_3 + HCOOCH_3(NaOCH_3) \longrightarrow$
(*c*) Cyclopentanone + diethyl carbonate (NaOEt) $\longrightarrow$
(*d*) $C_6H_5COCH_3 + (COOEt)_2(NaOEt) \longrightarrow$
(*e*) $2\ CH_3CH_2COOCH_3(NaOCH_3) \longrightarrow$
(*f*) $CH_3COCH_3 + HCOOEt(NaOEt) \longrightarrow$
(*g*) $CH_3COCH_3 + (COOMe)_2(NaOMe) \longrightarrow$
(*h*) $C_6H_5CH_2CN + CH_3COOEt(NaOEt) \longrightarrow$

* The basic catalyst (a sodium alkoxide) is shown in parentheses. In formulating the reactions, only the alkoxide anion (EtO^- or MeO^-) need be written in the equation.

D. RING CLOSURE REACTIONS INVOLVING VARIANTS OF ALDOL-LIKE CONDENSATIONS. THE DIECKMANN REACTION

When the two reacting species—the anionic carbon atom and the electrophilic carbonyl carbon atom—*are parts of the same molecule,* their reaction may serve to bring about a cyclization reaction (i.e., a ring closure). One of the commonest forms in which this reaction is encountered is the Dieckmann reaction, which is an intramolecular Claisen condensation. For example:

$H_2C(CH_2COOEt)$–H_2C–CH_2–$COOEt$ $\underset{}{\overset{EtO^-}{\rightleftharpoons}}$ $H_2C(\bar{C}HCOOEt)$–H_2C–CH_2–$COOEt$ $\rightleftharpoons$ H_2C—$CHCOOEt$, H_2C–CH_2–CO (ring) + EtO^-

$\rightleftharpoons$ {cyclopentanone-2-carboxylate anion, COOEt, $=\ddot{O}:$ $\longleftrightarrow$ COOEt, $-\ddot{O}:^-$} + EtOH

Final acidification of the reaction mixture yields the final ester, ethyl cyclopentanone-

2-carboxylate, in good yield. Other examples are the following:

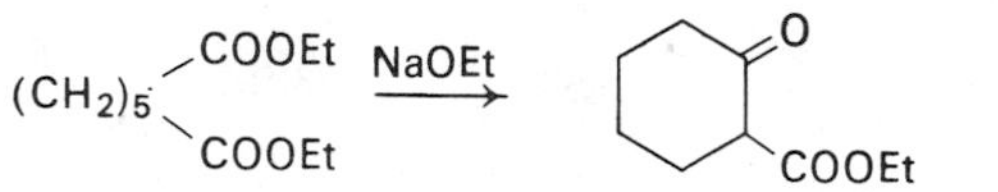

$(CH_2)_5(COOEt)_2$ —NaOEt→ ethyl cyclohexanone-2-carboxylate

diethyl pimelate

ethyl cyclohexanone-2-carboxylate

EtOOC, COOEt, CH_3, O —NaOEt→ EtOOC, O, O

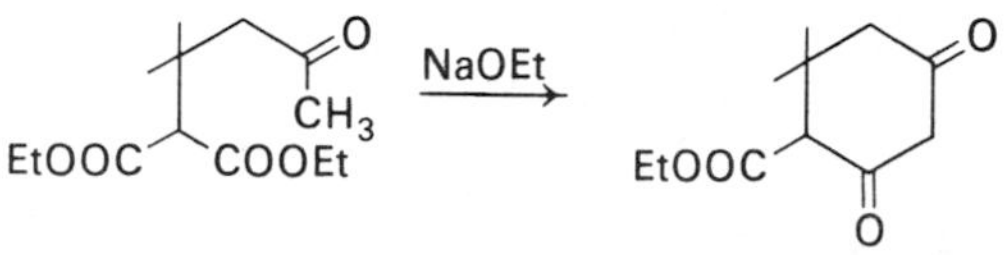

a variant in which the methyl group of $—COCH_3$ is the anionic attacking group

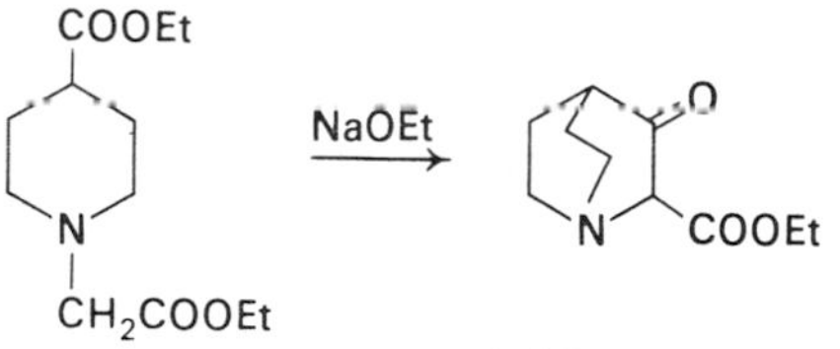

COOEt, N, CH_2COOEt —NaOEt→ N, COOEt, O

ethyl 3-quinuclidone-2-carboxylate

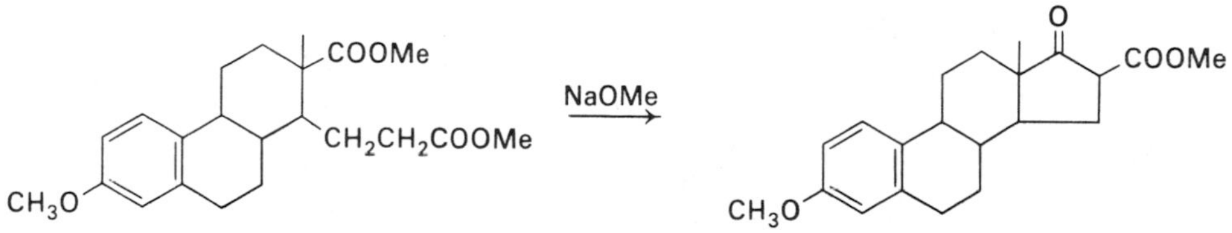

CH_3O, COOMe, CH_2CH_2COOMe —NaOMe→ CH_3O, O, COOMe

Exercise 4.6

Starting with the component(s) necessary for carrying out a Claisen or Dieckmann ester condensation, show how each of the following compounds could be prepared:

(*a*) $C_6H_5COCH(CH_3)COOEt$

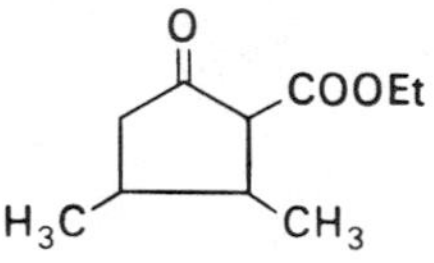

(*b*) $C_6H_5CH_2COCH(C_6H_5)COOEt$

(*c*) O, COOEt, H_3C, CH_3

(*d*) O, CHO

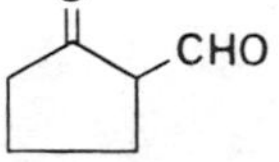

(*e*) O, $COCOOEt$

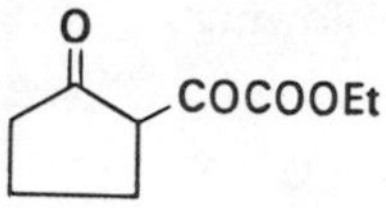

(*f*) C_6H_5, O, O, O, C_6H_5

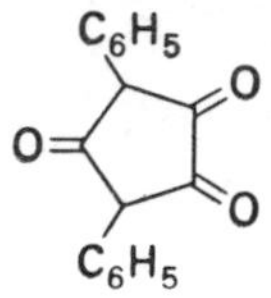

(*g*) O, $COOEt$, N, CH_3

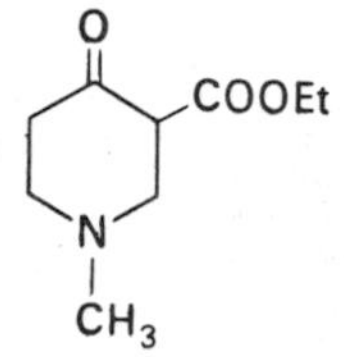

(*h*) $COOEt$, O, O, $COOEt$

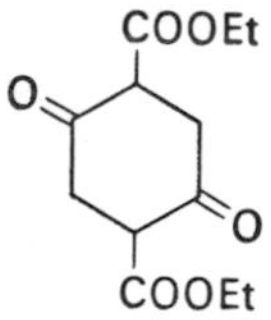

(*i*) $COOEt$, O, O, $COOEt$

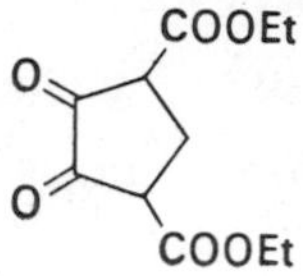

(*j*) $C_6H_5CH_2CHCOOEt$ (with CHO on the CH)

(*k*) $C_6H_5CH_2CHCOCOOEt$ (with $COOEt$ on the CH)

(*l*) O, O, O

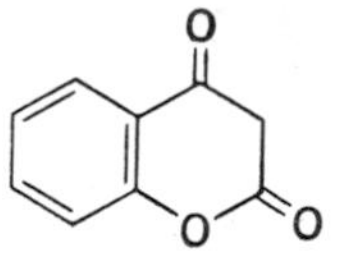

E. ADDITIONAL CLASSES OF ALDOL-LIKE CONDENSATIONS

(a) The Perkin reaction

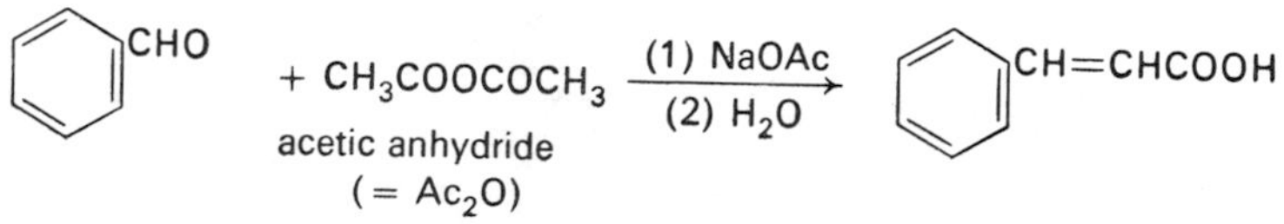

$$C_6H_5CHO + \underset{\text{acetic anhydride } (= Ac_2O)}{CH_3COOCOCH_3} \xrightarrow[(2)\ H_2O]{(1)\ NaOAc} C_6H_5CH{=}CHCOOH \quad \text{(E-1)}$$

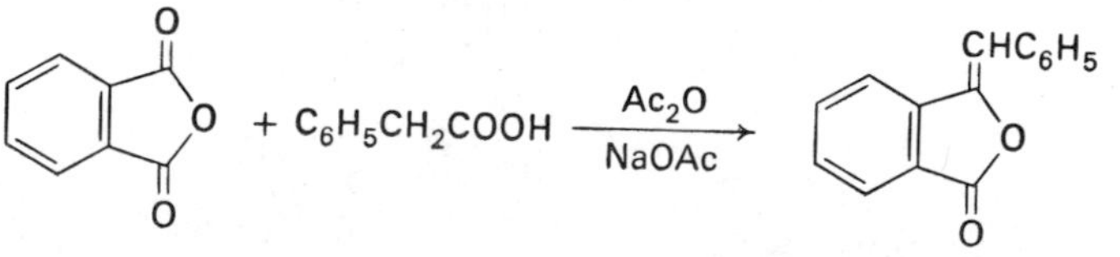

Phthalic anhydride + $C_6H_5CH_2COOH$ $\xrightarrow[NaOAc]{Ac_2O}$ benzylidenephthalide (CHC_6H_5, O, O) (E-2)

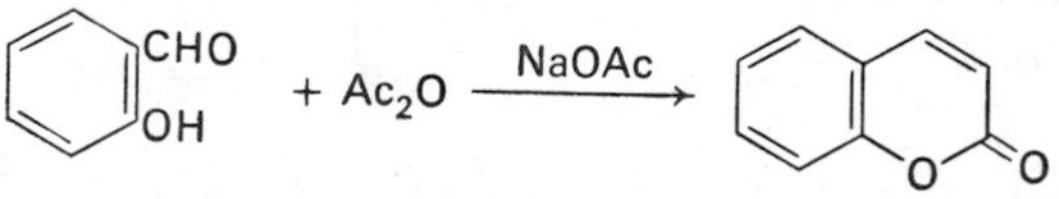

Salicylaldehyde (CHO, OH) + Ac_2O $\xrightarrow{NaOAc}$ coumarin (O, O) (E-3)

Exercise 4.7

Formulate the steps in the above condensations, Equations E-1–E-3, showing the generation of the nucleophilic reagent, its attack upon the carbonyl group, and a plausible scheme for completion of the overall reaction. In all of the examples formulated in these equations, the acetate ion is the basic reagent (that is, $B:^-$ in the general equations written earlier).

(b) The Stobbe condensation

$$\text{cyclohexanone} + \begin{array}{l}CH_2COOEt\\ |\\ CH_2COOEt\end{array} \xrightarrow{KOtBu} \text{cyclohexenyl}-\begin{array}{l}CHCOOEt^{*}\\ |\\ CH_2COOH\end{array} \qquad \text{(E-4)}$$

$$C_7H_{15}CHO + \begin{array}{l}CH_2COOEt\\ |\\ CH_2COOEt\end{array} \xrightarrow{KOtBu} C_7H_{15}CH{=}\overset{\displaystyle COOEt}{\overset{|}{C}}CH_2COOH \qquad \text{(E-5)}$$

Exercise 4.8

The Stobbe condensation is formulated as proceeding by way of (a) an initial aldol condensation, followed by (b) the formation of a lactone, and (c) final opening of the lactone to give the final product. Formulate Reaction E-4 in this way.

(c) Condensations involving malonic acid derivatives

Malonic acid (X = Y = OH) and its esters (X = Y = OR), as well as β-keto acids and their esters, possess a double-activated α-carbon atom

$$\begin{array}{l}Y{-}CO\\ \quad\quad\diagdown\\ \quad\quad\quad CH_2\\ \quad\quad\diagup\\ X{-}CO\end{array}$$

which is readily converted to the carbon anion by bases of moderate strength. The condensation of malonic acid with aldehydes can be carried out under mild conditions, often with the use of such relatively weak bases as pyridine and piperdine. Depending upon the reaction conditions, the initial condensation product may lose CO_2 to give as a final product the α-β-unsaturated acid:

$$\begin{array}{l}HOOC\\ \quad\diagdown\\ \quad\quad CH_2\\ \quad\diagup\\ HOOC\end{array} + RCHO \xrightarrow{\text{pyridine}} \left\{ RCH{=}C\begin{array}{l}\diagup COOH\\ \diagdown COOH\end{array} \right\} \xrightarrow{-CO_2} RCH{=}CHCOOH \qquad \text{(E-6)}$$

can be isolated — usual final product

This reaction, called the *Doebner reaction*, is often superior to the Perkin reaction. It proceeds under mild conditions and thus is applicable to aliphatic aldehydes that would not be stable under the usually vigorous conditions of the Perkin condensation: and it usually gives excellent yields.

---

* Note Equation E-7 and the sentence that follows it.

The related *Knoevenagel reaction* consists in the condensation of a malonic ester with a ketone or aldehyde in the presence of piperidine:

$$\text{cyclohexanone} + CH_2(COOEt)_2 \xrightarrow{\text{base}^*} \text{cyclohexylidene}{=}C(COOEt)_2 \rightleftarrows \text{cyclohexenyl}{-}CH(COOEt)_2 \qquad \text{(E-7)}$$

* basic catalyst = piperidine

Base-catalyzed double bond isomerization, as shown, can occur, and the product is often a mixture of the isomers.

Exercise 4.9

Show a practical means for the synthesis of the following compounds. Use a reaction of the general aldol type *at some stage* in the synthesis. Starting materials may be any compounds that you might reasonably expect to be available in a well-stocked storeroom. Compounds given by name should be looked up in a reference work or textbook:

(*a*) β-ionone from β-cyclocitral

(*b*) $CH_3CH(OH)CH_2CH_2OH$

(*c*) $C_6H_5CHBrCHBrCOOH$

(*d*) dicyclohexyl

(*e*) $(CH_3)_2CHCH_2COCH_3$

(*f*) $(CH_3)_2C{=}CHCOOH$

(*g*) $C_6H_5CH{=}CHCOOH{=}CHC_6H_5$

(*h*) 1-indanone, starting with benzaldehyde

(*i*) $C_6H_5CH_2CH(CH_3)COCH_2CH_3$

(*j*) $C_6H_5CH_2COCH_3$

F. THE MICHAEL REACTION AND RELATED REACTIONS

The addition of nucleophilic reagents to α,β-unsaturated carbonyl compounds (ketones, esters, nitriles) is a reaction of wide and useful application. It bears a kinship to the aldol condensation:

$$Y{:}^- + C{=}C{-}C{=}O \rightleftarrows \left\{ \left[Y{-}C{-}C{=}C{-}O \right]^- \leftrightarrow Y{-}C{-}\ddot{C}{-}C{=}O \right\} \xrightarrow{HA} Y{-}C{-}CH{-}C{=}O \qquad \text{(F-1)}$$

Compare with

$$Y{:} + C{=}O \rightleftarrows Y{-}C{-}O^- \overset{HA}{\rightleftarrows} Y{-}C{-}OH$$

Nucleophiles of many types are capable of such base-catalyzed addition; for example, amines:

$$(CH_3)_2NH + CH_2{=}CHCOCH_3 \longrightarrow (CH_3)_2NCH_2CH_2COCH_3 \qquad \text{(F-2)}$$

alcohols (alkoxides):

$$ROH + CH_2{=}CHCN \longrightarrow ROCH_2CH_2CN \qquad \text{(F-3)}$$

thiols (RSH; RS^-):

$$RSH + CH_2{=}CHCN \longrightarrow RSCH_2CH_2CN \qquad \text{(F-4)}$$

active methylene compounds (carbon anions):

$$(EtOOC)_2CH_2 + C_6H_5CH{=}CHCOOEt \longrightarrow C_6H_5\underset{\displaystyle CH(COOEt)_2}{\underset{|}{C}}HCH_2COOEt \qquad \text{(F-5)}$$

$$C_6H_5COCH_3 + C_6H_5CH{=}CHCOC_6H_5 \longrightarrow C_6H_5\underset{\displaystyle CH_2COC_6H_5}{\underset{|}{C}}HCH_2COC_6H_5 \qquad \text{(F-6)}$$

$$\text{2-methylenecyclohexanone } (=O,\ =CH_2) + CH_3NO_2 \longrightarrow \text{2-}(CH_2CH_2NO_2)\text{cyclohexanone} \qquad \text{(F-7)}$$

The latter of these (F-5–F-7) are designated as the *Michael reaction.* Examples F-2 and F-3 are often referred to as "Michael-like" reactions. There is no clear mechanistic distinction betweeen them.

Examples (including further useful transformations of products):

$$CH_2{=}CHCN + KCN \xrightarrow{EtOH} \underset{}{\overset{\displaystyle CN}{\overset{|}{C}}}H_2CH_2CN \xrightarrow{H_2O/H^+} HOOCCH_2CH_2COOH$$

$$\overset{}{C}H_2CH_2CN \text{ (cyanide adduct)} \xrightarrow{LiAlH_2} H_2NCH_2CH_2CH_2CH_2NH_2$$

$$CH_3NH_2 + CH_2{=}CHCOOCH_3 \longrightarrow CH_3N(CH_2CH_2COOCH_3)_2$$

$$CH_3N(CH_2CH_2COOCH_3)_2 \xrightarrow{\text{Dieckmann}} \text{1-}CH_3\text{-4-oxo-piperidine-3-COOMe}$$

$$(CH_3)_2C{=}CHCOCH_3 + CH_2(COOEt)_2 \xrightarrow{NaOEt} (CH_3)_2\underset{\substack{|\\CHCOOEt\\|\\COOEt}}{C}CH_2COCH_3 \xrightarrow{NaOEt}$$

[cyclic diketo ester: 5,5-dimethyl-4-carbethoxycyclohexane-1,3-dione] $\xrightarrow[(-CO_2)]{H_2O/H^+}$ [5,5-dimethylcyclohexane-1,3-dione]

dimedon

$$C_6H_5CH{=}CHCOOEt + C_6H_5CH_2COOEt \xrightarrow{NaOEt}$$

$$\begin{matrix} C_6H_5CHCH_2COOEt \\ | \\ C_6H_5CHCOOEt \end{matrix} \xrightarrow{\text{saponify}} \alpha,\beta\text{-diphenylglutaric acid}$$

[methyl vinyl ketone] + [2-methylcyclohexanone] $\xrightarrow[A]{tBuO^-}$ [2-methyl-2-(3-oxobutyl)cyclohexanone] $\xrightarrow[B]{NaOH}$ [bicyclic enone (octalone)]

(A Michael reaction, *A*, followed by a cyclic aldol condensation, *B*.)

$$C_6H_5CH{=}CHCOC_6H_5 + \text{[cyclopentadiene]} \xrightarrow{\text{piperidine}} C_6H_5\underset{\substack{|\\ \text{[cyclopentadienyl]}}}{C}HCH_2COC_6H_5$$

(Note that the $-CH_2-$ group of cyclopentadiene is "active" in the aldol and Michael condensations.)

Exercise 4.10

With the above examples as guides, formulate the following reactions. (In all, assume the use of an appropriate basic catalyst):

(*a*) $C_6H_5CH{=}CHCOCH_3 + CH_3CH(COOEt)_2 \longrightarrow$

(*b*) $CH_3CH{=}CHCOOEt + CH_3COCH_2COOEt \longrightarrow$

(*c*) $(C_6H_5CH_2)_2NH + CH_2{=}CHCN \longrightarrow$

(*d*) $CH_3OH + CH_3CH_2CH{=}CHCOOCH_3(NaOCH_3) \longrightarrow$

(*e*) $CH_3OCH_2CH_2COOEt + EtOH(NaOEt) \longrightarrow$

(*f*) [cyclohexanone] $+ C_6H_5COCH{=}CH_2 \longrightarrow$

(*g*) $C_6H_5COCH_2CH_2N(CH_3)_3{}^+I^- + CH_2(COOEt)_2 \longrightarrow$

(*h*) $CH_3CH{=}CHCOOEt + CH_3CH(COOEt)_2$ + trace of NaOEt $\longrightarrow$

(*i*) 2-methylenecyclohexanone + $CH_2(CN)(COOEt)$ $\longrightarrow$

(*j*) $C_6H_5CH{=}CHCOC_6H_5 + KCN \longrightarrow$

(*k*) When the reaction in (*h*) is carried out with the use of a 1 molar equivalent of NaOEt, the product is

$$CH_3CH(CH(COOEt)_2)-C(CH_3)HCOOEt.$$

Explain how this is formed. Is it reasonable to suppose that the methylmalonic ester adds as the fragments CH_3- and $-CH(COOEt)_2$?

G. THE MANNICH REACTION

Beta dialkylaminoketones can be prepared by a reaction that has the essential character of an aldol condensation. When a ketone is treated with an amine and an aldehyde in the presence of suitable acid-base catalysts, the following overall result (for example) is obtained:

$$C_6H_5COCH_3 + HCHO + Me_2\overset{+}{N}H_2\overset{-}{Cl} \longrightarrow C_6H_5COCH_2CH_2NMe_2{\cdot}HCl \qquad \text{(G-1)}$$

A simple and adequate description of the course of events that lead to this result is the following:

$$C_6H_5COCH_3 + B{:}^- \rightleftharpoons C_6H_5COCH_2{}^-{:} + BH \qquad \text{(G-2)}$$

$$MeNH_2 + HCHO \rightleftharpoons MeNCH_2OH \overset{H^+}{\rightleftharpoons} Me_2\overset{+}{N}{=}CH_2 + H_2O \qquad \text{(G-3)}$$

$$C_6H_5COCH_2{:}^- \longrightarrow CH_2{=}\overset{+}{N}Me_2 \rightleftharpoons C_6H_5COCH_2CH_2NMe_2 \qquad \text{(G-4)}$$

The anionic participant in the Mannich reaction need not be a ketone (such as $C_6H_5COCH_3$ in Example G-1). It may be a phenol, an aldehyde, a β-keto acid, a heterocyclic base or an acetylene. The following examples all represent reactions that may be included under the term Mannich reaction (or Mannich condensation):

$$\text{3,4-}(HO)_2C_6H_3CH_2CH_2NHCH_3 + HCHO \longrightarrow \text{6,7-dihydroxy-2-methyl-1,2,3,4-tetrahydroisoquinoline} \qquad \text{(G-5)}$$

$$\text{(indole)} + Me_2NH + HCHO \longrightarrow \text{(3-}CH_2NMe_2\text{-indole)} \qquad \text{(G-6)}$$

$$CH_3COCH_2COOH + Me_2NH + HCHO \longrightarrow CH_3COCH_2CH_2NMe_2 + CO_2 \qquad \text{(G-7)}$$

$$CH_3CHO + CH_3NH_2 + HCHO \longrightarrow \left\{ CH_3N\begin{matrix} CH_2CH_2CHO \\ CH_2CH_2CHO \end{matrix} \right\} \longrightarrow \text{(N-}CH_3\text{-tetrahydropyridine-3-}CHO) \qquad \text{(G-8)}$$

$$\text{(pyrrolidine)}NCH_2C{\equiv}CH + \text{(pyrrolidine, N–H)} + HCHO \longrightarrow \text{(pyrrolidine)}NCH_2C{\equiv}CCH_2N\text{(pyrrolidine)} \qquad \text{(G-9)}$$

$$(CH_3)_2CHNO_2 + Me_2NH + HCHO \longrightarrow (CH_3)_2C(NO_2)CH_2NMe_2 \qquad \text{(G-10)}$$

$$\text{(thiophene)} + Me_2NH + HCHO \longrightarrow \text{(2-}CH_2NMe_2\text{-thiophene)} \qquad \text{(G-11)}$$

Exercise 4.11

Write the detailed sequence of stages through which each of the reactions in (G-5–G-11) proceed. Identify the anionic participant that corresponds to $C_6H_5COCH_2{:}^-$ in (G-2) above.

The products of these reactions are commonly called "Mannich bases." They have many synthetic uses:

(a) They can be N-alkylated to yield the corresponding trialkylammonium compounds:

$$RCOCH_2CH_2NMe_2 + MeI \longrightarrow RCOCH_2CH_2\overset{+}{N}Me_3I^-$$

(b) The trialkylammonium compounds readily lose NR_3 to give the α,β-unsaturated carbonyl compounds:

$$RCOCH_2CH_2\overset{+}{N}Me_3I^- \xrightarrow[\text{or } B{:}^-]{\Delta} RCOCH{=}CH_2 + R_3N$$

(c) The carbonyl group undergoes addition reactions; for example, with Grignard reagents:

$$RCOCH_2CH_2NR'_2 + R''MgX \longrightarrow R-\underset{OH}{\overset{R''}{C}}CH_2CH_2NR'_2$$

(d) They can be used directly (in the quaternized form) as one of the addends in the Michael reaction:

$$RCOCH_2CH_2\overset{+}{N}Me_3 + CH_2(COOEt)_2 \xrightarrow{NaOEt} RCOCH_2CH_2CH(COOEt)_2 + NMe_3$$

(e) The $-\overset{+}{N}R_3$ grouping in some quaternized Mannich bases is displaceable in nucleophilic replacement reactions.

Exercise 4.12

The following reaction has been observed:

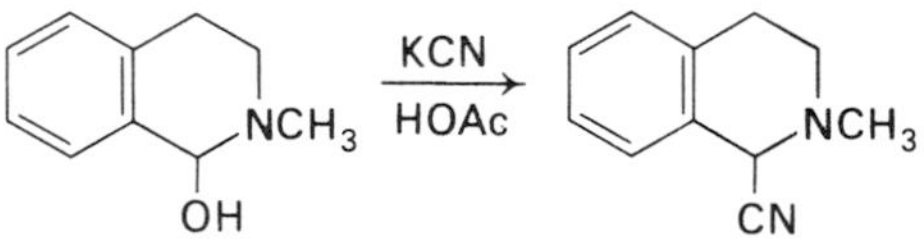

NCH3
OH
KCN
HOAc
NCH3
CN

How does this reaction relate to the Mannich reaction? Formulate it in detail, showing a typical Mannich reaction for comparison of the relevant steps.

Exercise 4.13

Write equations for the Mannich reactions and any other necessary steps that would lead to the following compounds:

(*a*) 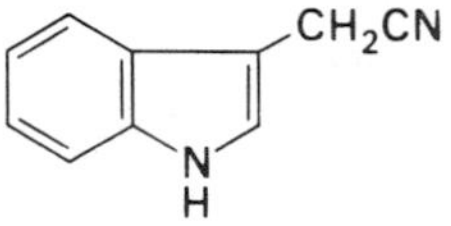

CH2CN
N
H

(*b*) 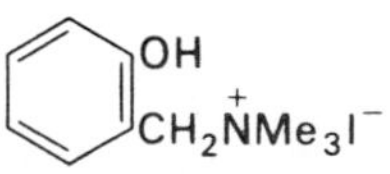

OH
CH2NMe3I⁻

(*c*) $CH_3-\underset{CH_3}{\overset{OH}{C}}CH_2CH_2NMe_2$

(*d*) 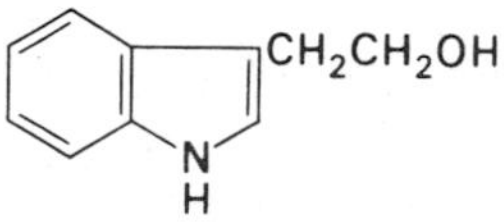

CH2CH2OH
N
H

(*e*) 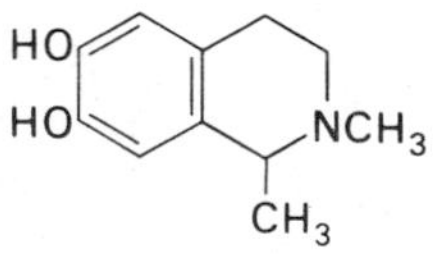

HO
HO
NCH3
CH3

(*f*) 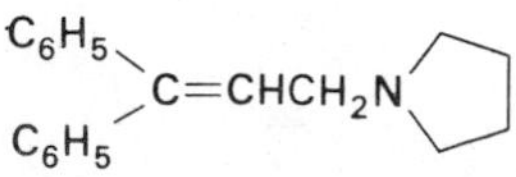

C6H5
C6H5
C=CHCH2N

(g) $\bigcirc NCH_2CH{=}CHCH_2N\bigcirc$ (pyrrolidine rings) (h) $CH_3-\overset{CH_3}{\underset{NH_2}{C}}CH_2NMe_2$

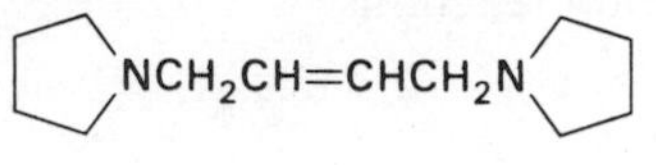

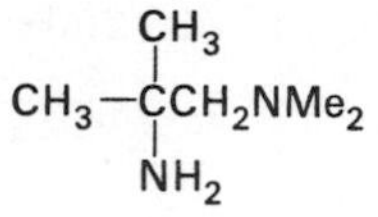

Comments on answers to some questions of exercise 4.13

(a) Since indole undergoes a Mannich reaction (Equation G-6), the compound

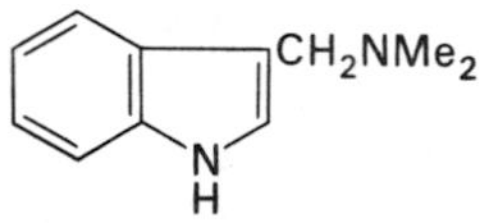

3-(CH_2NMe_2)indole (N–H)

is readily prepared. Conversion of this to the methiodide provides the $-NMe_3^+$ group, which is displaceable by CN^-:

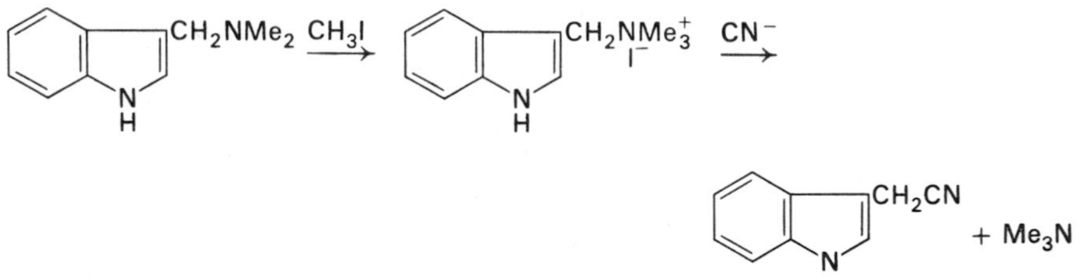

$$\text{indole-3-}CH_2NMe_2 \xrightarrow{CH_3I} \text{indole-3-}CH_2\overset{+}{N}Me_3\ I^- \xrightarrow{CN^-} \text{indole-3-}CH_2CN + Me_3N$$

(c) The presence of the $(CH_3)_2\overset{OH}{C}-$ grouping suggests the synthetic step:

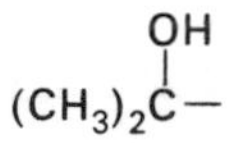

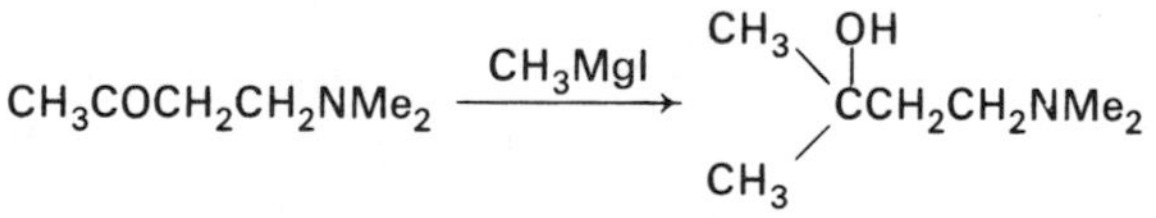

$$CH_3COCH_2CH_2NMe_2 \xrightarrow{CH_3MgI} (CH_3)_2\overset{OH}{C}CH_2CH_2NMe_2$$

The amino ketone is prepared by the Mannich condensation of acetone, formaldehyde, and dimethylamine.

(f) Note that cyclic secondary amines (piperidine, pyrrolidine, morpholine) are as useful in the Mannich reaction as are such amines as Me_2NH, Et_2NH, and so forth.

(g) Acetylenes are sufficiently acidic to act as the anionic reagent in a Mannich reaction (Equation G-9). It is also to be recalled that controlled catalytic hydrogenation of $C{\equiv}C$ gives *cis*- $HC{=}CH$.

PART **5**

Synthetic methods II. The use of Grignard reagents and other organometallic compounds in carbon-carbon bond formation

The formation of carbon-carbon bonds by the use of organometallic compounds is one of the most widely used and generally applicable of synthetic methods. The principal method involves the use of Grignard reagents, which are organomagnesium compounds formed by the direct reaction of magnesium with an alkyl or aryl halide in the presence of an ether (usually diethyl ether):

$$RX + Mg \xrightarrow{\text{ether}} RMgX$$

The symbol "RMgX" is a simplified expression but it serves adequately to represent the reagent in its synthetic applications. Other commonly used organometallic compounds, the behavior of which has much in common with that of Grignard reagents, are organolithium compounds (RLi), organocadmium compounds (R_2Cd or RCdX), and organozinc compounds (R_2Zn or RZnX).

GRIGNARD REAGENTS

The typical reactions of Grignard reagents can be expressed in the following general way:

$$RR'C{=}O + R''MgX \longrightarrow RR'C(R'')OMgX \xrightarrow{H_2O/H^+} RR'C(R'')OH + \text{Mg salts} \qquad (1)$$

(1) When R = R′ = H (for example, as in formaldehyde), the product is $R''CH_2OH$, a primary alcohol.
(2) When R′ = H and R = alkyl or aryl (as in an aldehyde), the product is $R{-}\underset{R''}{\underset{|}{C}}HOH$, a secondary alcohol.
(3) When R and R′ = alkyl or aryl (as in a ketone), the product is $R{-}\overset{OH}{\underset{R'}{C}}{-}R''$, a tertiary alcohol.
(4) When R = R′ = O (for example, carbon dioxide), the product is $R''COOH$, a carboxylic acid.
(5) When R = alkyl or aryl and R′ = alkoxy (as in an ester), the product is a tertiary alcohol, $R{-}\overset{OH}{\underset{R''}{C}}{-}R''$. (Note that with a tertiary alcohol both R″ groups are the same and are derived from R″MgX).
(6) Addition of R″MgX to RC≡N, followed by hydrolysis of $R{-}\overset{R''}{C}{=}NMgX$, gives a ketone, RCOR″.

The generality of the Grignard reactions is wide and subject to only a few limitations.

Esters, RCOOR′, react in two stages: in the first, addition of R″MgX to the carbonyl group and loss of MgX(OR′) leads to a ketone, RCOR″; this then reacts with R″MgX to yield the tertiary alcohol. It is possible, but not usually practicable, to halt the reaction at the ketone stage.

Differing degrees of reactivity of carbonyl groups sometimes permit selective attack at one center if a limited amount of Grignard reagent is used. In general (though much influenced by the degree of substitution in the vicinity of the carbonyl group) the order of reactivity of carbonyl groups is

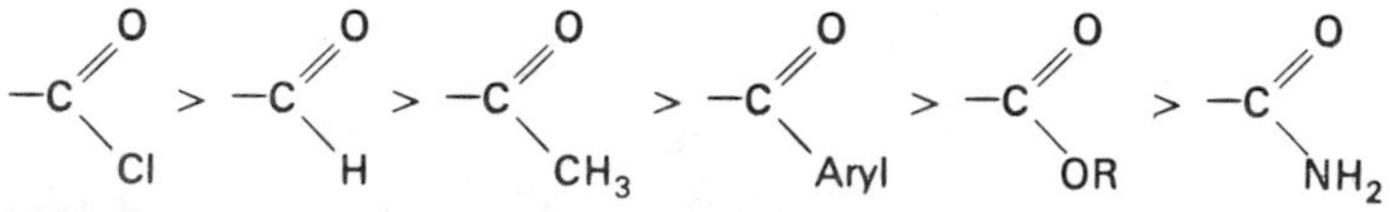

$$-C(=O)Cl > -C(=O)H > -C(=O)CH_3 > -C(=O)Aryl > -C(=O)OR > -C(=O)NH_2$$

Acid halides react very rapidly, the immediate product being the ketone. Under ordinary conditions this then adds RMgX to produce a tertiary alcohol as the final product. But if the amount of Grignard reagent is limited and the reaction is carried out at a low temperature, the intermediate ketone can often be isolated as the principal product.

The *Reformatsky reaction* is allied to the Grignard reaction. It involves the addition of an organozinc compound, prepared *in situ*, to a carbonyl compound. The general expression is the following:

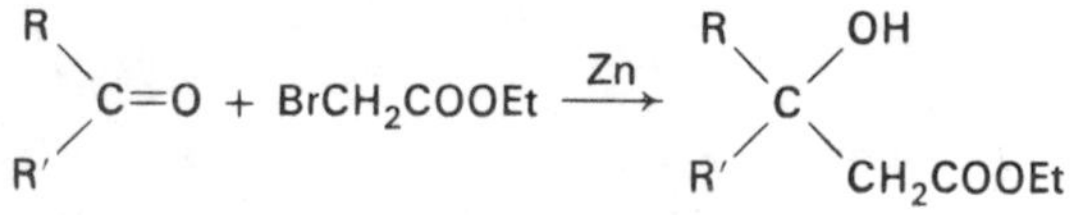

$$RR'C{=}O + BrCH_2COOEt \xrightarrow{Zn} RR'C(OH)CH_2COOEt \quad (via\ BrZnCH_2COOEt) \qquad (2)$$

Organolithium compounds are in many ways more versatile in their synthetic uses than Grignard reagents, but the many aspects of their chemistry will not be discussed here. In their addition to carbonyl compounds they resemble Grignard reagents, and in most cases the two are interchangeable. Simple organolithium compounds (methyllithium, butyllithium, phenyllithium) can be prepared by reaction between the alkyl or aryl halide with metallic lithium:

$$RCl + 2\,Li \longrightarrow RLi + LiCl \tag{3}$$

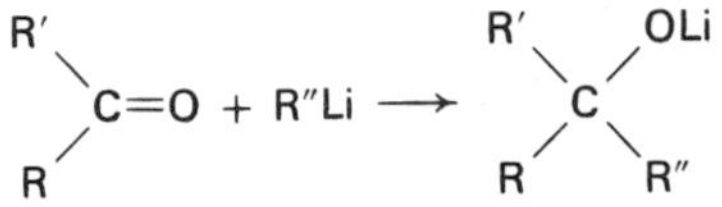

$$R'RC{=}O + R''Li \longrightarrow R'RC(OLi)R'' \tag{4}$$

Exercise 5.1

Show how a Grignard reaction can be used to prepare the following compounds:

(*a*) $CH_3CH_2CH(OH)CH_3$

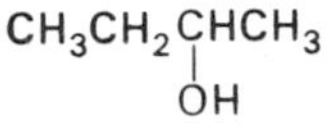

(*b*) $CH_3CH_2CH_2CH(CH_3)CH_2OH$

(*c*) $CH_3-C(OH)(CH_3)-CH_2CH_3$, from an ester

(*d*) $CH_3-C(OH)(CH_3)-CH_2CH_3$, from a ketone

(*e*) $CH_3CH(CH_3)CH_2CH_2COOH$

(*f*) $CH_3CH(CH_3)CH_2CH_2CH(OH)CH_2CH_2CH(CH_3)_2$

(*g*) $(CH_3)_3C-CH(OH)CH_2CH_3$

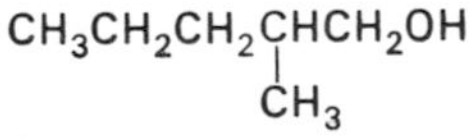

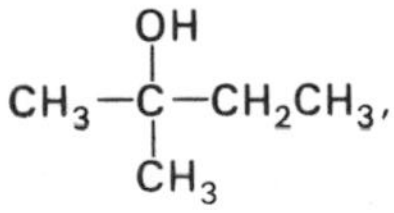

(*h*) $C_6H_5-C(OH)(CH_3)-CH_2CH_2NMe_2$

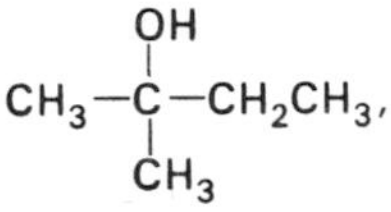

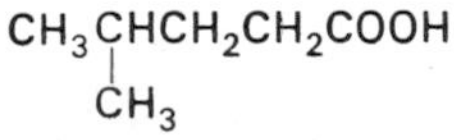

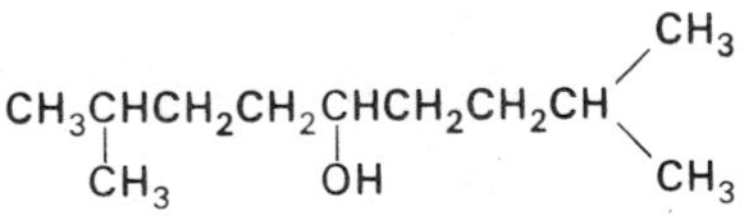

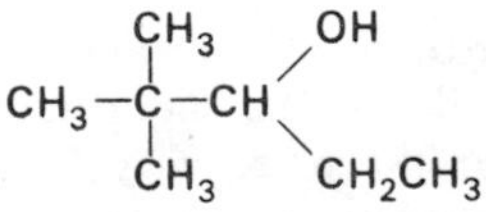

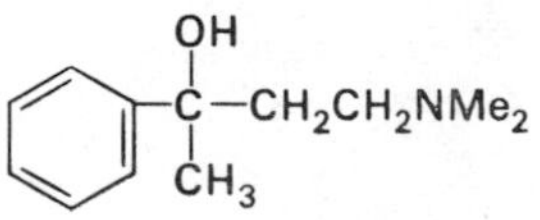

(*i*) $CH_3OCH_2CH_2\overset{\displaystyle OH}{\overset{|}{C}}HCH_2CH_3$

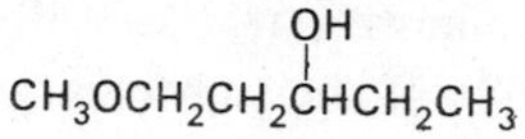

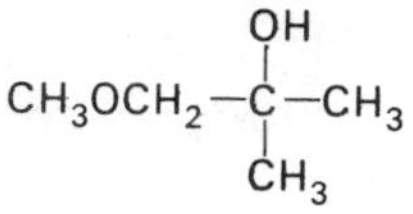

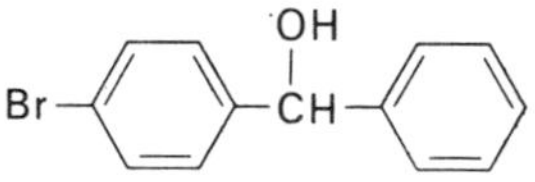

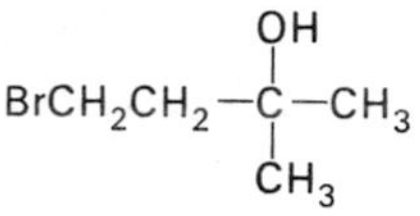

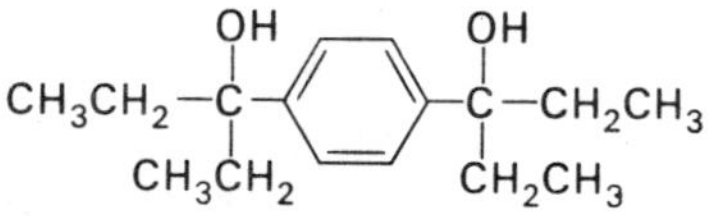

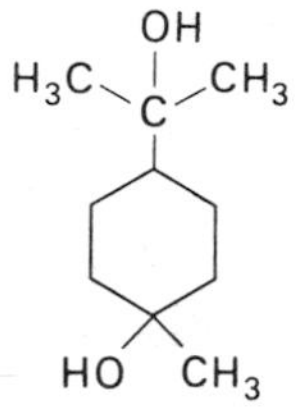

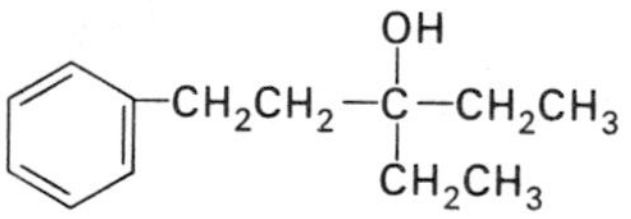

(*j*) $CH_3OCH_2-\underset{\displaystyle CH_3}{\underset{|}{\overset{\displaystyle OH}{\overset{|}{C}}}}-CH_3$

(*k*) $Br-C_6H_4-\overset{\displaystyle OH}{\overset{|}{C}}H-C_6H_5$

(*l*) $BrCH_2CH_2-\underset{\displaystyle CH_3}{\underset{|}{\overset{\displaystyle OH}{\overset{|}{C}}}}-CH_3$

(*m*) $CH_3CH_2-\underset{\displaystyle CH_3CH_2}{\underset{|}{\overset{\displaystyle OH}{\overset{|}{C}}}}-C_6H_4-\underset{\displaystyle CH_2CH_3}{\underset{|}{\overset{\displaystyle OH}{\overset{|}{C}}}}-CH_2CH_3$

(*n*) $H_3C-\overset{\displaystyle OH}{\overset{|}{C}}-CH_3$ attached to a cyclohexane ring bearing HO and CH_3 on the opposite carbon

(*o*) $C_6H_5-CH_2CH_2-\underset{\displaystyle CH_2CH_3}{\underset{|}{\overset{\displaystyle OH}{\overset{|}{C}}}}-CH_2CH_3$

(*p*) $CH_3CH_2-\underset{\displaystyle CH_2CH_3}{\underset{|}{\overset{\displaystyle OH}{\overset{|}{C}}}}-CH_2CH_2-\underset{\displaystyle CH_2CH_3}{\underset{|}{\overset{\displaystyle OH}{\overset{|}{C}}}}-CH_2CH_3$
starting with ethyl malonate and other necessary reagents

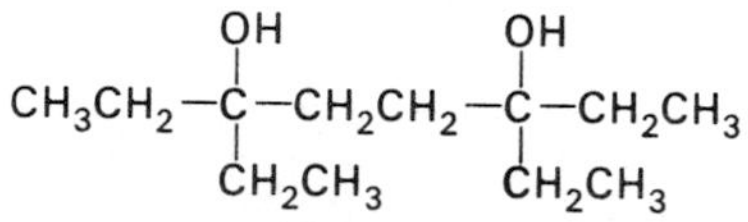

Exercise 5.2

With the use of any appropriate combinations of synthetic methods so far discussed (aldol and related reactions, and organometallic compounds), show how you could synthesize the following compounds. It is assumed that such other operations as oxidations, reductions, dehydrations, etc., will be employed when necessary. *Starting materials should include the one specified* and may include other compounds as required:

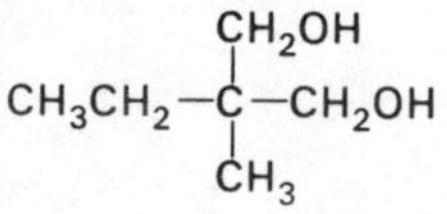

(*a*) $CH_3CH_2-\underset{\displaystyle CH_3}{\underset{|}{\overset{\displaystyle CH_2OH}{\overset{|}{C}}}}-CH_2OH$ from 2-methylbutanal (*b*) $C_6H_5CH_2CH_2-C(OH)(CH_3)-C_6H_5$ from benzaldehyde

(*c*) $C_6H_5CH=C(COC_6H_5)-C_6H_5$ from benzaldehyde

(*d*) $C_6H_5-C(CH_3)=CH_2$ from benzoic acid

(*e*) $C_6H_5-CH(CH_2C(OH)(CH_3)_2)CH_2C(CH_3)_2-OH$ from benzaldehyde

(*f*) $C_6H_5CH_2CH_2NH_2$ from benzaldehyde

(*g*) $(CH_3)_2CHCH=C(CH_3)_2$ from acetone

(*h*) 1-methylcyclohexene (CH_3 on ring) from pimelic acid

(*i*) $C_6H_5-C(OH)(CH_3)CH_2CH_2N$(pyrrolidine ring) from acetophenone

(*j*) 2-(CH_2CH_2COOH)cyclohexanone (ring C=O) from cyclohexanol

(*k*) 2,2,5,5-tetramethyltetrahydrofuran (H_3C, H_3C, O, CH_3, CH_3) from succinic acid

(*l*) 2-($CH_2CH_2COCH_3$)-2-(COOEt)cyclohexanone (ring C=O) from pimelic acid (*m*) [structure: 2-methyl-octahydroisoquinolin-6-one, N–CH_3] from $CH_2{=}CHCOOEt$ and $CH_2{=}CHCOCH_3$

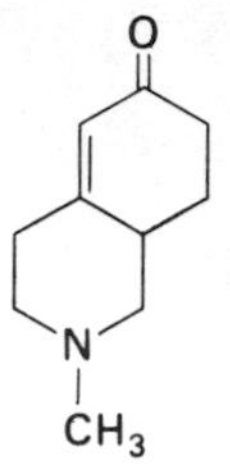

(*n*) [cyclopentane ring bearing CH_2CH_2OH and OH] from cyclopentanone

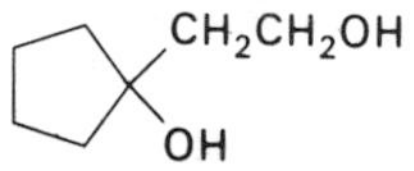

(*o*) $C_6H_5CH(CH_3)CH(OH)CH_2CH_3$ from acetophenone and $ClCH_2COOEt$

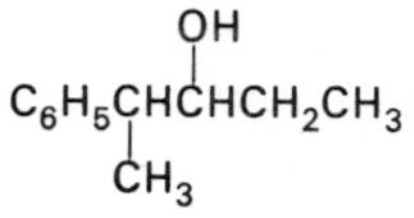

(*p*) $(C_6H_5)_2CHCH_2CH_2NMe_2$ from acetophenone

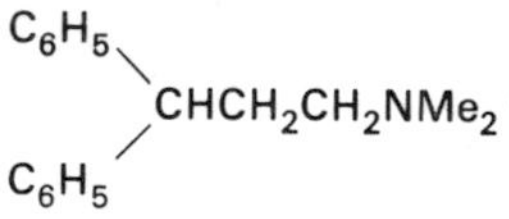

Partial answers to exercise 5.2

(*a*) $CH_3CH_2-CH(CH_3)CHO \xrightarrow[OH^-]{HCHO} CH_3CH_2-C(CH_3)(CH_2OH)-CH_2OH$

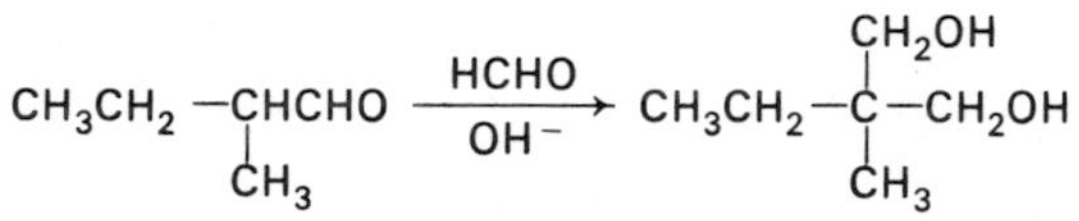

(*b*) $C_6H_5CHO \xrightarrow[\text{pyridine}]{CH_2(COOH)_2} C_6H_5CH{=}CHCOOH \xrightarrow[\text{(2) esterify}]{\text{(1) reduce } (H_2-Pt)}$

$C_6H_5CH_2CH_2COOEt \xrightarrow{C_6H_5MgBr} C_6H_5CH_2CH_2-C(OH)(C_6H_5)-C_6H_5$

(*c*) $C_6H_5CHO \xrightarrow[\text{(2) HCl}]{\text{(1) LiAlH}_4} C_6H_5CH_2Cl \xrightarrow{Mg} C_6H_5CH_2MgCl$

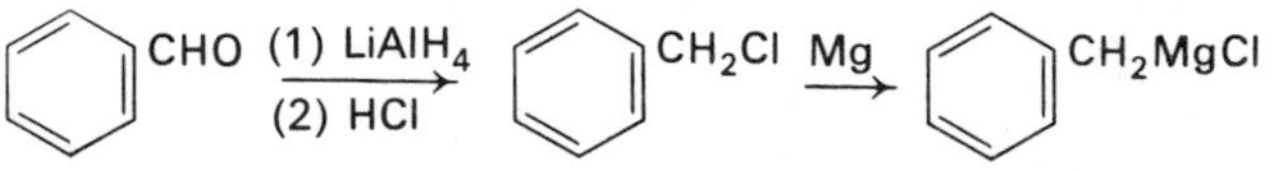

(*i*)

$C_6H_5CH_2MgCl + C_6H_5CHO \longrightarrow C_6H_5CH(OH)CH_2C_6H_5 \xrightarrow{CrO_3}$

$C_6H_5COCH_2C_6H_5$

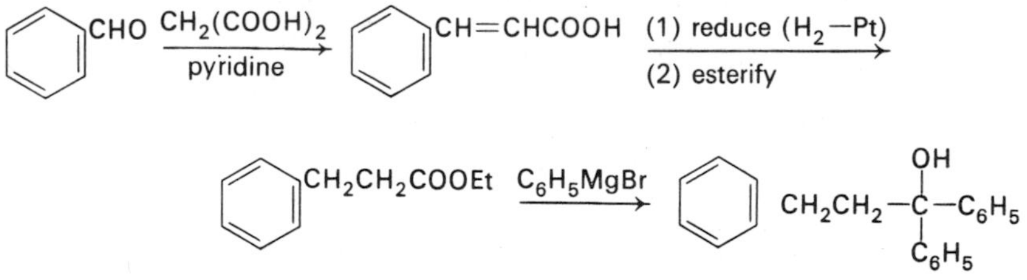

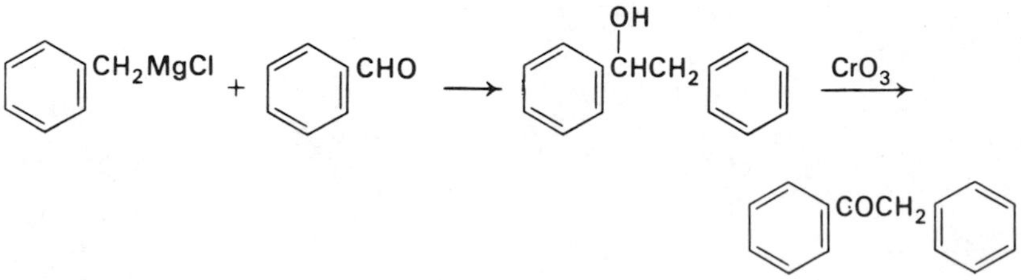

(*ii*)

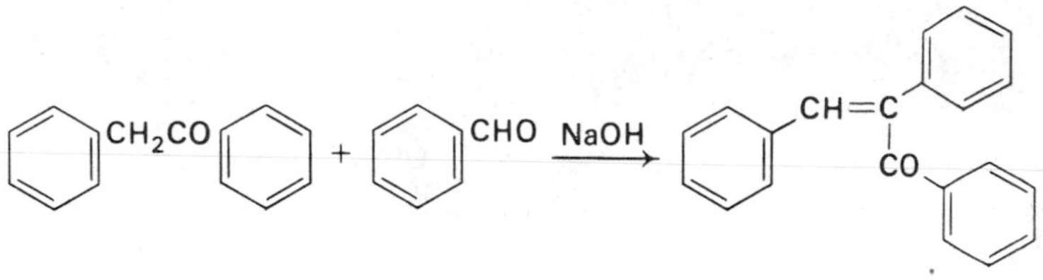

$$C_6H_5CH_2COC_6H_5 + C_6H_5CHO \xrightarrow{NaOH} C_6H_5CH{=}C(C_6H_5)COC_6H_5$$

(*iii*)

(*d*) $$C_6H_5COOH \xrightarrow[HCl]{EtOH} C_6H_5COOEt \xrightarrow{CH_3MgBr} C_6H_5C(CH_3)_2OH \xrightarrow[(-H_2O)]{\Delta,\ H^+} C_6H_5C(CH_3){=}CH_2$$

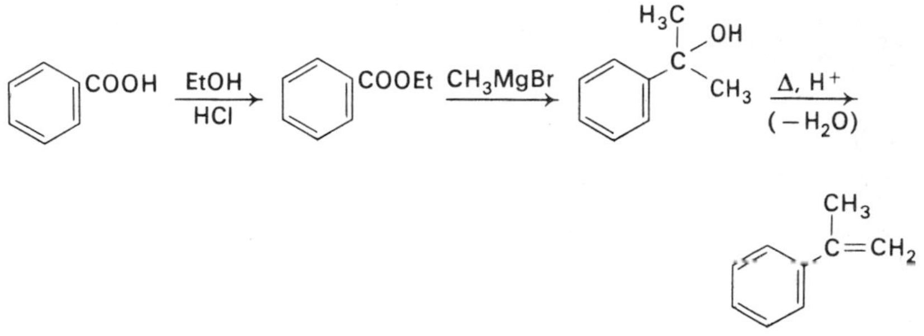

(*e*) $$C_6H_5CHO + CH_3COCH_3 \xrightarrow{OH^-} C_6H_5CH{=}CHCOCH_3$$

(*i*)

add $CH_2(COOEt)_2$ $$\xrightarrow[\text{condensation}]{\text{Michael}} C_6H_5{-}CH(CH(COOEt)_2)CH_2COCH_3$$

(*ii*)

saponify, decarboxylate $$\longrightarrow C_6H_5{-}CH(CH_2COOH)CH_2COCH_3$$

(*iii*)

esterify, then CH_3MgI $$\longrightarrow C_6H_5{-}CH(CH_2C(CH_3)_2OH)CH_2C(CH_3)_2{-}OH$$

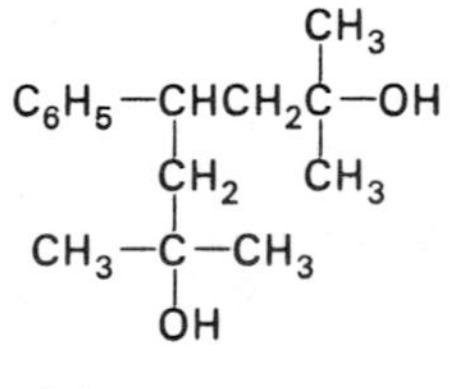

(*iv*)

(*f*) $$C_6H_5CHO + CH_3NO_2 \xrightarrow{NaOH} C_6H_5CH{=}CHNO_2 \xrightarrow{LiAlH_4} C_6H_5CH_2CH_2NH_2$$

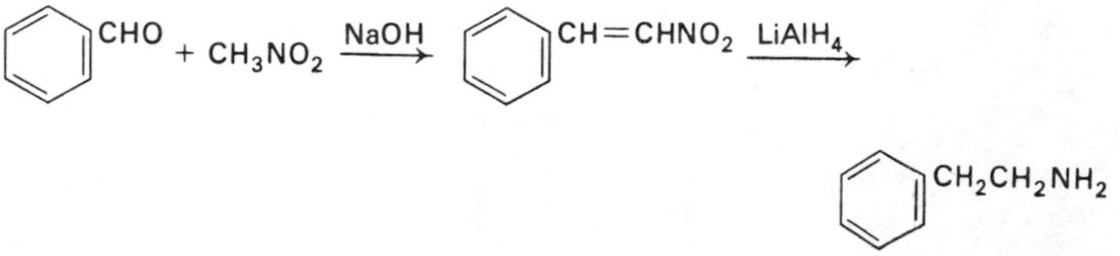

(*g*) acetone $\longrightarrow$ $(CH_3)_2C{=}CHCOCH_3$ $\xrightarrow[Pt]{H_2}$ $(CH_3)_2CHCH_2COCH_3$ $\xrightarrow{CH_3MgI}$

$(CH_3)_2CHCH_2C(OH)(CH_3){-}CH_3$ $\xrightarrow[(-H_2O)]{\Delta,\ H^+}$ $(CH_3)_2CHCH{=}C(CH_3)_2$

(*h*) diethyl pimelate (COOEt, COOEt) $\xrightarrow{NaOEt}$ 2-carbethoxycyclohexanone (O, COOEt) $\xrightarrow[(2)\ \Delta,\ -CO_2]{(1)\ saponify}$

cyclohexanone (O) $\xrightarrow{CH_3MgI}$ 1-methylcyclohexanol (CH_3, OH) $\xrightarrow[-H_2O]{\Delta,\ H^+}$ 1-methylcyclohexene (CH_3)

(*i*) $C_6H_5COCH_3$ + HCHO + pyrrolidine (N–H) $\longrightarrow$ $C_6H_5COCH_2CH_2N$(pyrrolidine) $\xrightarrow{EtMgBr}$

$C_6H_5(CH_3CH_2)C(OH)CH_2CH_2N$(pyrrolidine)

(*j*) cyclohexanol (OH) $\xrightarrow{CrO_3}$ cyclohexanone (O) $\xrightarrow{\text{pyrrolidine (N–H)}}$ 1-pyrrolidinocyclohexene (N)

$\xrightarrow[(2)\ H_2O/H^+\ \ (3)\ saponify]{(1)\ CH_2{=}CHCOOCH_3}$ 2-oxocyclohexyl (O) CH_2CH_2COOH

(*k*) pimelic acid (as above, in *h*) $\longrightarrow$ 2-carbethoxycyclohexanone (O, COOEt)

$\xrightarrow[NaOEt]{CH_2{=}CHCOCH_3}$ cyclohexanone (O) with $CH_2CH_2COCH_3$ and COOEt on C-2

PART **6**

Synthetic methods III. Carbon-carbon bond formation by C-alkylation and C-acylation of carbon anions

A. C-ALKYLATION OF MALONIC ESTERS AND β-KETO ESTERS

The general expression

$$\gt C:^- + RX \longrightarrow \gt C{-}R + :X^- \tag{1}$$

represents a reaction of great versatility and scope. It is most widely applied for the alkylation and acylation of the carbon anions formed from β-keto esters and esters of malonic acid, the anions of which are readily formed by the action of such bases as alkali metal alkoxides, sodamide and sodium hydride; or by metallic sodium.

1. Malonic esters

The simplest examples are found in the alkylation of malonic esters:

$$CH_2(COOEt)_2 + OEt^- \rightleftarrows {}^-{:}CH(COOEt)_2 + EtOH \tag{2}$$

$$(EtOOC)_2CH{:}^- + \underset{\displaystyle R}{\underset{|}{CH_2}}Br \longrightarrow (EtOOC)_2CHCH_2R + Br^- \tag{3}$$

A second alkylation can also be accomplished, for the monoalkylmalonic ester still possesses an active hydrogen atom. Thus, esters of the type

$$\begin{matrix} R' & & COOEt \\ & C & \\ R & & COOEt \end{matrix}$$

are readily prepared.

The alkylation reaction is not useful when RX is a tertiary halide, for dehydrohalogenation occurs. The practical uses of this reaction are many. The C-alkylated malonic esters can be

(a) saponified and decarboxylated to yield monocarboxylic acids:

$$\underset{R}{\overset{R'}{>}}C(COOEt)_2 \longrightarrow \underset{R}{\overset{R'}{>}}C(COOH)_2 \xrightarrow{-CO_2} \underset{R}{\overset{R'}{>}}CHCOOH \qquad (4)$$

(note that in the monoalkylated compounds, $R' = H$),

(b) condensed with urea to yield barbituric acids:

$$\underset{R}{\overset{R'}{>}}C\underset{COOEt}{\overset{COOEt}{<}} + \underset{NH_2}{\overset{NH_2}{>}}CO \xrightarrow{NaOEt} \underset{R}{\overset{R'}{>}}C\underset{CO-NH}{\overset{CO-NH}{<}}\!\!>CO \qquad (5)$$

(c) reduced with lithium aluminum hydride to yield 1,3-glycols:

$$\underset{R}{\overset{R'}{>}}C(COOEt)_2 \longrightarrow \underset{R}{\overset{R'}{>}}C\underset{CH_2OH}{\overset{CH_2OH}{<}} \qquad (6)$$

(d) transformed further by alterations of the —COOH group of $R_2CHCOOH$. Extensions of the reaction to yield C-alkylation products with additional functions expand its scope.* For example,

$$CH_2(COOEt)_2 + BrCH_2COOEt \xrightarrow{NaOEt} EtOOCCH_2CH\underset{COOEt}{\overset{COOEt}{<}} \xrightarrow[NaOEt]{RX} \qquad (7)$$

$$EtOOCCH_2\overset{R}{\overset{|}{C}}\underset{COOEt}{\overset{COOEt}{<}} \xrightarrow[(2)\ -CO_2]{(1)\ saponify} \begin{matrix} R \\ | \\ CHCOOH \\ | \\ CH_2COOH \end{matrix}$$

---

* *Note:* It will be recognized that the Michael reaction is also a C-alkylation process.

$$C_6H_5COCH_2Br + CH_2(COOEt)_2 \xrightarrow{NaOEt} C_6H_5COCH_2CH(COOEt)_2 \quad (8)$$

$$C_6H_5CH_2Cl + CH_2(COOEt)_2 \xrightarrow{NaOEt} C_6H_5CH_2CH(COOEt)_2 \xrightarrow[(2)\ -CO_2]{(1)\ \text{saponify}} \quad (9)$$

$$C_6H_5CH_2CH_2COOH \xrightarrow[(2)\ AlCl_3]{(1)\ PCl_5} \text{(1-indanone)}$$

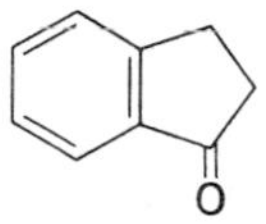

$$Br(CH_2)_nBr + CH_2(COOEt)_2 \xrightarrow{NaOEt} Br(CH_2)_nCH(COOEt)_2 \xrightarrow{NaOEt} \quad (10)$$

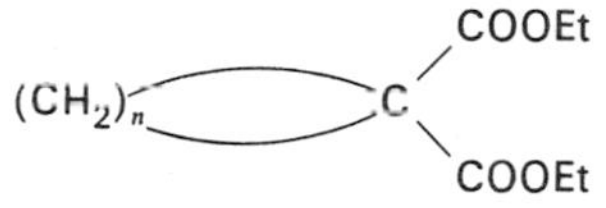

$$(CH_2)_n\text{(ring)}C(COOEt)_2 \quad \left(\text{for example, when } n = 2, \ \text{cyclopropane: } CH_2\text{–}CH_2 \text{ ring with } C(COOEt)_2\right)$$

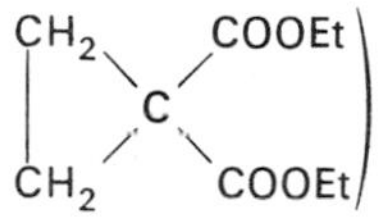

Exercise 6.1

Show how you could prepare the following compounds, using diethyl malonate as one reagent (additional steps may be needed—for instance, reduction of $-COOR$ to $-CH_2OH$ with $LiAlH_4$):

(*a*) $(CH_3)_2CHCH_2CH_2COOH$

(*b*) $CH_3CH(OH)CH_2CH_2CH_2OH$

(*c*) COOH

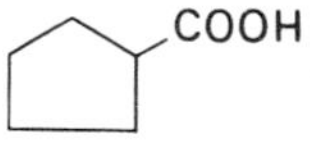

(*d*) $C_6H_5CH_2CH(CH_3)CH_2NH_2$

(*e*) $C_6H_5CH_2CH_2CH_2CH_2COOH$

(*f*) CH_3 CH_3

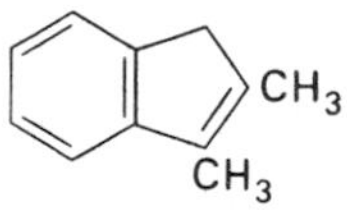

2. β-Keto esters and β-diketones*

The alkylation of β-keto esters follows the pattern described above for the alkylation of malonic ester. The $-CH_2-$ or $-CH(R)-$ group flanked by the two carbonyl

* β-Cyano esters and β-keto nitriles show similar behavior in the reactions described in this Section.

functions is readily transformed into the nucleophilic anion; for example, acetoacetic ester:

$$\begin{matrix} CH_3CO \\ \quad\quad \diagdown CH_2 \\ EtOOC \diagup \end{matrix} \xrightarrow{NaOEt} \begin{matrix} CH_3CO \\ \quad\quad \diagdown CH:^- \\ EtOOC \diagup \end{matrix} \xrightarrow{RX} \begin{matrix} CH_3CO \\ \quad\quad \diagdown CH{-}R \\ EtOOC \diagup \end{matrix} \tag{11}$$

One of the most useful applications of this kind of alkylation is in the preparation of ketones, for the *β*-keto esters can be saponified and decarboxylated:

$$CH_3CO\overset{COOEt}{\overset{|}{C}}HR \longrightarrow CH_3CO\overset{COOH}{\overset{|}{C}}HR \xrightarrow[-CO_2]{\Delta} CH_3COCH_2R \tag{12}$$

The alkylating agents (RX in (11)) can vary in character from simple primary and secondary alkyl halides to *α*-bromo-ketones and -esters, allylic and benzylic halides, etc. Thus, reactions of the following classes can be carried out:

$$CH_3COCH_2COOEt + C_6H_5COCH_2Br \longrightarrow CH_3CO\underset{CH_2COC_6H_5}{\underset{|}{C}}HCOOEt \xrightarrow[(2)\ -CO_2]{(1)\ \text{saponify}} CH_3COCH_2CH_2COC_6H_5 \tag{13}$$

$$CH_3COCH_2COOEt + BrCH_2COOEt \longrightarrow CH_3CO\underset{CH_2COOEt}{\underset{|}{C}}HCOOEt \xrightarrow[(2)\ -CO_2]{(1)\ \text{saponify}} \underset{\text{levulinic acid}}{CH_3COCH_2CH_2COOH} \tag{14}$$

$$CH_3CO\underset{CH_2COOEt}{\underset{|}{C}}HCOOEt + C_6H_5CH_2Cl \xrightarrow{NaOEt} CH_3CO\overset{CH_2C_6H_5}{\underset{CH_2COOEt}{\overset{|}{\underset{|}{C}}}}COOEt \xrightarrow[(2)\ -CO_2]{(1)\ \text{saponify}} CH_3CO\underset{CH_2COOH}{\underset{|}{C}}HCH_2C_6H_5 \tag{15}$$

$$CH_3COCH_2COOEt + BrCH_2CH_2Br \xrightarrow{NaOEt} CH_3CO\underset{CH_2CH_2Br}{\underset{|}{C}}HCOOEt \xrightarrow{\text{further}} CH_3CO\overset{COOEt}{\overset{|}{C}}HCH_2CH_2\overset{COOEt}{\overset{|}{C}}HCOCH_3 \tag{16}$$

3. Cyclic β-keto esters

Cyclic *β*-keto esters formed by Dieckmann condensations are also readily alkylated at the reactive $-CO-\underline{\overset{|}{C}H}COOEt$ position, and the reaction provides a means for the synthesis of a wide variety of compounds; for example:

$H_2C(CH_2COOEt)$–$H_2C(CH_2COOEt)$ $\xrightarrow{NaOEt}$ 2-carbethoxycyclopentanone $\xrightarrow[NaOEt]{RX}$ 2-R-2-carbethoxycyclopentanone (COOEt) $\xrightarrow[(2)\ -CO_2]{(1)\ saponify}$ 2-R-cyclopentanone (17)

This alkylation reaction was an important step in the early total synthesis of the sex hormone equilenin:

CH_3O-substituted tricyclic keto ester (COOEt, O) $\xrightarrow[CH_3I]{Na}$ CH_3O-substituted tricyclic keto ester (H_3C, COOEt, O) $\xrightarrow[\text{to equilenin}]{\text{further steps}}$ (18)

Other examples are

2-carbethoxycyclohexanone (O, COOEt) + $CH_2{=}CHCH_2Br$ $\xrightarrow{NaOEt}$ 2-allyl-2-carbethoxycyclohexanone (O, $CH_2CH{=}CH_2$, COOEt) (19)

2-carbethoxycyclohexanone (O, COOEt) + $BrCH_2COOEt$ $\xrightarrow{NaOEt}$ cyclohexanone (O, CH_2COOEt, COOEt) (20)

o-phenylenediacetic ester (CH_2COOEt, CH_2COOEt) $\xrightarrow{NaOEt}$ indanone (COOEt, =O) $\xrightarrow[CH_3I]{Na}$ indanone (H_3C, COOEt, =O) (21)

$CH_3COCH_3 + 2\ \underset{}{CH_2COOEt}$ (CN) $\longrightarrow$ $(CH_3)_2C(CHCOOEt)_2$ (each CHCOOEt bearing CN) $\xrightarrow{NH_3}$ (22)

dicyano glutarimide (CN, O, NH, O, CN) $\xrightarrow[CH_2I_2]{NaOEt}$ bicyclic imide (CN, O, NH, O, CN) $\xrightarrow[H^+]{H_2O}$ dimethylcyclobutane dicarboxylic acid (COOH, COOH)

Exercise 6.2

Devise syntheses for the following. At least one of the steps should be a C-alkylation of the kind described in the foregoing discussions:

(*a*) COOH, H_3C (3-methylcyclopentanecarboxylic acid)

(*b*) $CH_2{=}CHCH_2CH_2COCH_3$

(*c*) $C_6H_5CH_2CH{=}C(CH_3)CH_3$

(*d*) O, CH_2COCH_3 (2-acetonylcyclopentanone)

(*e*) CH_3, $CH{=}C(CH_3)CH_3$

(*f*) $CH_2CH{=}CH_2$, H_3C, O, H–N, =O, N–H, O

(*g*) O, CH_3 (3-methylcyclopent-2-enone)

(*h*) $HOOCCH_2CH_2CH(COOH)CH_2CH_2COOH$

(*i*) O, $CH_2CH_2CH_2$, O

(*j*) CH_3, O, CH_3, CH_3, O, CH_3 (tetramethylcyclopentanedione)

(*k*) COC_6H_5, COOH

(*l*) $CH_3{-}C(CH_3){=}CHCH_2CH_2COCH_3$

Notes about answers to exercises 6.2

(*a*) The "reverse" pathway for the preparation of 3-methyl-cyclopentanecarboxylic acid is the following:

3-methylcyclopentane-COOH $\xleftarrow{-CO_2}$ 3-methylcyclopentane-1,1-(COOH)$_2$ $\xleftarrow{\text{saponify}}$ 3-methylcyclopentane-1,1-(COOEt)$_2$

↑ NaOEt $(+CH_2(COOEt)_2)$

$CH_3CH(CH_2COOH)COOH$ $\xrightarrow{LiAlH_4}$ $CH_3CH(CH_2CH_2OH)CH_2OH$ $\xrightarrow[\text{or } PBr_3]{HBr}$ $CH_3CH(CH_2CH_2Br)CH_2Br$

↑ $-CO_2$

$CH_3{-}C(COOH)_2CH_2COOH$ $\xleftarrow{\text{saponify}}$ $CH_3C(COOEt)_2CH_2COOEt$ { from $CH_2(COOEt)_2$ and CH_3I; then $BrCH_2COOEt$

(*e*) The compound *cannot* be prepared by a route involving the step

$$\text{2-(COOEt)cyclohexanone} + (H_3C)_2C{=}CHBr \not\rightarrow \text{2-[CH=C(CH}_3)_2\text{]-2-(COOEt)cyclohexanone}$$

for the vinyl bromide is not reactive in the displacement (alkylation) reaction. The route that could be taken is (in reverse order):

$$\text{1-}CH_3\text{-2-[CH=C(CH}_3)_2\text{]cyclohexene} \xleftarrow[(-2\,H_2O)]{H^+} \text{1-HO-1-}CH_3\text{-2-[}CH_2C(OH)(CH_3)CH_3\text{]cyclohexane} \xleftarrow{CH_3MgI} \text{2-(}CH_2COCH_3\text{)cyclohexanone}$$

and 2-acetonylcyclohexanone can be prepared by way of the alkylation of ethyl cyclohexanone-2-carboxylate with bromoacetone. A variant of this route is given in the Answers Section.

(*k*) It is clear that no direct substitution reaction can give the product (*o*-benzoylbenzoic acid). This is most simply derived by the following oxidation:

$$\text{1-}C_6H_5\text{-indene} \xrightarrow{CrO_3} o\text{-}(COC_6H_5)C_6H_4COOH$$

The 1-phenylindene is readily prepared from 1-indanone as follows:

$$\text{1-indanone} \xrightarrow{C_6H_5MgBr} \text{1-HO-1-}C_6H_5\text{-indane} \xrightarrow{-H_2O} \text{1-}C_6H_5\text{-indene}$$

The synthetic problem is now reduced to the preparation of 1-indanone, which can be accomplished by a route starting with the alkylation of ethyl malonate with benzylchloride (Equation 9).

B. C-ALKYLATION WITH THE USE OF ENAMINES

The direct alkylation of anions derived from ketones is not generally practicable because of the possibility of polyalkylation. For example, the following reactions (in which $B\!:^-$ is a strong base)

$$CH_3COCH_3 + B\!:^- \rightleftarrows CH_3COCH_2\!:^- + B\!:\!H \qquad (23)$$

$$CH_3COCH_2{:}^- + RCH_2X \longrightarrow CH_3COCH_2CH_2R + X^- \tag{24}$$

are valid as they are written, but under the actual conditions of the experiment the further reactions

$$CH_3COCH_2CH_2R \overset{B:^-}{\rightleftarrows} {}^-{:}CH_2COCH_2CH_2R \xrightarrow{RCH_2X} RCH_2CH_2COCH_2CH_2R \tag{25}$$

and

$$CH_3COCH_2CH_2R \overset{B:^-}{\rightleftarrows} CH_3CO\ddot{\bar{C}}HCH_2R \xrightarrow{RCH_2X} CH_3COCH(CH_2R)CH_2R \tag{26}$$

can and usually will occur to produce polyalkylation products.

When polyalkylation is not undesirable (as in the following reaction)

$$C_6H_5COCH_3 + CH_3I \xrightarrow{NaNH_2} C_6H_5COC(CH_3)_3 \tag{27}$$

or when the products of the polyalkylation can be separated conveniently, the reaction has been used; but these uses are exceptional.

If it were possible to (1) generate a single anionic (carbanion) center, (2) alkylate it, and thus (3) produce a product that would no longer yield an anion, it is clear that *mono*alkylation could be achieved. This can be done in the following way.

Treatment of a ketone (cyclohexanone will be used as the example) with a secondary amine gives rise to an unsaturated amine, formed in the following way:

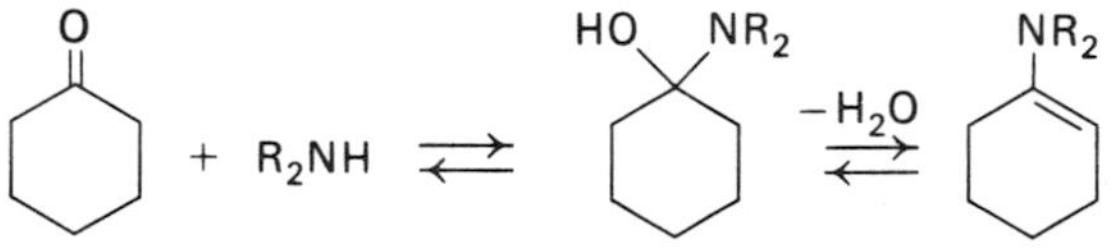

$$\text{cyclohexanone (C=O)} + R_2NH \rightleftarrows \text{1-hydroxy-1-}NR_2\text{-cyclohexane (HO, }NR_2) \overset{-H_2O}{\rightleftarrows} \text{1-}NR_2\text{-cyclohexene} \tag{28}$$

The *enamine* can provide electrons to an attacking electrophile in the following way:

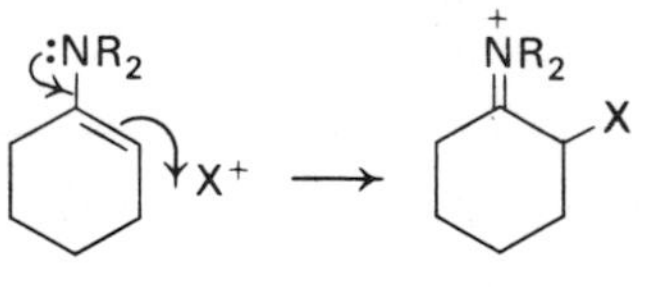

$$\text{1-}{:}NR_2\text{-cyclohexene} + X^+ \longrightarrow \text{2-X-cyclohexylidene-}\overset{+}{N}R_2 \tag{29}$$

Hydrolysis of this product readily yields the substituted ketone:

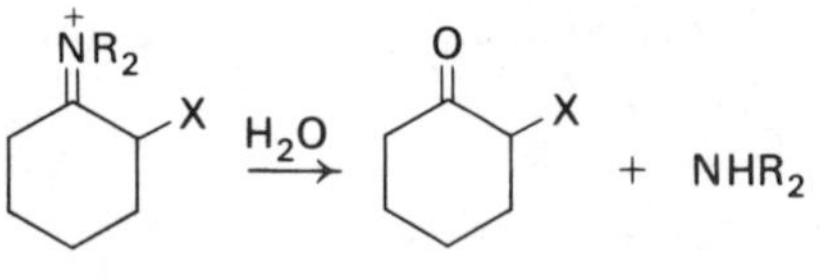

$$\text{2-X-cyclohexylidene-}\overset{+}{N}R_2 \xrightarrow{H_2O} \text{2-X-cyclohexanone (C=O)} + NHR_2 \tag{30}$$

Conversion of the α-substituted ketone into the enamine proceeds in the following way:

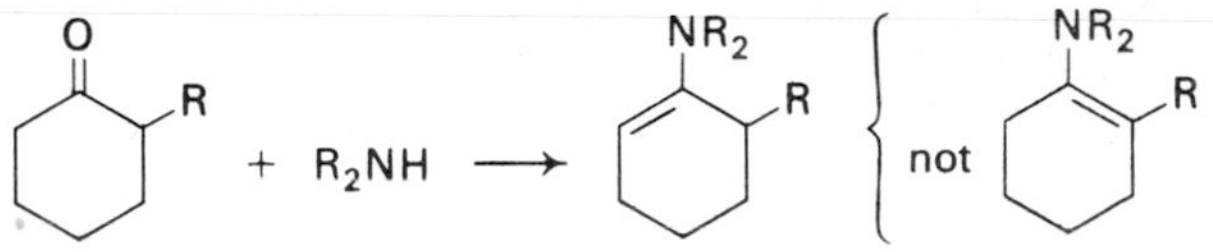

$$\text{2-R-cyclohexanone} + R_2NH \longrightarrow \text{1-}NR_2\text{-6-R-cyclohexene} \quad \left\{ \text{not } \text{1-}NR_2\text{-2-R-cyclohexene} \right. \qquad (31)$$

Thus, symmetrical (that is, 2,6-) dialkylation can be accomplished:

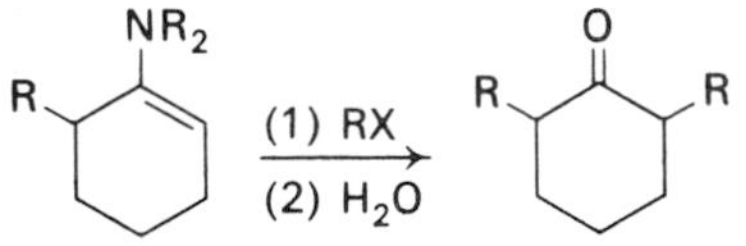

$$\text{1-}NR_2\text{-6-R-cyclohexene} \xrightarrow[(2)\ H_2O]{(1)\ RX} \text{2,6-di-R-cyclohexanone}$$

Exercise 6.3

Show how the enamine method can be used in the preparation of the following. (In some cases additional steps, not involving enamines, are necessary.)

(*a*) $CH_3COCH_2CH_2CH{=}CH_2$

(*b*) $CH_3CH_2COCH(CH_3)CH_2COCH_3$

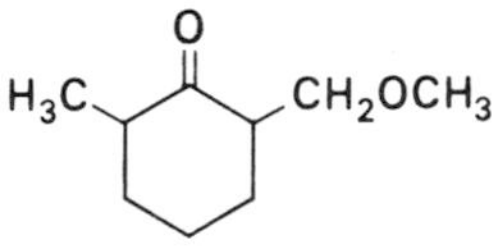

(*c*) 2-(H_3C)-6-(CH_2OCH_3)-cyclohexanone

(*d*) 1-CH_3-2-($CH_2C_6H_5$)-cyclopentene

(*e*) 3-CH_3-2-(C_6H_5)-cyclohex-2-en-1-one

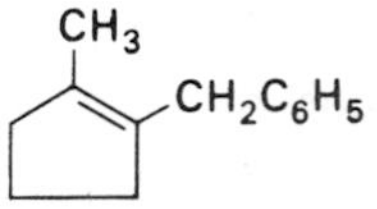

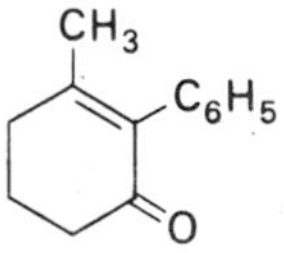

(*f*) $(CH_3)_2CHCOCH_2CH_2CH{=}CH_2$

(*g*) 1-($CH_2C_6H_5$)-cyclohexene

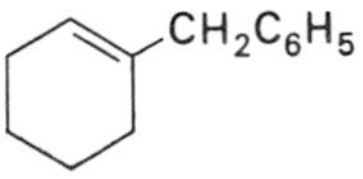

Partial answer to exercise 6.3

(*a*) Enamine from acetone ($CH_3C(=CH_2){-}NR_2$) and allyl bromide as alkylating agent.
(*c*) Alkylation of 2-methylcyclohexanone enamine using CH_3OCH_2Cl.
(*d*) Alkylate cyclopentanone enamine with benzyl chloride. Then add CH_3MgI to the resulting 2-benzylcyclopentanone. Then dehydrate the resulting tertiary alcohol.

Enamines can also be used as the anionic (nucleophilic) reagent in a Michael reaction:

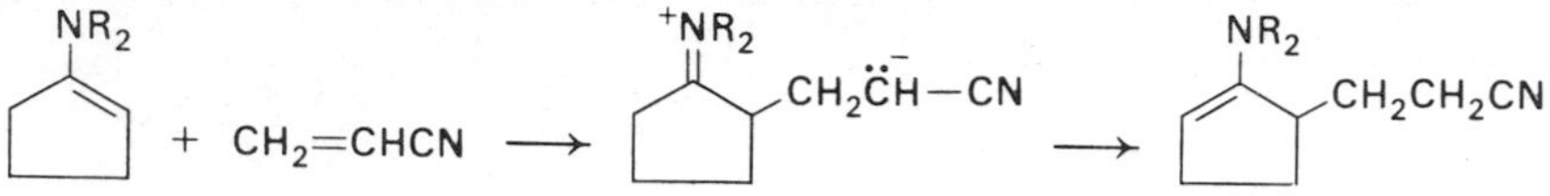

$$\text{NR}_2\text{-cyclopentene} + CH_2{=}CHCN \longrightarrow {}^{+}NR_2{=}\text{cyclopentane-}CH_2\ddot{C}H^{-}{-}CN \longrightarrow NR_2\text{-cyclopentene-}CH_2CH_2CN \quad (32)$$

Exercise 6.4

Show how the following compound could be prepared from simple starting materials:

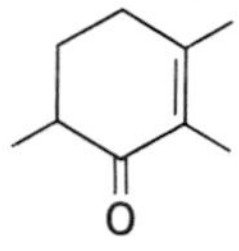

Answer to exercise 6.4

The immediate precursor of this compound can be the diketone, which by a cyclic aldol condensation gives the final desired product:

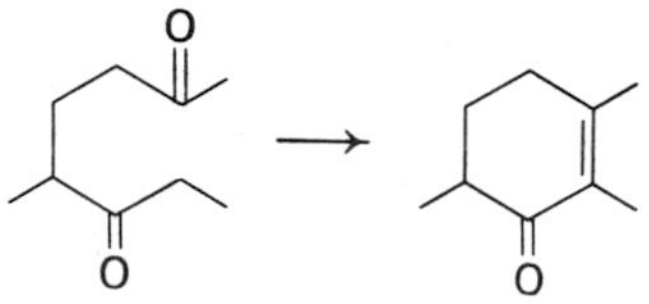

This diketone can be obtained by the alkylation of diethyl ketone in the following way,

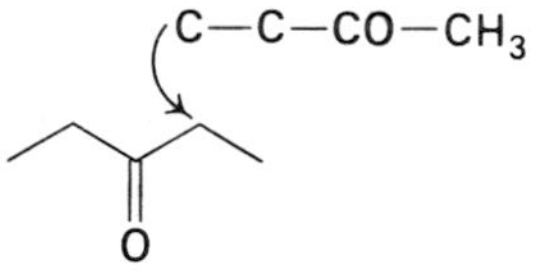

$$C{-}C{-}CO{-}CH_3$$

a result that can be achieved by the Michael condensation of methyl vinyl ketone with the enamine derived from diethyl ketone:

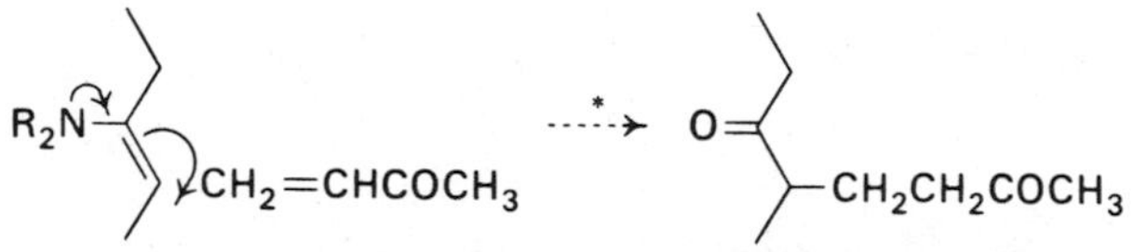

$$R_2N{-}\text{enamine} + CH_2{=}CHCOCH_3 \overset{*}{\dashrightarrow} O{=}\text{C}{-}CH_2CH_2COCH_3 \quad (33)$$

---

* The dotted arrow will often be used to represent a series of steps that can be supplied by the student, which include such unexceptional manipulations as hydrolysis, saponification, etc.

Exercise 6.5

Using a procedure similar to that in Exercise 6.4., and starting with 3-methylbutanal as one reagent, show how

[structure: 4-isopropylcyclohex-2-enone] could be prepared.

C. C-ACYLATION

In principle, the anions whose *C-alkylation* has been described above can be *C-acylated* in an analogous way:

$$-\overset{|}{\underset{|}{C}}:^- + R\!-\!C(=O)\!-\!X \longrightarrow -\overset{|}{\underset{|}{C}}\!-\!C(=O)\!-\!R + X:^- \qquad (34)$$

The Claisen and Dieckmann reactions (where X = OR) are C-acylations; they have been described in Part 4.

C-Acylation of the active methylene group of malonic esters or β-keto esters leads to the formation of α-acyl-β-keto esters (Equation 35) or acylmalonic esters (Equation 36):

$$(CH_3CO\ddot{C}HCOOEt)^- + RCOCl \longrightarrow CH_3CO\overset{COR}{\overset{|}{C}}HCOOEt \qquad (35)$$

$$(:CH(COOEt)_2)^- + RCOCl \longrightarrow RCOCH(COOEt)_2 \qquad (36)$$

These α-acyl-β-keto esters and acylmalonic esters can in principle be cleaved by bases (for example, NaOH) to yield β-diketones or β-keto esters, as in the following general expressions:

$$RCO\!-\!CH(COR')(COOEt) \xrightarrow{\text{cleave at } a} RCOCH_2COR'$$
$$\xrightarrow{\text{cleave at } b} RCOCH_2COOEt \qquad \xrightarrow{\text{cleave at } c} R'COCH_2COOEt \qquad (37)$$

$$RCOCH(COOEt)_2 \longrightarrow RCOCH_2COOEt \quad \text{or} \quad RCOOH + CH_2(COOEt)_2 \qquad (38)$$

It is apparent, however, that because several carbonyl groups are present, numerous possible courses (Equation 37) are available for these cleavage reactions and that the experimental procedures are liable to lead to mixtures or to unsatisfactory yields of desired products. Although it is possible in particular cases to specify the conditions needed to carry out a given reaction successfully, it is beyond our present scope to enter into such details. There are, however, certain devices that are of general application in synthetic manipulations of this kind.

The cleavage of acylmalonic esters by bases (for example NaOH, NaOEt) usually results in loss of the acyl group

$$\mathrm{RCOCH(COOEt)_2 \xrightarrow{EtO^-} RCOOEt + (:CH(COOEt)_2)^-} \qquad (39)$$

Consequently, the following desirable result

$$\mathrm{RCOCH(COOEt)_2 \longrightarrow RCOCH_2COOEt} \qquad (40)$$

sometimes cannot be achieved. There is, however, a way around this difficulty by providing for the removal of one or two —COOR groups without the use of a basic reagent. Such an end can be attained by using a malonic ester in which one COOR group (or both—see Equation 45) is susceptible either to hydrolysis under *acidic* conditions *or to removal by other means.* There are three useful ways of doing this.

1. A *t*-butyl ester is readily hydrolyzed by acid. Thus, the following series of reactions is a practicable one:*

$$\mathrm{Et\overset{+}{O}Mg{:}\overset{-}{C}H(COOEt){-}COO\mathit{t}Bu \xrightarrow{RCOCl} RCOCH(COO\mathit{t}Bu)(COOEt) \xrightarrow{H^+} RCOCH_2COOEt + CO_2 + CH_2{=}CMe_2} \qquad (41)$$

The final step is carried out with the use of *p*-toluenesulfonic acid in refluxing benzene.

2. A tetrahydropyranyl (THP) ester can be used:

$$\mathrm{H_2C(COOH)(COOEt) + \text{dihydropyran} \xrightarrow{H^+} H_2C(COO\text{-THP})(COOEt)} \quad \text{(THP)} \qquad (42)$$

$$\mathrm{CH_2(COO(THP))(COOEt) \xrightarrow[\text{(2) RCOCl}]{\text{(1) \textit{via} —MgOEt salt}} RCOCH(COO(THP))(COOEt) \xrightarrow{H_2O/H^+} RCOCH(COOH)(COOEt) \xrightarrow{\Delta} RCOCH_2COOEt} \qquad (43)$$

* The use of the alkoxymagnesium derivative of the malonic ester, rather than a sodium or potassium derivative, is the preferred way of carrying out this acylation.

3. A benzyl ester can be used. Hydrogenolysis (with hydrogen and a palladium catalyst) removes the benzyl group without affecting other parts of the molecule:

$$\mathrm{RCOCH}\begin{matrix}\diagup \mathrm{COOCH_2C_6H_5} \\ \diagdown \mathrm{COOEt}\end{matrix} \xrightarrow[\mathrm{Pd}]{\mathrm{H_2}} \mathrm{RCOCH}\begin{matrix}\diagup \mathrm{COOH} \\ \diagdown \mathrm{COOEt}\end{matrix} + \mathrm{C_6H_5CH_3} \qquad (44)$$

It will be noted that if the malonic ester contains *two* *t*-butyl, THP, or benzyl groups, the method is a valuable route to methyl ketones:

$$\mathrm{RCOCH}\begin{matrix}\diagup \mathrm{COO(THP)} \\ \diagdown \mathrm{COO(THP)}\end{matrix} \xrightarrow{\mathrm{H_2O/H^+}} \mathrm{RCOCH}\begin{matrix}\diagup \mathrm{COOH} \\ \diagdown \mathrm{COOH}\end{matrix} \xrightarrow[-2\,\mathrm{CO_2}]{\Delta} \mathrm{RCOCH_3} \qquad (45)$$

Exercise 6.6

Assuming the availability of the half-ester of malonic acid, $HOOCCH_2COOEt$, devise syntheses for the following:

(*a*) $C_6H_5COCH_2COOEt$

(*b*) $CH_3CH_2CH_2COCH_3$

(*c*) $CH_2{=}CHCH_2COCH_2COOEt$

(*d*) $CH_2{=}CHCH_2CH_2COCH(CH_3)_2$

(*e*) $CH_3C_6H_4COCH_2CH_2CH(CH_3)_2$

Answer to exercise 6.6(*e*)

In reverse order of steps, the product can be prepared as follows:

$$\mathrm{CH_3C_6H_4COCH_2CH_2CH(CH_3)_2} \xleftarrow[(2)\ -\mathrm{CO_2}]{(1)\ \text{saponify}} \mathrm{CH_3C_6H_4COCH(CH_2CH(CH_3)_2)COOEt}$$

$$\uparrow \mathrm{(CH_3)_2CHCH_2Br/NaOEt}$$

$$\mathrm{CH_3C_6H_4COCH}\begin{matrix}\diagup \mathrm{COOH} \\ \diagdown \mathrm{COOEt}\end{matrix} \xrightarrow{-\mathrm{CO_2}} \mathrm{CH_3C_6H_4COCH_2COOEt}$$

$$\uparrow \mathrm{H_2O/H^+}$$

$$\mathrm{CH_3C_6H_4COCH}\begin{matrix}\diagup \mathrm{COO{-}THP} \\ \diagdown \mathrm{COOEt}\end{matrix} \xleftarrow{\mathrm{CH_3C_6H_4COCl}} \left\{\mathrm{CH}\begin{matrix}\diagup \mathrm{COO{-}THP} \\ \diagdown \mathrm{COOEt}\end{matrix}\right\} \mathrm{EtOMg}$$

The acylation of cyclohexanone (by way of its enamine, as discussed in Section 6.B) leads to 2-acylcyclohexanones. These are smoothly cleaved by alkali to give the ring-opened keto acids:

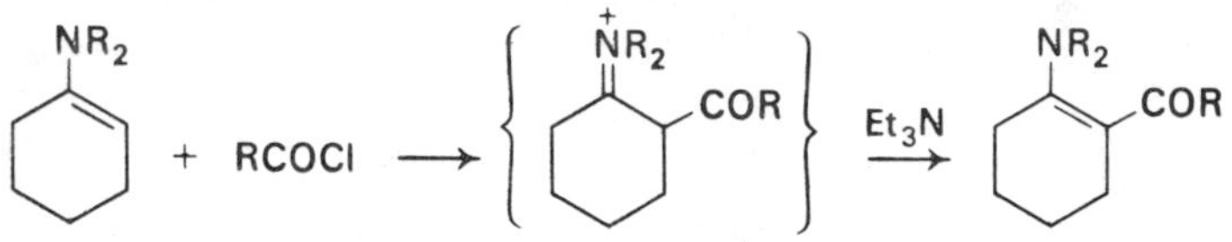

$$\text{1-}(NR_2)\text{cyclohexene} + RCOCl \longrightarrow \left\{ \text{2-COR-1-}(\overset{+}{N}R_2\text{=})\text{cyclohexane} \right\} \xrightarrow{Et_3N} \text{2-COR-1-}(NR_2)\text{cyclohexene} \qquad (46)$$

$$\text{2-COR-1-}(NR_2)\text{cyclohexene} \underset{H^+}{\xrightarrow{H_2O}} \text{2-COR-cyclohexanone} \qquad (47)$$

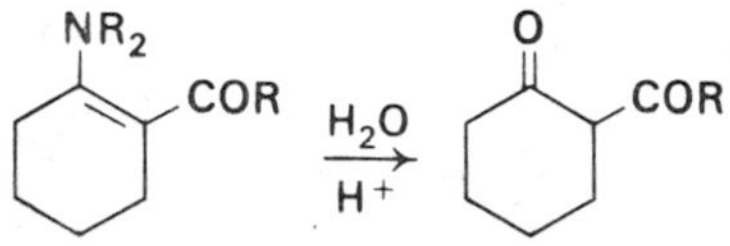

$$\text{2-COR-cyclohexanone} \underset{(KOH)}{\xrightarrow{H_2O}} HOOC\text{–}(CH_2)_5\text{–}COR \equiv RCO(CH_2)_5COOH \qquad (48)$$

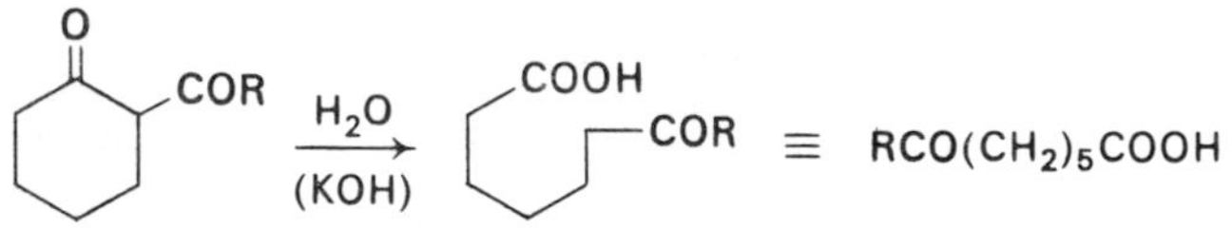

(*Note:* Chloroformic esters can also be used (that is, in Equation 46, RCOCl = ClCOOEt).)

When substituted cyclohexanones are used, the corresponding substituted keto acids are obtained:

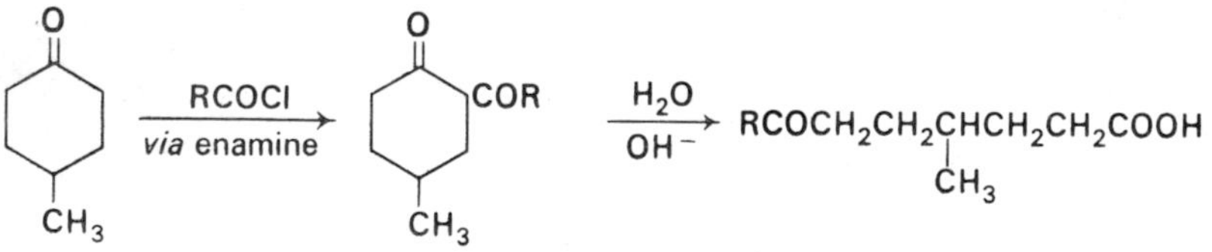

$$\text{4-}CH_3\text{-cyclohexanone} \underset{via\ enamine}{\xrightarrow{RCOCl}} \text{2-COR-4-}CH_3\text{-cyclohexanone} \underset{OH^-}{\xrightarrow{H_2O}} RCOCH_2CH_2\underset{\displaystyle CH_3}{\underset{|}{C}}HCH_2CH_2COOH \qquad (49)$$

Exercise 6.7

By use of the enamine acylation reaction, show how the following compounds can be prepared:

(*a*) 2-($COCH_2CH_3$)cyclopentanone

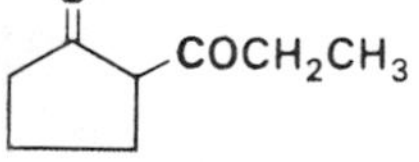

(*b*) 2-($COCH(CH_3)_2$)cyclohexanone

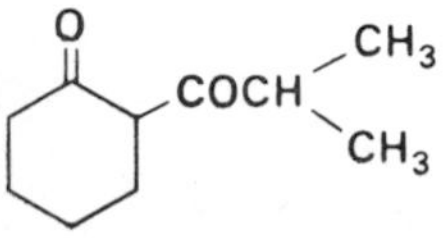

(*c*) $CH_3CH_2CH_2COCH_2CH_2CH_2CH_2CH_2COOH$

(*d*) $HOOC(CH_2)_5COCH_2CH_2CH_2CO(CH_2)_5COOH$

(*e*) $CH_3CH_2CH_2CO\underset{|}{C}HCOOEt$
CH_2CH_3

(*f*) CH_3
$C_6H_5CO\underset{|}{C}HCOCH_2CH_3$

(*g*) $CH_3CH_2CO\underset{|}{C}HCOCH_2C_6H_5$
CH_3

(*h*) $COOH$
$CH_2CH_2CH_2COCH_2CH_3$

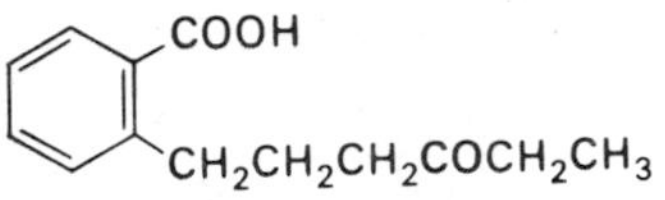

(*i*) CH_2COCH_3
CH_2CH_2COOH

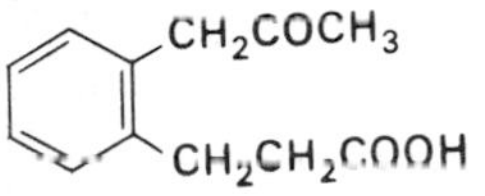

Notes about exercise 6.7

(*d*) Dicarboxylic acids (as the acid chlorides) serve as bifunctional acylating agents, reacting with enamines at both —COCl groups.

(*e*) The acylating agent is ClCOOEt.

(*h*) 1-Tetralone forms the expected enamine:

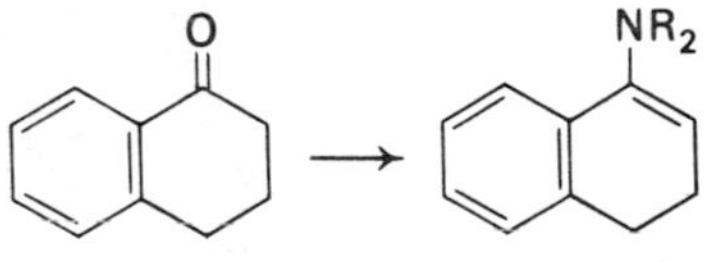

O → NR_2

(*i*) 2-Tetralone forms the enamine:

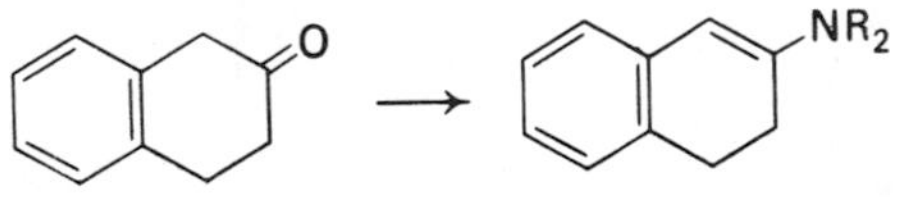

O → NR_2

PART 7

The Wittig synthesis of carbon-carbon bonds

The Wittig synthesis is a carbon-carbon bond-forming reaction that leads directly to the formation of olefinic compounds. Its general expression is the following:

$$\gt C{=}O + \begin{matrix} R \\ \diagdown \\ C{=}PR''_3 \\ \diagup \\ R' \end{matrix} \longrightarrow \gt C{=}C\begin{matrix} R \\ \diagup \\ \\ \diagdown \\ R' \end{matrix} + R''_3PO \qquad (1)$$

where the isolated carbonyl group represents an aldehyde or ketone.

The phosphorane, $R_2C{=}PR_3$, is prepared in the following way. Addition of an alkyl halide (or other reactive halide) to triphenylphosphine (R″ in (1) is usually phenyl) proceeds to give the expected quaternary phosphonium salt.*

$$RCH_2Br + (C_6H_5)_3P \longrightarrow R_2CH_2\overset{+}{P}(C_6H_5)_3Br^- \qquad (2)$$

The abstraction of a proton by a base leads to the alkylidene-triphenylphosphorane:

$$RCH_2\overset{+}{P}(C_6H_5)_3 \xrightarrow{B:^-} R\overset{..-}{C}H{-}\overset{+}{P}(C_6H_5)_3 \longleftrightarrow RCH{=}P(C_6H_5)_3$$

* It will be recognized by the student that phosphorus corresponds to nitrogen in the number of electrons in the external (valence) shells. In phosphorus, however, these five electrons (two 3*s*, three 3*p*) occupy the *M* shell; phosphorus has the capacity to utilize the 3*d* orbitals and thus is not confined to an octet ($3s^23p^6$) in bond formation. In this respect, R_4N^+ corresponds to R_4P^+. But while PCl_5 is known, NCl_5 is not.

Exercise 7.1

Could a corresponding nitrogen compound, with a contributing structure $RCH{=}NR_3$, be formed in the same way? Why?

The synthetic uses of the Wittig reaction (in Equation 1) are many. A particular advantage is the fact that the position of the double bond that is introduced is unambiguous. For examples, contrast Equations 3–4 and 5–6:

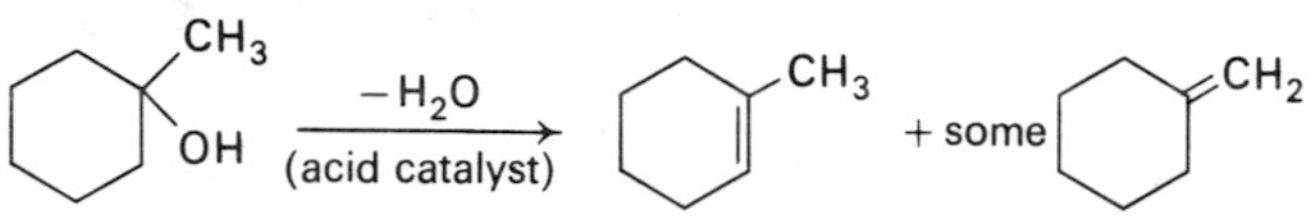

1-methylcyclohexanol $\xrightarrow[\text{(acid catalyst)}]{-H_2O}$ 1-methylcyclohexene (CH_3) + some methylenecyclohexane (CH_2) (3)

but:

cyclohexanone ($=O$) + $(C_6H_5)_3P{=}CH_2$ $\longrightarrow$ methylenecyclohexane ($=CH_2$) only (4)

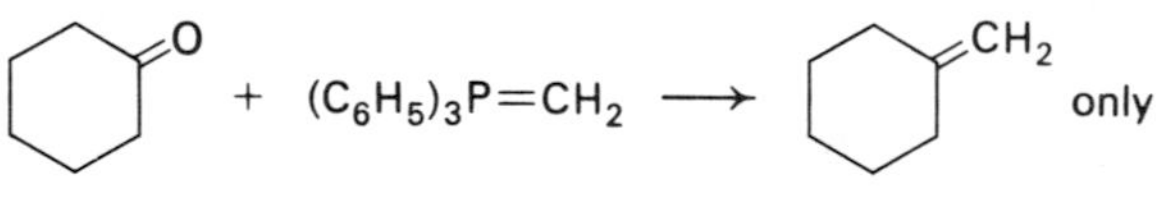

$$(CH_3)_2CHCH_2\overset{CH_3}{\overset{|}{C}}HOH \xrightarrow[\text{(acid catalyst)}]{-H_2O} (CH_3)_2C{=}CHCH_2CH_3 + (CH_3)_2CHCH{=}CHCH_3 \quad (5)$$

mixture depends upon experimental conditions

but:

$$(CH_3)_2CHCHO + (C_6H_5)_3P{=}CHCH_3 \longrightarrow (CH_3)_2CHCH{=}CHCH_3 \text{ only} \quad (6)$$

Bifunctional Wittig reagents offer unique opportunities for elaborate syntheses. The synthesis of the important polyene hydrocarbon, squalene, has been accomplished as follows:

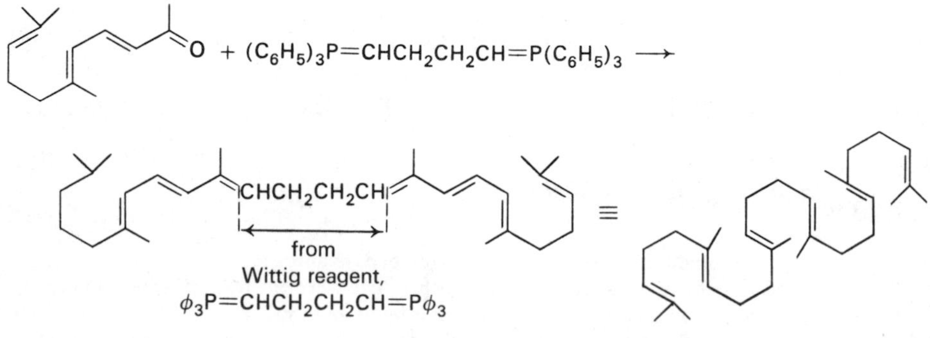

$$\text{(ketone)}{=}O + (C_6H_5)_3P{=}CHCH_2CH_2CH{=}P(C_6H_5)_3 \longrightarrow \quad (7)$$

(squalene) $=CHCH_2CH_2CH=$ ... $\equiv$ (squalene)

from Wittig reagent, $\phi_3P{=}CHCH_2CH_2CH{=}P\phi_3$

Ring formation can also be accomplished:

$$\text{o-}C_6H_4(CHO)_2 + (C_6H_5)_3P{=}CH(CH_2)_nCH{=}P(C_6H_5)_3 \xrightarrow{n=1} \text{benzocycloheptatriene}; \quad \xrightarrow{n=2} \text{benzocyclooctatetraene} \tag{8}$$

Halides other than alkyl halides can be used, and they provide for great versatility in the method:

$$(C_6H_5)_3P + BrCH(CH_3)COOEt \xrightarrow{a} \xrightarrow{b} (C_6H_5)_3P{=}C(CH_3)COOEt \tag{9}$$

$$(C_6H_5)_3P + ClCH_2OCH_3 \xrightarrow{a} \xrightarrow{b} (C_6H_5)_3P{-}CHOCH_3 \tag{10}$$

$$(C_6H_5)_3P + BrCH_2COOEt \xrightarrow{a} \xrightarrow{b} (C_6H_5)_3P{=}CHCOOEt \tag{11}$$

In the above equations a = formation of the quaternary salt and b = removal of a proton to give the neutral phosphorane. The use of the reagents formed as in (10) and (11) are illustrated in Equations 12–15:*

$$(C_6H_5)_3P{=}CHOCH_3 + C_6H_{10}{=}O \longrightarrow C_6H_{10}{=}CHOCH_3 \tag{12}$$

The product of (12) is a vinyl ether, and undergoes ready hydrolysis (with acid) to form the aldehyde

$$C_6H_{10}{=}CHOCH_3 \xrightarrow[H^+]{H_2O^\dagger} C_6H_{11}CHO \tag{13}$$

$$(C_6H_5)_3P{=}CHCOOEt + C_6H_5CHO \longrightarrow C_6H_5CH{=}CHCOOEt \tag{14}$$

---

* The Wittig reaction is not always predictably stereospecific. Both *cis* and *trans* olefinic linkages are formed; the ratio of isomers is altered if the experimental conditions are altered and is dependent upon the structures of the reacting species.

† *Query:* Formulate this hydrolysis in detail, and account for the fact that water does not add in the reverse manner to give 1-(methoxymethyl)cyclohexanol ($C_6H_{10}(OH)CH_2OCH_3$)

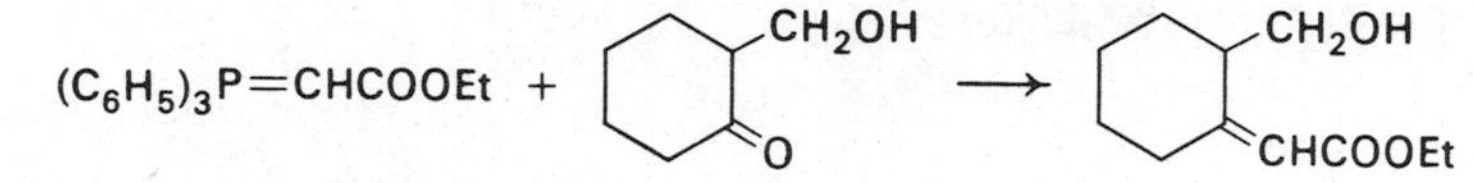

(C6H5)3P=CHCOOEt +
CH2OH
O
CH2OH
CHCOOEt

(15)

An extension of the use of α-bromo esters (Equations 9 and 11) is the use of γ-bromocrotonic esters, such as $BrCH_2CH{=}CHCOOEt$.

Exercise 7.2

Show how you would synthesize the following, using a Wittig reagent prepared from the halide shown below each, and a carbonyl compound of requisite structure. (In some cases additional steps will be required.)

(*a*) 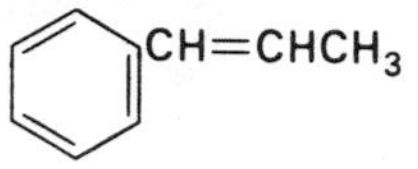

CH=CHCH3

(CH_3CH_2Br)

(*b*) 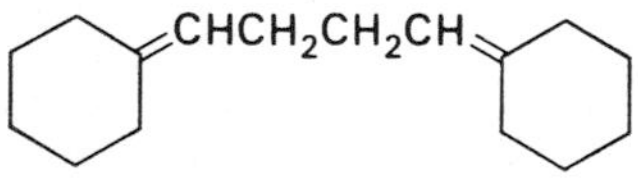

CHCH2CH2CH

($BrCH_2CH_2CH_2CH_2Br$)

(*c*) 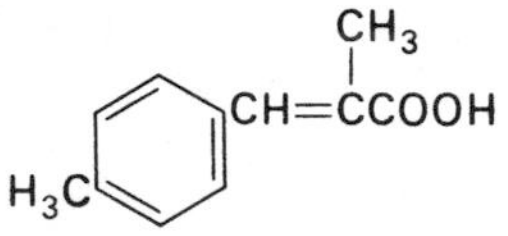

CH3
CH=CCOOH
H3C

($BrCH(CH_3)COOEt$)

(*d*) 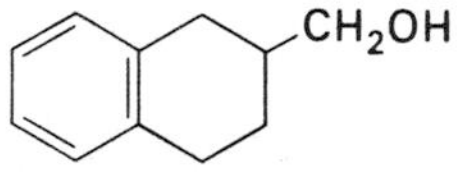

CH2OH

($ClCH_2OCH_3$)

(*e*) 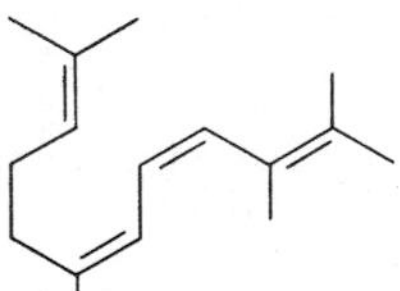

($(CH_3)_2C{=}C(CH_3)CH_2Br$)

(*f*) 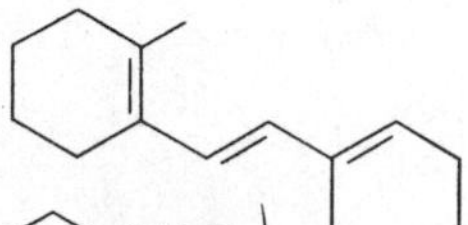

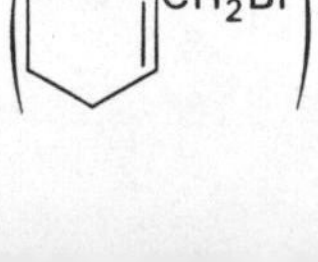

CH2Br

Answers to exercise 7.2

In each synthesis, the carbon-carbon bond-forming step can be reconstructed as shown in Equation 1.

In the parts of this exercise, the carbonyl compound required in the C=C bond-forming (Wittig) step need not be separately synthesized.

Exercise 7.3

Select the necessary reagents and show how the following compounds could be prepared by use of the Wittig reaction:

(*a*) Starting with cyclohexanone, prepare

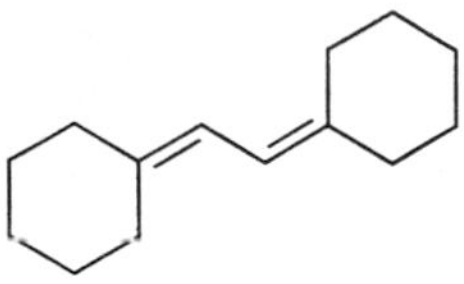

(*b*) Starting with geranyl bromide and crocetin dialdehyde, prepare lycopene. (Look up the structures of the named compounds.)

Answer to exercise 7.3(*a*)

The final step in the synthesis is the Wittig reaction

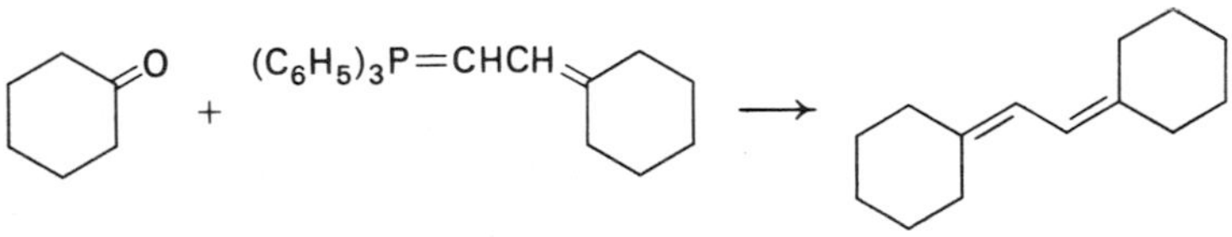

$(C_6H_5)_3P{=}CHCH{=}$

It remains to prepare

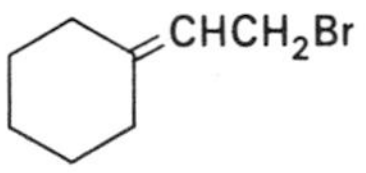

$CHCH_2Br$

Prepare this by reaction of the corresponding alcohol with PBr_3. The alcohol is prepared by $LiAlH_4$ reduction of the ester. The ester is obtained from the reaction of cyclohexanone with the Wittig reagent prepared from $BrCH_2COOEt$.

PART 8

Nucleophilic displacement reactions in synthesis

A. INTRODUCTION

The nucleophilic displacement reaction is one of wide scope and general application in the synthesis of organic compounds. The general expression

$$X{:} + \;\rangle C{-}Y \longrightarrow X{-}C\langle\; + {:}Y \qquad (1)$$

describes reactions of many kinds, which differ in (a) the nature of the nucleophile, X: ; and (b) the nature of the displaced atom or group, :Y.

The *nucleophile* (X: in Equation 1) possesses an unshared pair of electrons, and is in most cases basic in character. The range of basicity is wide; nucleophiles include

(a) hydroxide and alkoxide ions
(b) ammonia and amines
(c) cyanide ions, CN^-
(d) halide ions, especially I^-
(e) mercaptans (RSH) and their anions (RS^-)
(f) water, alcohols, ethers
(g) carboxylic acid anions ($RCOO^-$)

(h) carbon anions (see Part 6)
(i) a wide variety of miscellaneous substances, including thiourea (NH_2CSNH_2), thiocyanate ion (SCN^-), thiosulfate ion ($S_2O_3^{--}$), and others.

The *displaced* group (usually called the "leaving group"), represented by :Y in Equation 1, is also a nucleophile (it contains an unshared electron pair), but its nucleophilicity is very low. Thus, although Equation 1 is in principle reversible, the practical applications of the reaction are those in which the superior nucleophile, X:, displaces the weak nucleophile, :Y, in what is for practical synthetic purposes a reaction complete to the right (Equation 1).

Most useful leaving groups (:Y) are the *anions* (*conjugate bases*) *of strong or moderately strong acids.* Thus, typical leaving groups are:

(j) Cl^-, Br^-, I^- (as in RI, RBr, RCl)
(k) anions of sulfonic or sulfuric acids, RSO_2O^-, $ROSO_2O^-$ (as in $R'OSO_2R$)
(l) H_2O (as in ROH_2^+)
(m) NR_3 (as in $R'{-}NR_3^+$)
(n) variants of (l), that is, R_2O (as in $R'{-}\overset{+}{O}\begin{smallmatrix} R \\ R \end{smallmatrix}$).

Many of the important details (that is, the stereochemistry) of the nucleophilic displacement reaction are beyond the intent of this workbook and will not be discussed here. Attention will be directed principally to application of the reaction in synthesis.

B. PREPARATION OF ALKYL HALIDES

A widely used reaction is the conversion of alcohols into alkyl halides. The latter are of particular utility in the Grignard reaction, and thus provide access to an almost unlimited variety of organic compounds (as is outlined in Part 5). The general route is

$$RCH_2OH \longrightarrow RCH_2Br \longrightarrow RCH_2MgBr \longrightarrow \begin{cases} \text{alcohols} \\ \text{carboxylic acids} \\ \text{ketones} \end{cases} \tag{2}$$

The conversion of alcohols into alkyl halides is represented by the reaction of a primary alcohol with HBr:

$$RCH_2OH + HBr \rightleftharpoons RCH_2OH_2^+ + Br^- \tag{3}$$

$$RCH_2OH_2^+ + Br^- \longrightarrow RCH_2Br + H_2O \tag{4}$$

The conversion can also be made by the use of PBr_3 or (for alkyl iodides) $P + I_2$.

This reaction is quite general; the ease of replacement of —OH by —Br* is greatest in tertiary alcohols, in which the reaction occurs almost at once at ordinary temperature. Secondary and primary alcohols require longer times and higher temperature but give satisfactory yields. Allylic and benzylic alcohols react rapidly.

Exercise 8.1

Write equations for the reactions required for the following transformations:†

(*a*) $CH_3CH_2CH_2OCOCH_3 \longrightarrow CH_3CH_2CH_2Br$

(*b*) $(CH_3)_3COH \longrightarrow (CH_3)_3CCl$

(*c*) $C_6H_5CH_2OCH_3 \longrightarrow C_6H_5CH_2Br$

(*d*) $CH_2{=}CHCH_2OH \longrightarrow CH_2{=}CHCH_2Br$

(*e*) $C_6H_5CHO \longrightarrow C_6H_5{-}\underset{\displaystyle Br}{\underset{|}{C}}HCH_2C_6H_5$

(*f*) $(CH_3)_3COH \longrightarrow (CH_3)_3C{\cdot}COOH$

(*g*) $CH_3CH_2CH_2OH \longrightarrow CH_3CH_2CH_2CH_2Br$

(*h*) $CH_2{=}CHCH_2OH \longrightarrow CH_2{=}CHCH_2COOH$

† In some of these problems additional steps (such as Grignard reactions, saponification, reduction, oxidation) will be required at some point in the series.

C. DISPLACEMENT REACTIONS OF ALKYL HALIDES

The nucleophilic displacement reaction is used for many transformations of which the following are typical examples.

(a) Ethers. A general expression is

$$RO^- + R'CH_2Br \longrightarrow ROCH_2R' + Br^- \tag{5}$$

This reaction, known as the "Williamson ether synthesis," is most effective with primary alkyl halides, allylic halides and benzylic halides:

$$Na^+CH_3O^- + CH_3CH_2CH_2Br \longrightarrow CH_3CH_2CH_2OCH_3 + NaBr \tag{6}$$

* In most cases the bromo compounds are the preferred halides; they are readily prepared, are more reactive than chlorides, and, in preparation, purification, and preservation, offer some advantages over iodides.

$$Na^+CH_3CH_2O^- + CH_2{=}CHCH_2I \longrightarrow CH_2{=}CHCH_2OCH_2CH_3 + NaCl \qquad (7)$$

$$\text{methyl } \beta\text{-D-glucopyranoside} \xrightarrow[Ag_2O]{CH_3I} \text{(2,3,4,6-tetra-}O\text{-methyl derivative)} + AgI \qquad (8)$$

methyl β-D-glucopyranoside

$$C_6H_{11}O^-Na^+ + C_6H_5CH_2Br \longrightarrow C_6H_{11}OCH_2C_6H_5 + NaBr \qquad (9)$$

There are limitations to the generality of the Williamson ether synthesis. Tertiary halides give little or no ether when treated with sodium alkoxides. The principal reaction is the dehydrohalogenation of the halide:

$$\text{1-chloro-1-methylcyclohexane} + CH_3O^- \longrightarrow \text{1-methylcyclohexene} \text{ not } \text{1-methoxy-1-methylcyclohexane} \qquad (10)$$

$$(CH_3)_3CBr + CH_3O^- \longrightarrow (CH_3)_2C{=}CH_2, \text{ not } (CH_3)_3COCH_3 \qquad (11)$$

Secondary halides often give poor yields of ethers, some dehydrohalogenation occurring:

$$CH_3CH_2CHBrCH_3 + CH_3CH_2O^- \longrightarrow CH_3CH{=}CHCH_3 + CH_3CH_2CH(OCH_2CH_3)CH_3 \qquad (12)$$

It is to be noted that the same reaction, occurring intramolecularly, is involved in the formation of cyclic ethers. In particular, the ring closure of 1,2-halohydrins is a convenient route to expoxides:

$$HOCH_2CH_2Br \xrightarrow{OH^-} \{^-OCH_2CH_2Br\} \longrightarrow H_2C\overset{O}{—}CH_2 \qquad (13)$$

$$\text{trans-2-chlorocyclohexanol} \xrightarrow{NaOH} \text{cyclohexene oxide} + NaCl \qquad (14)$$

Exercise 8.2

Write equations for the reaction required for a practical synthesis of the following ethers:

(*a*) $(CH_3)_2CHOCH_2CH_3$

(*b*) 1-methoxy-1-methylcyclohexane (ring bearing CH_3 and OCH_3)

(*c*) $CH_2{=}CHCH_2OCH_3$

(*d*) $C_6H_5CH_2OCH_2C_6H_5$

(*e*) $CH_3HC\overset{O}{-}CHCH_3$ (*cis*)

(*f*) $C_6H_5OCH_2CH{=}CH_2$

(*g*)* naphthalene–OCH_3

(*h*) $C_6H_5OCH_2COOH$

(*i*) $(CH_3)_3COCH_2C_6H_5$

(*j*) $C_6H_5OCH_2CH_2OC_6H_5$

(*k*) ring with O; substituents CH_3CH_2, OCH_3, H, H, H, H, CH_3O, OCH_3

(*l*) $CH_3{-}\underset{OCH_3}{CH}COOCH_3$

*(g) requires a modification of the Williamson synthesis, in which an aryloxide (e.g., phenoxide, $C_6H_5O^-$) is used instead of an alkoxide. The Williamson synthesis differs from the latter only in the conditions used; since phenols are acidic compounds, their ionization is essentially complete in strongly alkaline aqueous solution. Thus, the reaction called for in (g) can be accomplished by treatment of an alkaline solution of β-naphthol with methyl iodide (or, as described further on in this Part, in Section D, with dimethyl sulfate.

(b) Amines. The nucleophilic nitrogen atom in ammonia and amines can displace halide ion from alkyl halides in the following way:

$$RCH_2Br + NH_3 \longrightarrow RCH_2\overset{+}{N}H_3\ Br^- \qquad (15)$$

$$R_2CH_2Br + R'NH_2 \longrightarrow RCH_2\overset{+}{N}H_2R'\ Br^- \qquad (16)$$

$$RCH_2Br + (CH_3)_3N \longrightarrow RCH_2\overset{+}{N}(CH_3)_3\ Br^- \qquad (17)$$

$$RCH_2Br + C_5H_5N \longrightarrow C_5H_5\overset{+}{N}CH_2R\ Br^- \qquad (18)$$

It will be noticed that the products of these reactions are ammonium salts. The detailed description of the reaction shows this:

$$R_3N\!:\longrightarrow \underset{\displaystyle R}{\underset{|}{CH_2}}\!-\!Br \longrightarrow \left.R_3\overset{+}{N}\!-\!\underset{\displaystyle R}{\underset{|}{CH_2}}\right\} Br^- \tag{19}$$

When the amine is tertiary (as in Equations 17 and 18), the product, a quarternary ammonium salt, is stable and is usually formed smoothly (without complication) and in good yield. When the amine is ammonia, RNH_2 or RNH, the product is a protonated amine:

$$R_2\overset{\displaystyle H}{\overset{|}{N}}\!: + \underset{\displaystyle R'}{\underset{|}{CH_2Br}} \longrightarrow R_2\overset{+}{N}(H)\!-\!CH_2R' + Br^- \tag{20}$$

Consequently, in the presence of excess R_2NH, proton exchange occurs, and the free amine is liberated and can react further with alkyl halide:

$$R_2\overset{+}{N}HCH_2R' + R_2NH \rightleftharpoons R_2NCH_2R' + R_2\overset{+}{N}H_2 \tag{21}$$

$$R_2NCH_2R' + R'CH_2Br \longrightarrow R_2\overset{+}{N}(CH_2R')_2 + Br^- \tag{22}$$

It will be seen, therefore, that the reactions of Equations 15, 16, and 20 do not represent the final result; further reaction according to Equations 21 and 22 can and usually will occur to give a mixture of amines. As a result, the monoalkylation of ammonia, RNH_2, and R_2NH with an alkyl halide is not a generally practicable procedure, for mixtures will be formed.

The process known as "exhaustive methylation," usually carried out with methyl iodide, leads as the final result to the trimethylammonium compound:

$$RCH_2NH_2 + \text{excess } CH_3I + \text{a base} \xrightarrow[\text{above}]{\text{as in 19–22}} RCH_2\overset{+}{N}(CH_3)_3I^- \tag{23}$$

The purpose of the added base (for example, sodium carbonate) is to provide for the deprotonation of the ammonium salts $R\overset{+}{N}H_2CH_3$, etc. so that the reactions corresponding to Equation 22 ($-CH_2R' = CH_3$) can proceed.

Exercise 8.3

Write the equations for the following reactions, showing principal products only:

(*a*) $CH_3N(C_2H_5)_2$ + CH_2Br ⟶

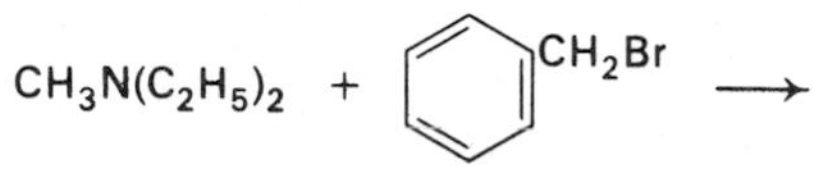

(*b*) N + $BrCH_2COOEt$ ⟶

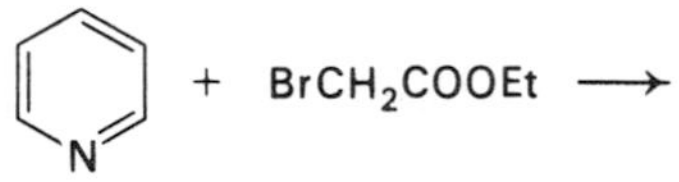

(*c*) CH_3O $N(CH_3)_2$ + CH_3OSO_2 Br ⟶

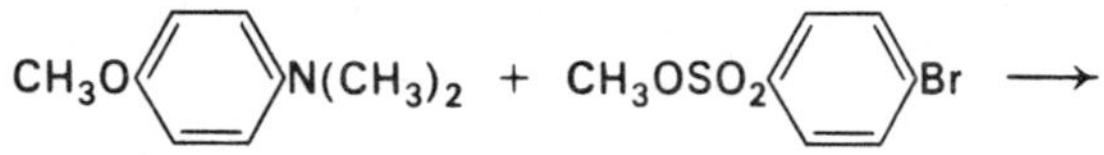

(*d*) $(CH_3)_2NH + CH_2{=}CHCH_2Cl$ ⟶

(*e*) 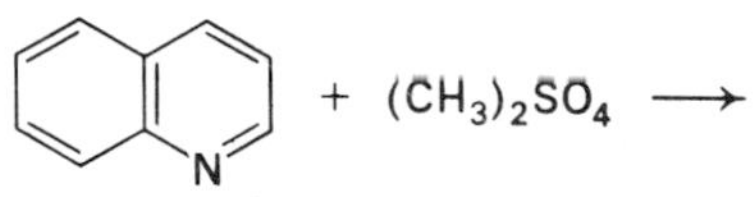 N + $(CH_3)_2SO_4$ ⟶

(*f*) H_3C $NHCH_3$ + C_2H_5Br ⟶

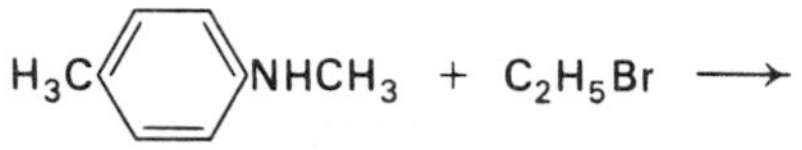

(*g*) NCH_3 + CH_3I ⟶

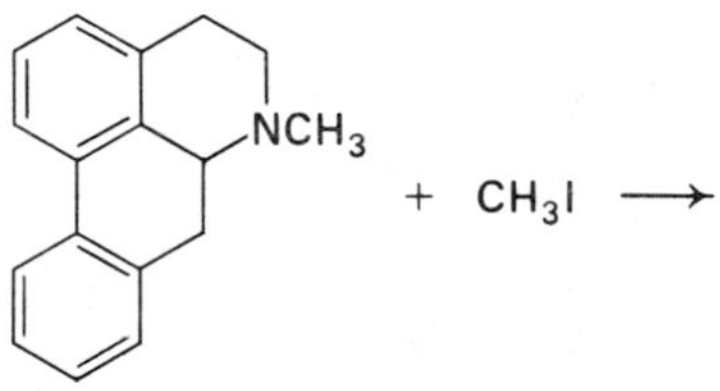

(*h*) 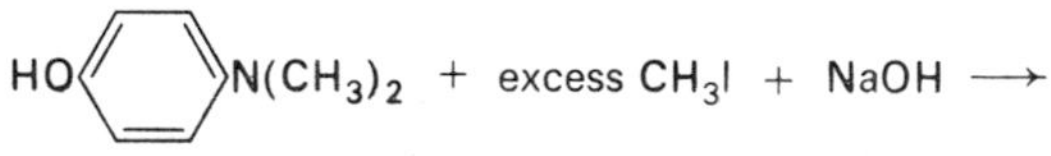 HO $N(CH_3)_2$ + excess CH_3I + NaOH ⟶

(*i*) NCH_3 + C_2H_5I ⟶

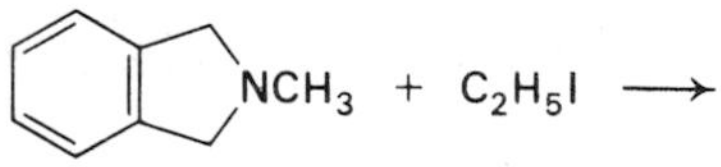

(*j*)* $CH_2N(CH_3)_2$ N H + CH_3I ⟶

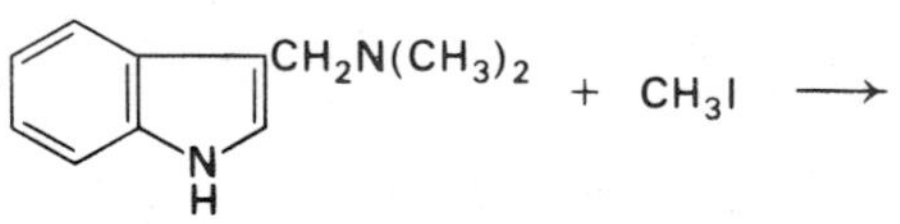

*The principal product of Reaction *j* is a salt having the composition $C_{20}H_{22}N_3I$. Can you give the structure of this salt, and explain how it is formed?

(c) Nitriles. The displacement of halogen by the nucleophilic cyanide ion, CN^-, provides an effective synthesis of nitriles:

$$CH_3CH_2CH_2Br + KCN \longrightarrow CH_3CH_2CH_2CN + KBr \tag{24}$$

$$BrCH_2CH_2Br + 2\,KCN \longrightarrow \begin{matrix} CH_2CN \\ | \\ CH_2CN \end{matrix} + 2\,KBr \tag{25}$$

$$CH_2{=}CHCH_2Br + KCN \longrightarrow CH_2{=}CHCH_2CN \tag{26}$$

$$p\text{-}CH_3OC_6H_4CH_2Cl + KCN \longrightarrow p\text{-}CH_3OC_6H_4CH_2CN \tag{27}$$

Nitriles can be hydrolyzed to carboxylic acids:

$$RCH_2CN + H_2O(H^+) \longrightarrow RCH_2COOH(+NH_4{}^+) \tag{28}$$

or reduced with lithium aluminum hydride (or catalytically) to primary amines:

$$RCH_2CN \xrightarrow[\text{(2) } H_2O/H^+]{\text{(1) } LiAlH_4} RCH_2CH_2NH_2 \tag{29}$$

If instead of water, ethanol or methanol is used in the reaction corresponding to Equation 28, *alcoholysis*, instead of hydrolysis, occurs:

$$RCH_2CN \xrightarrow[(H_2SO_4)]{MeOH} RCH_2COOCH_3 \tag{30}$$

Exercise 8.4

Devise syntheses for the following compounds, using the starting material shown and any other necessary reagents:

(*a*) $(CH_3)_2C{=}CHCH_2OH \dashrightarrow (CH_3)_2C{=}CHCH_2CH_2NH_2$

(*b*) $C_6H_5CH_2OH \dashrightarrow C_6H_5C(CN)(CH_2CH_2COOCH_3)_2$

(*c*) $CH_2{=}CHCH_2Cl \dashrightarrow CH_2{=}CHCH_2COOCH_2CH{=}CH_2$

(*d*) $HOCH_2CH_2OH \dashrightarrow H_2NCH_2CH_2CH_2CH_2NH_2$

(*e*) phthalan (C_6H_4 fused ring: CH_2–O–CH_2) $\dashrightarrow$ 2-indanone (C_6H_4 fused ring: CH_2–CO–CH_2)

Answers to exercises 8.4

(*b*) The final step is a Michael addition of $C_6H_5CH_2CN$ to $CH_2{=}CHCOOCH_3$.
(*e*) The final product is formed by saponification and decarboxylation of

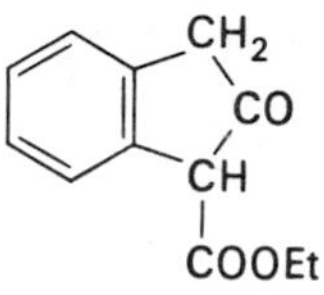

[Structure: benzene ring fused to a five-membered ring of CH_2, CO and CH; COOEt on the CH]

This keto ester is the product of the (Dieckmann) ring closure of

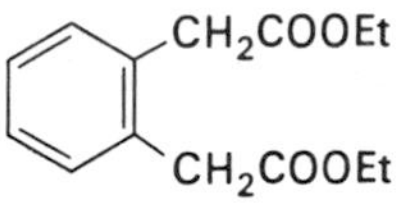

[Structure: benzene ring bearing two ortho CH_2COOEt groups]

The formation of the ester is left to the reader. The requisite *o*-bis-(bromomethyl)-benzene is formed by the action of concentrated HBr upon the ether given as the starting material.

(d) Miscellaneous further examples. The general reaction

$$X{:}^- + \underset{\displaystyle R}{\overset{|}{C}}H_2{-}Y \longrightarrow X{-}\underset{\displaystyle R}{\overset{|}{C}}H_2 + {:}Y^- \tag{31}$$

has now been illustrated by examples in which the nucleophile $X{:}^-$ is OR^-, ammonia, or an amine, CN^- (and, as outlined in Part 6, carbon anions of several kinds). The wide generality of this reaction is such that a variety of nucleophiles of other kinds are capable of taking part as $X{:}^-$ in Equation 31.

The hydrolysis of alkyl halides (Equation 32)

$$RCH_2Br + OH^- \longrightarrow RCH_2OH + Br^- \tag{32}$$

bears an obvious close resemblance to the Williamson synthesis (in which RO^- would be used instead of OH^-). It has not been treated here in detail inasmuch as the principal emphasis has been on converting alcohols into corresponding halides, and then into other compounds. Nevertheless, the reaction in (32) is a generally useful one when the alkyl halide is available as the starting compound in a synthesis.

Another class of nucleophilic reagents is that of certain sulfur compounds. The reaction of alkyl halides with mercaptans in the presence of alkali is the direct counterpart of the Williamson synthesis of ethers

$$RSH + OH^- \rightleftarrows RS^- + H_2O \tag{33}$$

$$RS^- + R'CH_2Br \longrightarrow RSCH_2R' + Br^- \tag{34}$$

and is subject to comparable limitations.

An effective procedure for preparing mercaptans is to use thiourea, NH_2CSNH_2, as the nucleophile. This displays its nucleophilic character in the following way:

$$NH_2-\overset{:NH_2}{\overset{|}{C}}=S \quad \underset{R}{\underset{|}{CH_2}}-Br \longrightarrow \left. NH_2-\overset{^+NH_2}{\overset{\|}{C}}-SCH_2R \right\} Br^- \tag{35}$$

a thiouronium salt

Hydrolysis of the thiouronium compound leads to its smooth conversion into the mercaptan:

$$\left\{ RCH_2SC \begin{matrix} \overset{+}{N}H_2 \\ NH_2 \end{matrix} \right\} X^- \xrightarrow[NaOH]{H_2O} RCH_2S^- + N{\equiv}CNH_2 + X^- \tag{36}$$

The nucleophilic character of the sulfite ion, SO_3^{--}, is displayed at the sulfur, not the oxygen, atom:

$$O=\underset{O^-}{\underset{|}{\overset{O^-}{\overset{|}{S}}}}: \quad \underset{R}{\underset{|}{CH_2}}Br \longrightarrow RCH_2-SO_3^- + Br^- \tag{37}$$

The product (after acidification) is an alkanesulfonic acid.

The nucleophilic character of the anions of carboxylic acids is greater than what might have been expected, because these anions are very weak bases. But the reaction

$$RC \begin{matrix} \nearrow O \\ \searrow O^- \end{matrix} \quad \underset{R'}{\underset{|}{CH_2}}X \longrightarrow RC \begin{matrix} \nearrow O \\ \searrow OCH_2R' \end{matrix} + X^- \tag{38}$$

is often a useful one. Its principal application is in the identification and characterization of carboxylic acids. The halide ($R'CH_2X$ in Equation 38) is chosen so that the resulting esters ($RCOOCH_2R'$ in Equation 38) are crystalline solids whose melting points are used as characterizing properties. A number of such halides are useful; a typical reagent is *p*-nitrobenzyl bromide. The reaction of this highly reactive halide with acetate ion is the following:

$$CH_3COO^- + p\text{-}NO_2C_6H_4CH_2Br \longrightarrow p\text{-}NO_2C_6H_4CH_2OOCCH_3 + Br^- \tag{39}$$

The reaction is quite general, and the melting points of the *p*-nitrobenzyl esters of most of the common (both aliphatic and aromatic) acids are recorded in handbooks.

Exercise 8.5

Write the equations for the preparation of each of the following products by a route including a nucleophilic displacement reaction; add any additional necessary steps:

(*a*) $CH_3COOCH_2C_6H_5$

(*b*) $CH_2{=}CHCH_2SCH_2C_6H_5$

(*c*) $C_6H_5N(CH_3)CH_2C_6H_5$

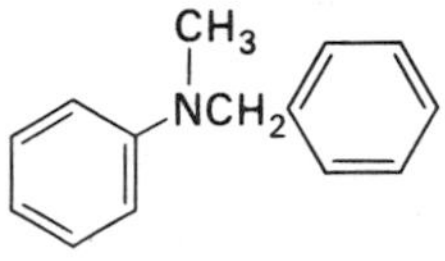

(*d*) $CH_3COOCH_2CH_2OCOCH_3$

(*e*) thiane (six-membered ring containing S)

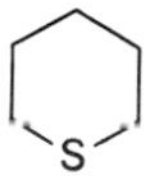

(*f*) $C_6H_5{-}C_6H_4COCH_2OCOCH(CH_3)_2$

(*g*) $C_6H_5CH_2SO_3H$

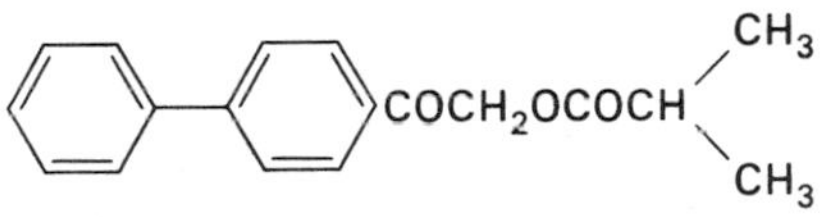

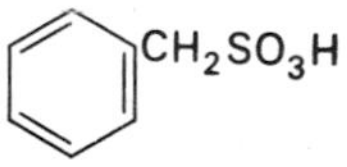

(*h*) $(CH_3)_2CHCH_2SC({=}NH)NH_2$

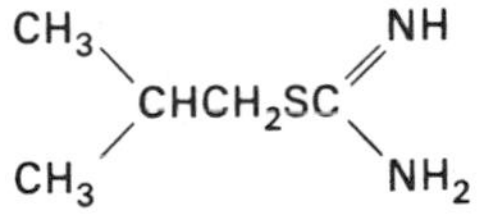

(*i*) $(CH_3)_3S^+I^-$

(*j*) $C_5H_5\overset{+}{N}{-}OCH_3\ \ I^-$

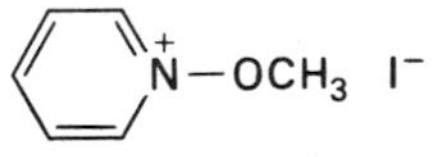

Notes about exercise 8.5

(*a*) This is directly comparable to Equation 39.

(*e*) $Br(CH_2)_5Br + SH^- \longrightarrow HS(CH_2)_5Br \xrightarrow{OH^-}$ thiane (six-membered ring containing S)

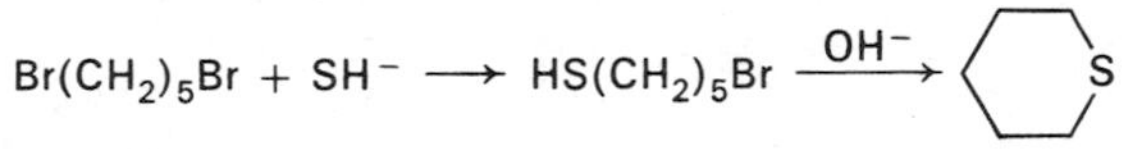

Query: How could you prepare (six-membered ring)N—CH_3?

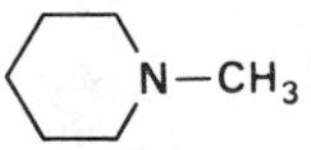

(*f*) This is another variant of the procedure of which Equation 38 is an example. This is prepared by the reaction of the salt of $(CH_3)_2CHCOOH$ with a halogen-containing compound, RCH_2Br. What is R in this example?

(*i*) $(CH_3)_2S$ is a nucleophile, identical in kind (but not in degree) with $(CH_3)_3N$.
(*j*) Treatment of pyridine with hydrogen peroxide converts it into the N-oxide:

$$C_5H_5N \xrightarrow{H_2O_2} C_5H_5\overset{+}{N}{-}O^-$$

Would this be a nucleophile? Where is the nucleophilic center? What would be the structure of the salt formed by the reaction of pyridine N-oxide with HCl?

D. DISPLACEMENT REACTIONS OF SULFONIC ACID ESTERS

Reaction of an alcohol in the presence of a base (such as sodium hydroxide or pyridine) with a sulfonyl chloride yields an ester of the sulfonic acid. The choice of the sulfonyl chloride is usually dictated by accessibility; *p*-toluenesulfonyl chloride and methanesulfonyl chloride are readily available and inexpensive and are commonly used:

$$CH_3OH + CH_3C_6H_4SO_2Cl \xrightarrow{NaOH} CH_3OSO_2C_6H_4CH_3 \qquad (40)$$

$$(CH_3)_2CHOH + CH_3SO_2Cl \xrightarrow{NaOH} CH_3SO_2OCH(CH_3)_2 \qquad (41)$$

In the following discussion the general designation RSO_2— will be used for the sulfonyl group. In addition to esters of this kind, which are prepared as a part of a synthetic process, the methyl ester of sulfuric acid, dimethyl sulfate ($CH_3OSO_2OCH_3$), is an expensive commercial product and is widely employed as a methylating agent.

Sulfonic and sulfuric esters are effective alkylating agents and can be regarded for practical purposes as equivalent to the corresponding bromides or iodides.

Thus, the following pairs of reactions (using the typical nucleophiles CH_3O^- and CN^- in the examples) represent essentially equivalent alternatives. The "leaving groups" Br^- and $CH_3SO_2O^-$, despite the fact that they are chemically quite unrelated, are substantially the same in their behavior (refer again to Section 8.A):

$$CH_3Br + CH_3O^- \longrightarrow CH_3OCH_3 + Br^- \qquad (42)$$

$$CH_3OSO_2CH_3 + CH_3O^- \longrightarrow CH_3OCH_3 + CH_3SO_2O^- \qquad (43)$$

$$C_6H_5CH_2Cl + CN^- \longrightarrow C_6H_5CH_2CN + Br^- \qquad (44)$$

$$C_6H_5CH_2OSO_2CH_3 + CN^- \longrightarrow C_6H_5CH_2CN + CH_3SO_2O^- \qquad (45)$$

It will be apparent that this procedure can provide many advantages over the route: alcohol $\longrightarrow$ alkyl halide $\longrightarrow$ product. For one thing the stereochemistry of the displacement reaction is under better control in the sulfonic ester route. This is because in the series starting with the optically active alcohol

$$\underset{\mathrm{OH}}{\mathrm{R\overset{*}{C}HR'}} + \mathrm{HBr} \xrightarrow[A]{} \mathrm{R{-}\underset{Br}{CH}{-}R'} \xrightarrow[B]{\mathrm{X:^-}} \mathrm{R\overset{X}{C}H'} + \mathrm{Br^-} \tag{46}$$

inversion of configuration (at *) takes place in step *A*, but the stereochemical result may in some cases not be "clean;" that is, a certain (and often unpredictable) amount of racemization can occur. A second inversion takes place in step *B*.

In the sulfonic ester route, the step

$$\overset{\mathrm{OH}}{\mathrm{R\underset{*}{C}HR'}} \xrightarrow[\text{pyridine}]{\mathrm{CH_3SO_2Cl}} \overset{\mathrm{OSO_2CH_3}}{\mathrm{R\underset{*}{C}HR'}}$$

occurs with no change in configuration at the starred C's; inversion occurs only during replacement:

$$\overset{\mathrm{OSO_2CH_3}}{\mathrm{RCHR'}} + \mathrm{X:^-} \longrightarrow \mathrm{R{-}\underset{X}{CH}{-}R'} + \mathrm{CH_3SO_2O^-} \tag{47}$$

Sulfonic esters may be substituted advantageously for corresponding halides in nearly all reactions in which a nucleophilic displacement is carried out. They can be used, for example, in the C-alkylation reactions discussed in Part 6, Section A.

The ready availability of dimethyl sulfate, $CH_3OSO_2OCH_3$, makes it a valuable methylating agent; for example*

$$\mathrm{C_6H_5O^-(Na^+)} + \mathrm{CH_3OSO_2OCH_3} \longrightarrow \mathrm{C_6H_5OCH_3} + \mathrm{CH_3OSO_2O^-(Na^+)} \tag{48}$$

$$\mathrm{C_5H_5N} + \mathrm{CH_3OSO_2OCH_3} \longrightarrow \mathrm{C_5H_5\overset{+}{N}{-}CH_3} + \mathrm{CH_3OSO_2O^-} \tag{49}$$

$$\mathrm{C_6H_5COO^-(Na^+)} + \mathrm{CH_3OSO_2OCH_3} \longrightarrow \mathrm{C_6H_5COOCH_3} + \mathrm{CH_3OSO_2O^-} \tag{50}$$

It will be recognized that these reactions are the exact counterparts of those that were described above, in which CH_3I was used.

---

* Both $—CH_3$ groups of $CH_3OSO_2OCH_3$ can be utilized if the proper experimental conditions are used. In most cases of its use in the laboratory, only one $—CH_3$ is utilized, as in (48).

Exercise 8.6

Suggest a mechanism (i.e., the stepwise course) for each of the following reactions:

(*a*) $C_6H_4(CH_2OH)_2$ (1,2-) + $CH_3C_6H_4SO_2Cl$ $\xrightarrow{NaOH}$ $C_6H_4(CH_2)_2O$ (cyclic ether)

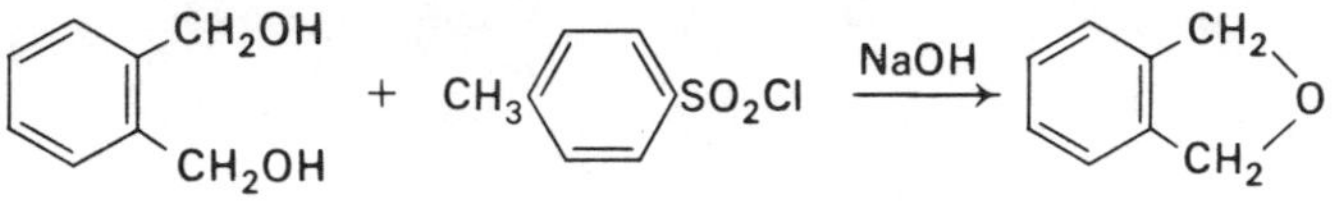

(*b*) cyclohexane with $OCOCH_3$ and OSO_2CH_3 (trans) $\xrightarrow[NaOAc]{EtOH}$ cyclohexane-fused 1,3-dioxolane with C bearing CH_3 and OCH_2CH_3

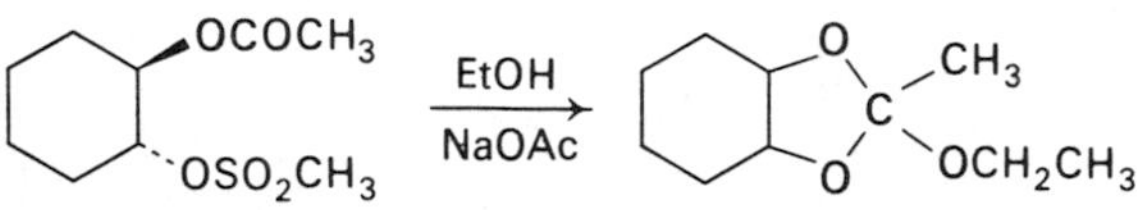

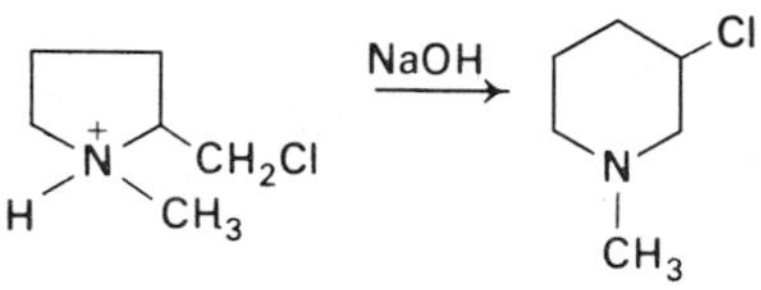

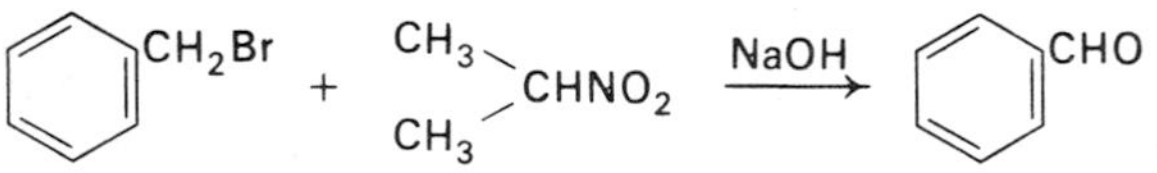

(*c*) pyrrolidinium ($\overset{+}{N}$ bearing H and CH_3) with CH_2Cl $\xrightarrow{NaOH}$ 3-chloro-1-methylpiperidine (Cl, N–CH_3)

(*d*) $C_6H_5CH_2Br$ + $(CH_3)_2CHNO_2$ $\xrightarrow{NaOH}$ C_6H_5CHO

(*e*)* $C_6H_5CH_2OSO_2C_6H_5$ + $(CH_3)_2S{=}O$ $\longrightarrow$ C_6H_5CHO

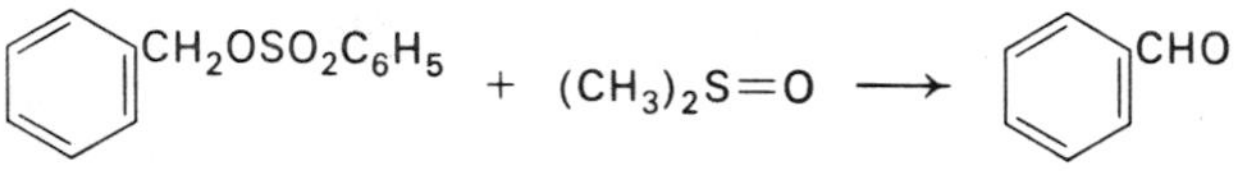

*(*e*) The first step is a nucleophilic displacement reaction, with $(CH_3)_2S{=}O$ acting as the nucleophile (at oxygen). The dimethyl sulfoxide is converted into dimethyl sulfide, Me_2S.

PART 9

Aromatic substitution reactions

A. ELECTROPHILIC SUBSTITUTION

Electrophilic substitution into the benzene or substituted benzene ring can be discussed in terms of three principal reactions:

halogenation,

$$C_6H_6 + Br_2 \xrightarrow{FeBr_3} C_6H_5Br + HBr \quad (1)$$

nitration,

$$C_6H_6 + HNO_3 \xrightarrow{H_2SO_4} C_6H_5NO_2 + H_2O \quad (2)$$

acylation,

$$C_6H_6 + CH_3COCl \xrightarrow{AlCl_3} C_6H_5COCH_3 + HCl \quad (3)$$

All of these may be viewed in terms of a common mechanism: the attack upon the benzene ring (a nucleophile) by an electrophilic reagent.

$$C_6H_6 + X^+ \longrightarrow [C_6H_6X]^+ \xrightarrow{-H^+} C_6H_5X \qquad (4)$$

The *nucleophile* in this reaction is the aromatic compound: the *electrophile* is the underlined electron-deficient species, formed in the following ways:

$$Br{-}BrFeBr_3 = \underline{\overset{\delta+}{Br}}\cdots Br \cdots \overset{\delta-}{FeBr_3} \qquad (5)$$

$$HNO_3 + H_2SO_4 \rightleftharpoons H_2ONO_2^+ + HSO_4^- \rightarrow \underline{NO_2^+} + H_2O \qquad (6)$$

$$CH_3\overset{O}{\overset{\|}{C}}{-}ClAlCl_3 = \underline{CH_3{-}\overset{O}{\overset{\|}{C}}}{}^{\delta+}\cdots Cl \cdots \overset{\delta-}{AlCl_3} \qquad (7)$$

B. ACTIVATING AND DIRECTING EFFECTS OF SUBSTITUENTS

When the aromatic ring is substituted, the substituent (Y in Equation 8) affects (a) the rate of the substitution reaction and (b) the position at which electrophilic attack occurs.

$$Y{-}C_6H_5 + X^+ \xrightarrow{\text{as in (4)}} Y{-}C_6H_4{-}X \qquad (8)$$

If Y furnishes electrons to the ring, it increases the nucleophilic character of the ring and increases the rate of the reaction. Thus, *activating substituents are typically groups or atoms that possess unshared electrons* (or, as in the methyl group, an inductive electron release); for example:

$$R\ddot{O}{-} \quad R_2\ddot{N}{-} \quad H\ddot{O}{-} \quad {}^{-}\!:\ddot{O}{-} \quad \overrightarrow{CH_3}{-}$$

It would be expected, and is in fact observed, that the electron-donating capacity of the group Y would be directly related to the base strength of the corresponding compound HY. Thus,

base strength	$^{-}OH > HNR_2 > ROH \sim HOH$
activating ability	$^{-}O{-}Ar > R_2N{-}Ar > RO{-}Ar \sim HO{-}Ar$

Acylation of O— and N— leads to diminished activating ability. The substituents

$$CH_3CO{-}O{-}Ar \quad \text{and} \quad CH_3CO\overset{R}{\overset{|}{N}}{-}Ar$$

are activating but much less so than, respectively, HO— and RNH—.

The CH_3O— group activates the aromatic ring to electrophilic attack. Does CH_3O— also have a *directing* influence? That is, does the nitration (for example) of anisole lead to *o*-, *m*-, or *p*-nitroanisole?

The fact is that only* the *p*- and *o*- compounds are formed. The reason for the *o*-*p*-directing effect of these activating groups will not be dwelt upon here; it can be found in a textbook of organic chemistry.

Electron-withdrawing groups are deactivating and *m*-directing; indeed nitrobenzene is inert to acylation (Equation 3) and is brominated with difficulty.

$$C_6H_5NO_2 \xrightarrow[H_2SO_4]{HNO_3} m\text{-}C_6H_4(NO_2)_2 \qquad (9)$$

$$C_6H_5COOCH_3 \xrightarrow[H_2SO_4]{HNO_3} m\text{-}O_2NC_6H_4COOCH_3 \qquad (10)$$

Relative activating and directing effects of two substituents on an aromatic ring lead to substitution of the incoming group at a position controlled by the more effective electron-donating group; thus, nitration or bromination of the following compounds will occur at the position(s) marked:

$p\text{-}CH_3OC_6H_4NO_2$ $\quad$ $m\text{-}CH_3C_6H_4NO_2$ $\quad$ $p\text{-}CH_3C_6H_4NHCOCH_3$ $\quad$ $p\text{-}H_2NC_6H_4OCH_3$

Halogen (for example, Br) is *deactivating* but nevertheless *o-p-directing*. Thus:

$$C_6H_5Br \xrightarrow[H_2SO_4]{HNO_3} p\text{-}BrC_6H_4NO_2 + o\text{-}BrC_6H_4NO_2 \qquad (11)$$

slow compared with nitration of benzene

* A small amount of the *m*- compound may be formed, but this is relatively minute. For reasons outside the scope of this brief summary discussion the major product is *p*-nitroanisole.

Exercise 9.1

Each of the following compounds will give largely a single monosubstitution product upon nitration. Indicate where the substitution will occur:

(*a*) 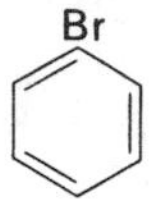 Br

(*b*) $COCH_3$

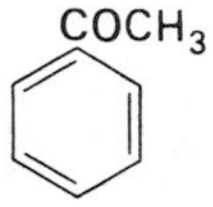

(*c*) 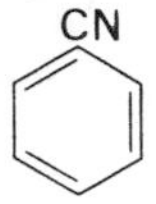 CN

(*d*) OCH_3, OCH_3

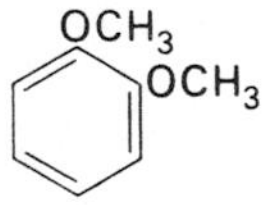

(*e*) NO_2, CH_3

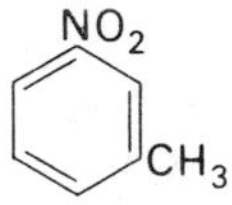

(*f*) 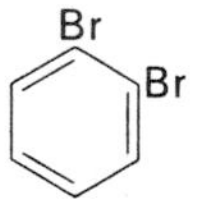 Br, Br

(*g*) $CH(NO_2)_2$

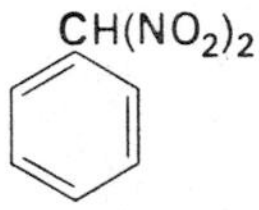

(*h*) 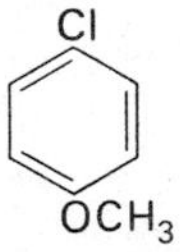 Cl, OCH_3

(*i*) $N(COCH_3)_2$

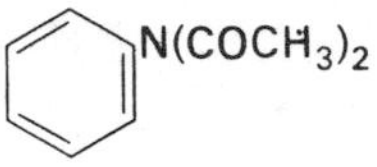

(*j*) OCH_3, $N(COCH_3)_2$

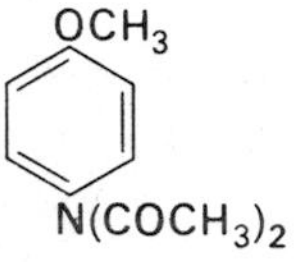

(*k*) 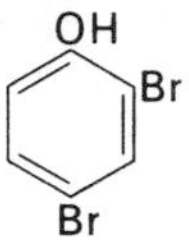 OH, Br, Br

(*l*) NO_2, NO_2

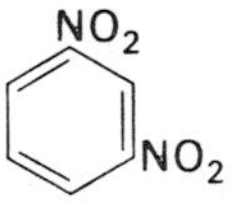

(*m*) NO_2, CN

(*n*) $N(CH_3)_2$, OCH_3

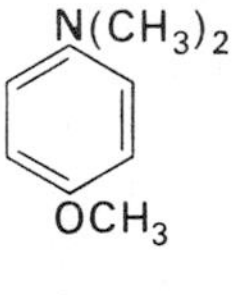

(*o*) $\overset{+}{N}(CH_3)_3$

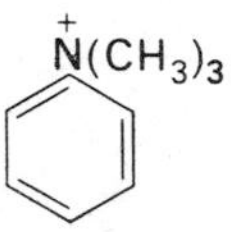

(*p*) NO_2, NO_2

(*q*) $\overset{+}{O}$

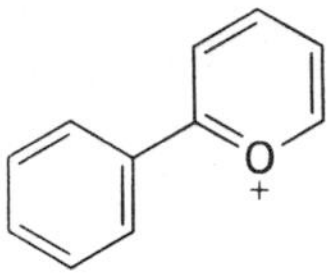

(*r*) NO_2

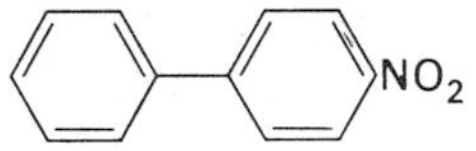

(*s*) $NHCOCH_3$, $OCOCH_3$

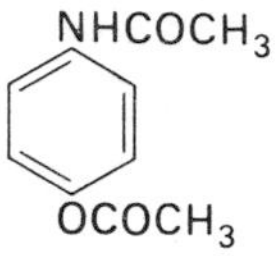

(*t*) $\overset{+}{S}(CH_3)_2$

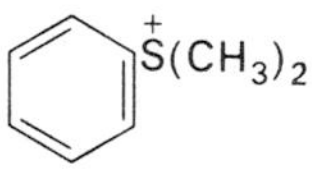

(*u*) CF_3

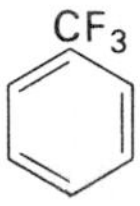

(*v*) CF_3, CH_3

(*w*) CH_2CH_2COOH

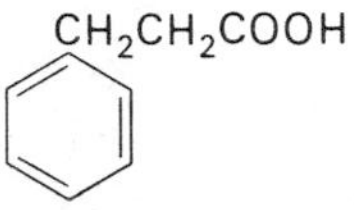

(*x*) $COOCH_3$, Br

Exercise 9.2

Starting with any *monosubstituted* benzene, show how each of the following could be prepared:*

(*a*) 1-CH_3, 4-NO_2 benzene

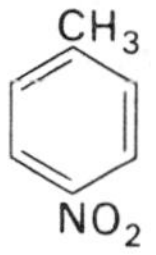

(*b*) 1-CH_3, 4-Br benzene

(*c*) 1-CH_3, 2-NO_2, 4-$COCH_3$ benzene

(*d*) 1,4-dibromobenzene (Br, Br)

(*e*) 1-$NHCOCH_3$, 4-Br benzene

(*f*) 1-$CH{=}CHCOOH$, 3-NO_2 benzene

(*g*) 1-CH_3, 2-NO_2, 4-NO_2 benzene

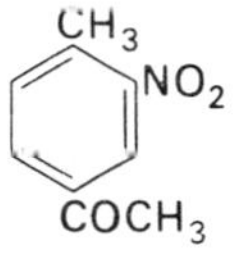

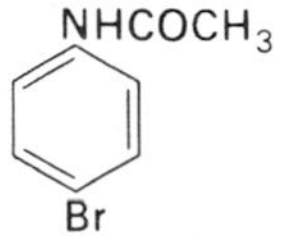

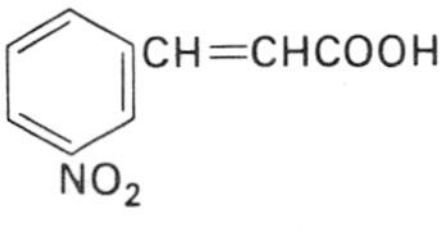

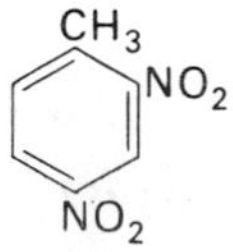

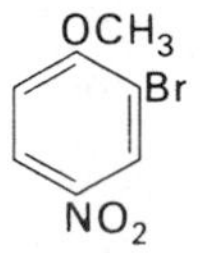

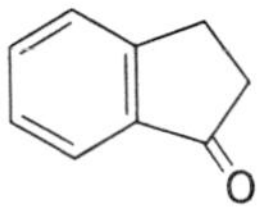

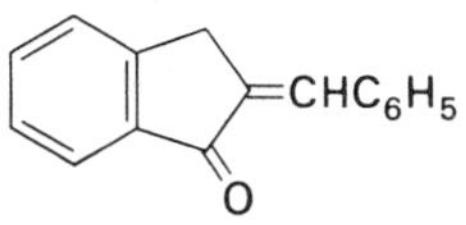

(*h*) 1-OCH_3, 2-Br, 4-NO_2 benzene

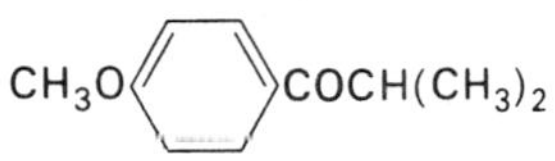

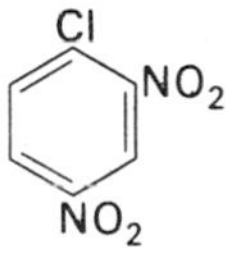

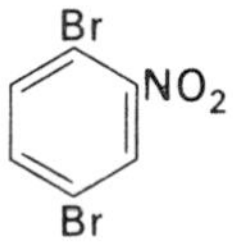

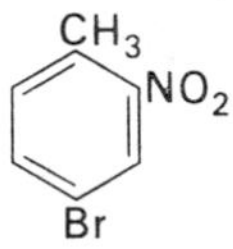

(*i*) O

(*j*) 2-($=CHC_6H_5$)indan-1-one

(*k*) CH_3O–C_6H_4–$COCH(CH_3)_2$ (para)

(*l*) 1-Cl, 2-NO_2, 4-NO_2 benzene

(*m*) 1-Br, 2-NO_2, 4-Br benzene

(*n*) 1-CH_3, 2-NO_2, 4-Br benzene

(*o*) CH_3O–C_6H_4–$COCH_2CH_2$–C_6H_5 (para)

* For the purpose of the Exercise the arbitrary assumption may be made that when

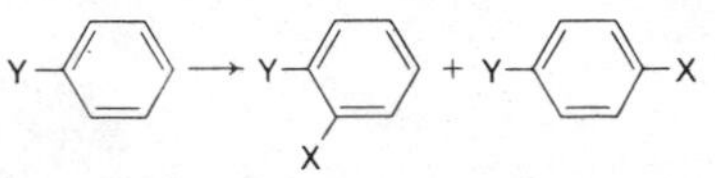

$$Y\text{–}C_6H_5 \longrightarrow Y\text{–}C_6H_4\text{–}X\ (meta) + Y\text{–}C_6H_4\text{–}X\ (para)$$

the *p*-substituted compound may be regarded as the principal product.

Notes about Exercise 9.2

1. Strongly deactivating substituents (for example, NO_2) are usually introduced late in a sequential synthesis; their presence normally makes electrophilic substitution reactions very slow or impossible:

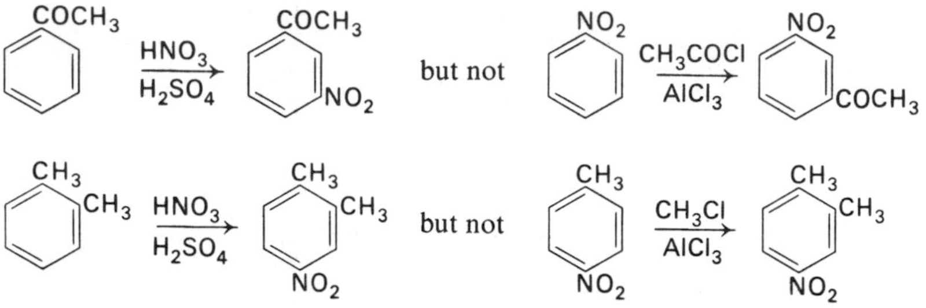

$COCH_3$-benzene $\xrightarrow[H_2SO_4]{HNO_3}$ $COCH_3$, NO_2-benzene but not NO_2-benzene $\xrightarrow[AlCl_3]{CH_3COCl}$ NO_2, $COCH_3$-benzene

CH_3, CH_3-benzene $\xrightarrow[H_2SO_4]{HNO_3}$ CH_3, CH_3, NO_2-benzene but not CH_3, NO_2-benzene $\xrightarrow[AlCl_3]{CH_3Cl}$ CH_3, CH_3, NO_2-benzene

2. Successive introduction of strongly electron-attracting groups makes further substitution difficult or impossible.

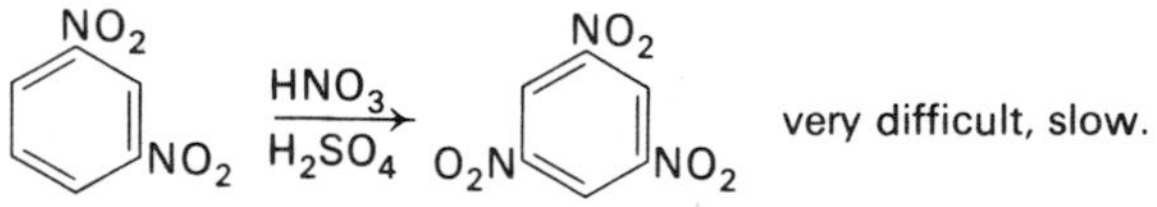

NO_2, NO_2-benzene $\xrightarrow[H_2SO_4]{HNO_3}$ O_2N, NO_2, NO_2-benzene very difficult, slow.

But:

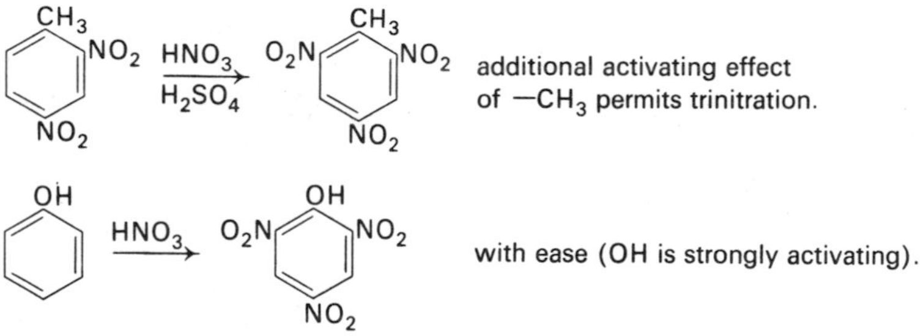

CH_3, NO_2, NO_2-benzene $\xrightarrow[H_2SO_4]{HNO_3}$ CH_3, O_2N, NO_2, NO_2-benzene additional activating effect of $-CH_3$ permits trinitration.

OH-benzene $\xrightarrow{HNO_3}$ OH, O_2N, NO_2, NO_2-benzene with ease (OH is strongly activating).

Electrophilic substitution leading to ring formation

Aromatic substitution with closure of a ring is observed in many forms. Most of these are directly or mechanistically related to the electrophilic acylation reaction (the Friedel-Crafts reaction in Equation 3); that is, they represent the nucleophilic attack of the aromatic nucleus upon an electrophilic (electron-deficient) carbon atom.

In the conventional Friedel-Crafts reaction, acylation occurs in the following manner:

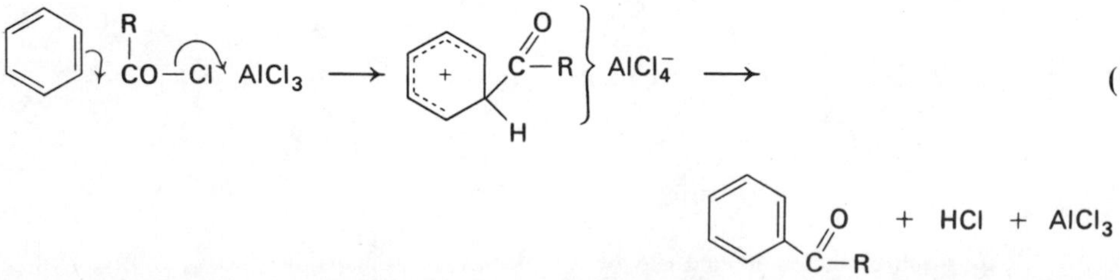

$$C_6H_6 + R{-}CO{-}Cl \; AlCl_3 \longrightarrow \left\{ [C_6H_6(H)(C(=O){-}R)]^+ \right\} AlCl_4^- \longrightarrow C_6H_5{-}C(=O){-}R + HCl + AlCl_3 \qquad (12)$$

When the acyl function is part of a substituent attached to the ring, cyclization occurs:

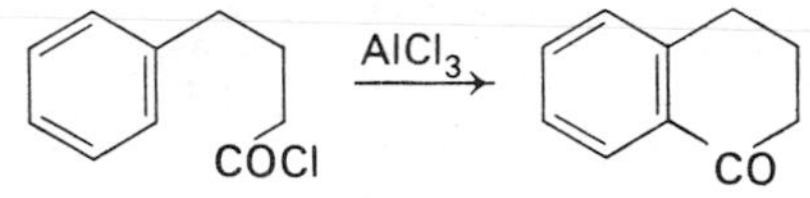

AlCl$_3$

COCl → CO

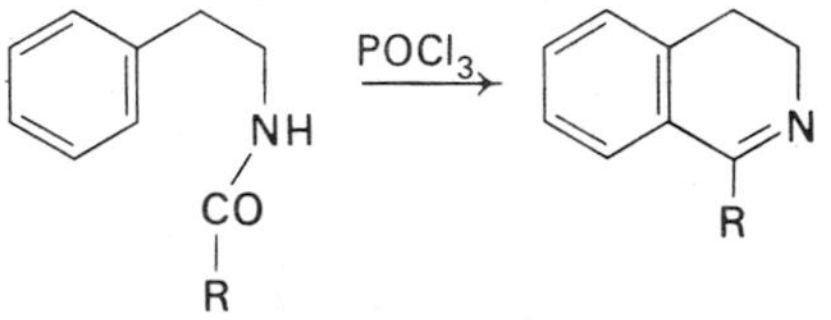

POCl$_3$

NH, CO, R → N, R

the Bischler-Napieralski synthesis of isoquinolines

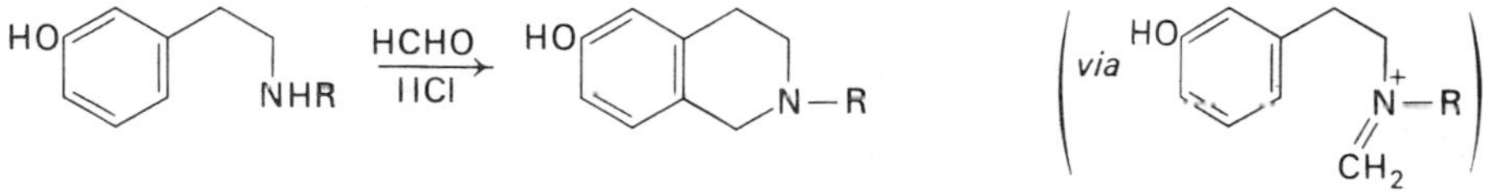

HO, NHR $\xrightarrow[HCl]{HCHO}$ HO, N—R (*via* HO, $\overset{+}{N}$—R, CH_2)

the Mannich reaction

Examples

(1) How could CH_3O [indanone structure] be prepared?

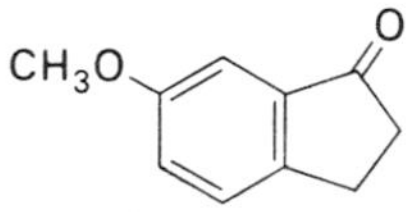

The final step is an intramolecular Friedel-Crafts cyclization reaction:

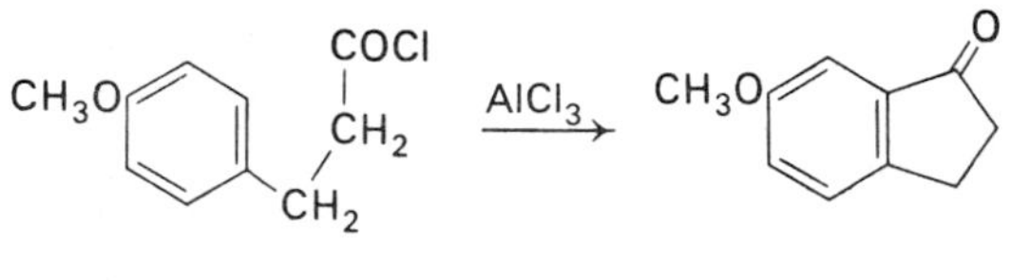

CH_3O–C_6H_4–CH_2CH_2COCl $\xrightarrow{AlCl_3}$ CH_3O [indanone]

The *p*-methoxyhydrocinnamic acid is prepared from *p*-methoxybenzaldehyde (anisaldehyde) as follows:

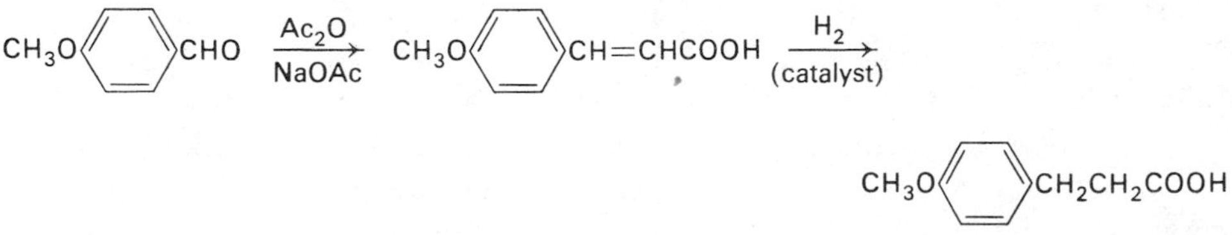

CH_3O–C_6H_4–CHO $\xrightarrow[NaOAc]{Ac_2O}$ CH_3O–C_6H_4–CH=CHCOOH $\xrightarrow[\text{(catalyst)}]{H_2}$ CH_3O–C_6H_4–CH_2CH_2COOH

(2) How could H_3C [tetralone structure] be prepared?

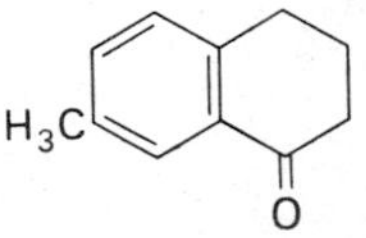

The final step is the ring closure, as in Example 1, of 4-(*p*-tolyl)-butanoic acid, prepared as follows:

H_3C–C_6H_4–H + $H_2C(CO)$–$H_2C(CO)$ >O $\xrightarrow{AlCl_3}$ H_3C–C_6H_4–CO–CH_2–CH_2COOH $\xrightarrow[\text{CO to } CH_2]{\text{reduce}}$

H_3C–C_6H_4–$CH_2CH_2CH_2COOH$ $\xrightarrow{SOCl_2}$ H_3C–C_6H_4–$CH_2CH_2CH_2COCl$ $\xrightarrow{AlCl_3}$ H_3C-tetralone (C=O)

C. NUCLEOPHILIC SUBSTITUTION

Halogenated benzenes (bromobenzene, chlorobenzene) are quite inert to the attack of nucleophilic reagents, and remain unaffected by treatment with solutions of sodium hydroxide, sodium alkoxides and amines. Chlorobenzene reacts with sodium hydroxide only under severe experimental conditions (high temperature and pressure); under those conditions the reaction (which yields phenol) may proceed by a mechanism other than direct substitution of Cl by attack of hydroxide ion.

When a *para* (or *ortho*) nitro group is present, the halogen atom is replaced with comparative ease:

$C_6H_5Br \xrightarrow[NaOMe]{MeOH}$ little or no reaction

$p\text{-}BrC_6H_4NO_2 \xrightarrow[NaOMe]{MeOH} p\text{-}MeOC_6H_4NO_2 + NaBr$

Two nitro groups (*o*- and *p*- to the halogen) greatly increase the ease with which this substitution occurs; 2,4,6-trinitrobromobenzene reacts very rapidly with nucleophilic reagents.

The explanation for the unreactivity of bromobenzene is that the transition state for attack of hydroxide ion appears to be one of high activation energy, for the aromatic structure is replaced by one of lower stability. If the negatively charged intermediate were replaced by one in which additional delocalization (resonance stabilization) were afforded by a substituent capable of accommodating the negative charge, it would be expected that a lower activation energy would be required and the reaction rate would increase.

This is in fact accomplished by the presence of the *p*-nitro group:

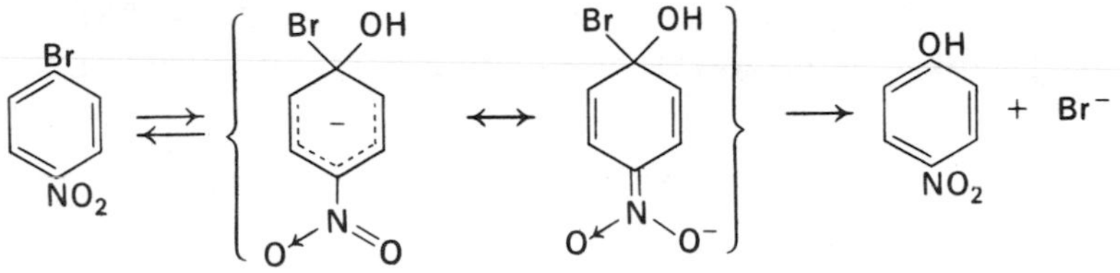

It is clear that a second nitro group at the position *ortho* to Br provides additional opportunity for charge delocalization:

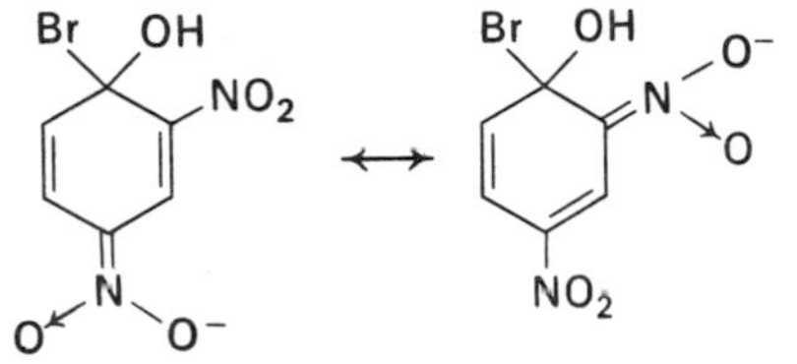

and the replacement of the halogen will be much faster.

It can be seen that the nucleophilic replacement of halogen from the benzene ring is quite different from the nucleophilic displacement of halogen at a saturated carbon atom (outlined in Part 8). Indeed, it will be observed that nucleophilic displacement by attack at the rear side of the ring carbon atom holding the bromine atom is impossible, for because of the planarity of the aromatic nucleus the nucleophile cannot approach in this way.

Aromatic nucleophilic replacement in the manner described above is a useful synthetic reaction. One of its well known applications is in the preparation of 2,4-dinitrophenylhydrazine, a valuable reagent for the characterization of aldehydes and ketones:

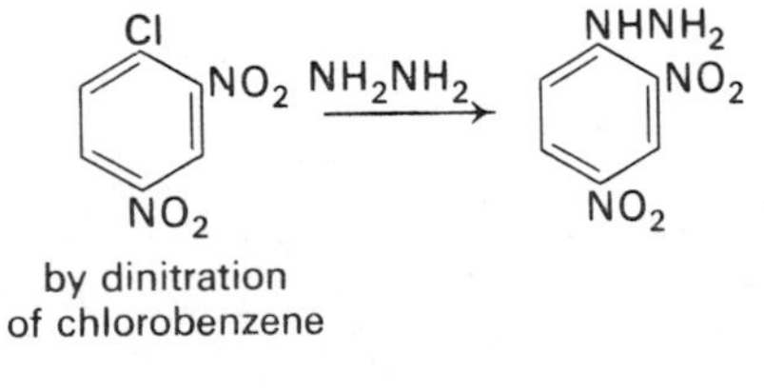

by dinitration
of chlorobenzene

Another important application of the reaction is in the structure proof of proteins (polypeptides). A terminal $—NH_2$ group reacts with 2,4-dinitrofluorobenzene in the following way:

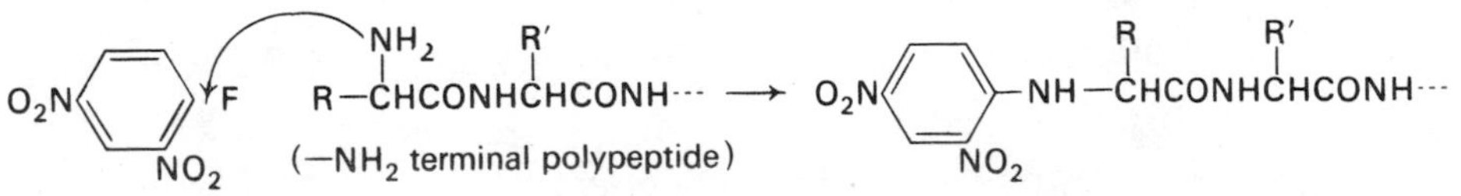

($—NH_2$ terminal polypeptide)

Hydrolysis of this product yields the N-2,4-dinitrophenylamino acid, which can be identified, usually by comparison with the known compound.

The formation of nitro-, dinitro- and trinitro- phenols and aniline derivatives is readily accomplished in this way; some examples are to be found in Exercise 9.3.

Exercise 9.3

Show how the following compounds can be prepared by a nucleophilic aromatic substitution reaction:

(*a*) OCH_3, NO_2

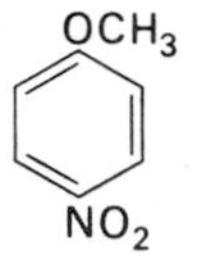

(*b*) OH, NO_2, NO_2

(*c*) NHOH, NO_2, NO_2

(*d*) —NH— NO_2

(*e*) H_3C–N–CH_3, Br, NO_2

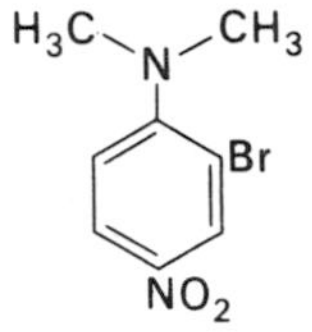

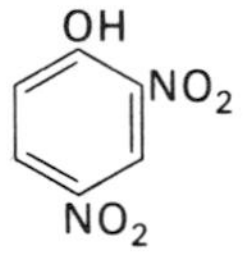

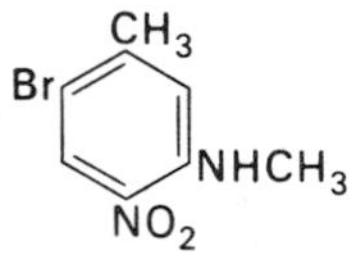

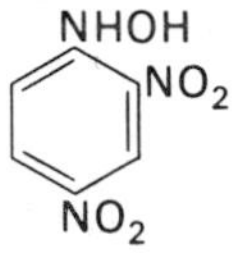

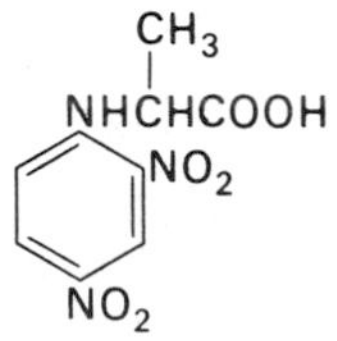

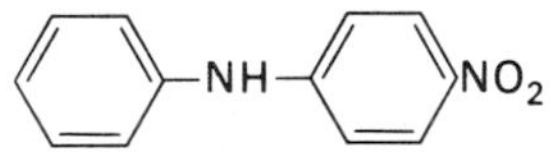

(*f*) CH_3, Br, $NHCH_3$, NO_2

(*g*) CH_3, NHCHCOOH, NO_2, NO_2

(*h*) O_2N $NHCH_2CH_2NH$ NO_2

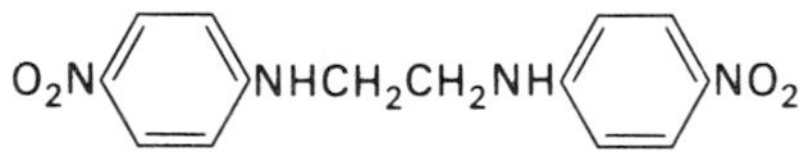

D. TRANSFORMATIONS OF AROMATIC SUBSTITUENTS

In the synthesis of substituted benzene derivatives it is often impracticable or impossible to carry out the desired synthesis by a simple substitution reaction. For example, *p*-dinitrobenzene cannot be made by direct nitration; *m*-bromotoluene cannot be efficiently made by bromination of toluene; *m*-dibromo- or 1,3,5-tribromobenzene cannot be made by direct bromination of benzene.

There are a variety of devices that can be used for the introduction of substituents in positions that are not accessible by direct substitution. Expressed first of all in general terms, these include:

(a) Transformation of one substituent into another. Examples of this are:

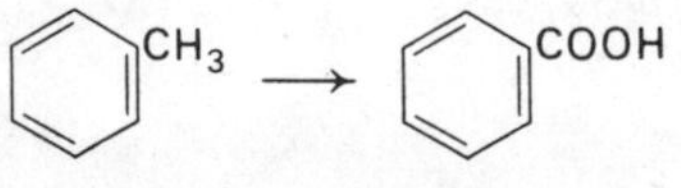

$$C_6H_5CH_3 \longrightarrow C_6H_5COOH \quad \text{oxidation} \qquad (13)$$

$C_6H_5COCH_3 \longrightarrow C_6H_5NHCOCH_3 \longrightarrow C_6H_5NH_2$ (14)

Beckmann rearrangement, *via* oxime

$C_6H_5NO_2 \longrightarrow C_6H_5NH_2$ reduction (15)

$C_6H_5COOH \longrightarrow C_6H_5NH_2$ Hofmann or Curtius reaction (16)

$C_6H_5NH_2 \longrightarrow C_6H_5\overset{+}{N}_2Cl^- \longrightarrow C_6H_5X$ (17)

diazotization X = halogen, CN, OH, H, aryl, etc.

(b) Employment of a "directing" group and its removal at a later stage. For example:

NHCOCH$_3$/CH$_3$ → NHCOCH$_3$/NO$_2$/CH$_3$ → NH$_2$/NO$_2$/CH$_3$ —(1) diazotize, (2) H_3PO_2→ NO$_2$/CH$_3$ (18)

(c) Combinations of (a) and (b).

CH$_3$ → CH$_3$/NO$_2$ → CH$_3$/NHCOCH$_3$ → CH$_3$/Br/NHCOCH$_3$ → (19)

CH$_3$/Br/NH$_2$ —*via* diazonium salt→ CH$_3$/Br/Br

It will often be recognized that a desired compound may be *one* of several products of a direct substitution reaction. The aim of the synthetic chemist, however, is to prepare a *specific* compound of *unambiguous* structure in satisfactory yield and uncontaminated with other compounds.

E. ALTERATION OF $-NH_2$

One of the most versatile of aromatic substituents is the amino group. It is easily introduced by the two steps: nitration, and reduction of $-NO_2$:*

$$C_6H_5OCH_3 \xrightarrow[H_2SO_4]{HNO_3} p\text{-}O_2NC_6H_4OCH_3 \xrightarrow[HCl]{Sn} p\text{-}H_2NC_6H_4OCH_3 \dashrightarrow \text{further steps as shown in chart} \qquad (20)$$

Diazotization of the amino group provides a diazonium salt, the versatility of which is shown by the following chart of the transformations it undergoes:

CHART SHOWING TRANSFORMATIONS OF A DIAZONIUM SALT

$$C_6H_5NH_2 \xrightarrow{HCl/HNO_2} C_6H_5N_2^+Cl^-$$

- (a) H_3PO_2 → C_6H_6
- (b) H_2O → C_6H_5OH
- (c) CuCN → C_6H_5CN
- (d) C_6H_5OH → $C_6H_5-N{=}N-C_6H_4-OH$
- (e) Cu_2Br_2 → C_6H_5Br
- (f) Cu_2Cl_2 → C_6H_5Cl
- (g) KI → C_6H_5I

One of the most useful properties of the diazonium group is the ease of its removal; that is, its replacement by H (Reaction *a* in the chart). This reaction permits one to use the *o-p*-directing power of $-NHCOCH_3$.† Examples are the following:

$$C_6H_5Br \longrightarrow p\text{-}BrC_6H_4NO_2 \longrightarrow p\text{-}BrC_6H_4NH_2 \longrightarrow p\text{-}BrC_6H_4NHCOCH_3 \xrightarrow{Br_2} \qquad (21)$$

$$2,4\text{-}Br_2C_6H_3NHCOCH_3 \xrightarrow[(2)\ HCl/HNO_2\ (3)\ H_3PO_2]{(1)\ H_2O/H^+} m\text{-}C_6H_4Br_2$$

---

* The reduction of $-NO_2$ to $-NH_2$ can be accomplished with a number of reagents. The use of Sn/HCl is chosen here as a typical example.

† The strongly activating property of $-NH_2$ is usually moderated by acetylation to $-NHCOCH_3$. Note that $-NHCOCH_3$ dominates $-CH_3$ and $-Br$ in directing power.

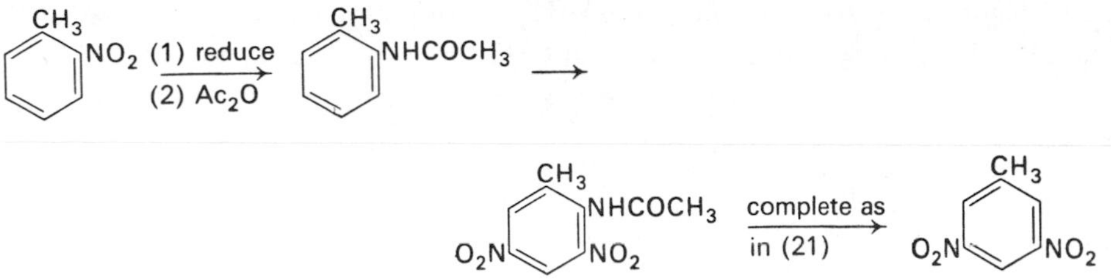

CH_3, NO_2 —(1) reduce; (2) Ac_2O→ CH_3, $NHCOCH_3$ →

CH_3, $NHCOCH_3$, O_2N, NO_2 —complete as in (21)→ CH_3, O_2N, NO_2 (22)

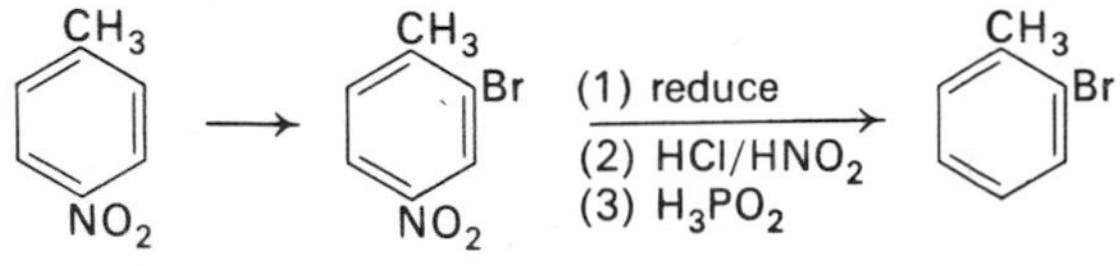

CH_3, NO_2 → CH_3, Br, NO_2 —(1) reduce; (2) HCl/HNO_2; (3) H_3PO_2→ CH_3, Br (23)

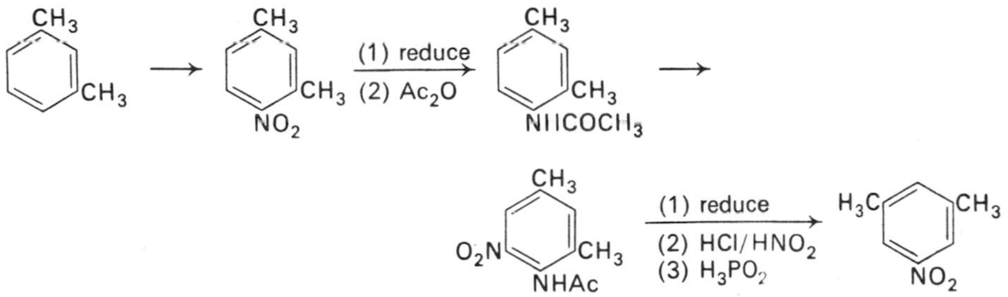

CH_3, CH_3 → CH_3, CH_3, NO_2 —(1) reduce; (2) Ac_2O→ CH_3, CH_3, $NHCOCH_3$ → (24)

CH_3, O_2N, CH_3, NHAc —(1) reduce; (2) HCl/HNO_2; (3) H_3PO_2→ H_3C, CH_3, NO_2

Exercise 9.4

Starting with any *monosubstituted* benzene, show how the following could be prepared (in each, the replacement of $-NH_2$ by $-H$ should be used as one of the steps):*

(*a*) *m*-nitrobromobenzene
(*b*) *m*-bromoanisole
(*c*) *m*-chlorobromobenzene
(*d*) 2,6-dibromoaniline
(*e*) 1,3,5-tribromobenzene
(*f*) *m*-chloroaniline
(*g*) 3,5-dibromotoluene
(*h*) 3,5-dibromoaniline
(*i*) 3,5-dibromonitrobenzene
(*j*) *o*-bromobenzonitrile
(*k*) *m*-bromobenzonitrile

* Mononitration or bromination of anisole, acetanilide, bromobenzene, toluene, chlorobenzene may be assumed here to give only the *p*-nitro compound.

Replacement of $-NH_2$ (*via* $-N_2^+Cl^-$) by halogen or cyano (see chart, above) provides access to a wide variety of compounds. It will be recalled that the cyano group is a versatile functional group, for it can be altered in a number of ways (see Part 8).

By means of the replacement of the $—N_2^+$ group by Cl or CN syntheses such as the following can be performed (some intermediate steps are omitted):

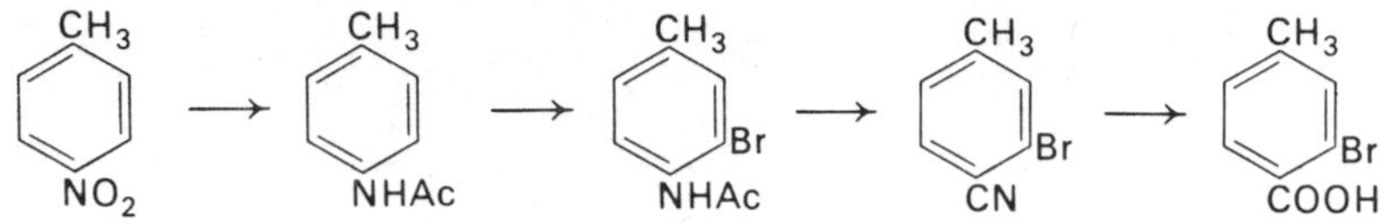

(25)

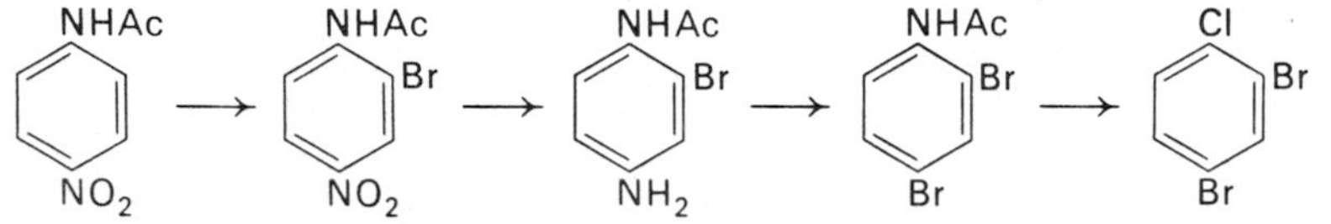

(26)

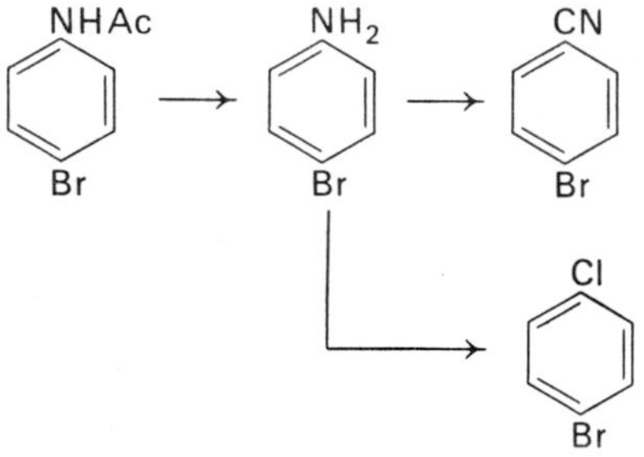

(27)

Exercise 9.5

Starting with any *monosubstituted* benzene, devise practical syntheses for the following:

(*a*) *p*-toluic acid
(*b*) 2,4-dibromobenzonitrile
(*c*) 2-bromo-4-methylacetophenone
(*d*) *o*-bromobenzoic acid
(*e*) *p*-bromobenzylamine
(*f*) *p*-bromoisopropylbenzene
(*g*) 2,5-diethylbenzoic acid
(*h*) 3,4-diaminochlorobenzene
(*i*) 1,2,3-tribromobenzene
(*j*) 1,2,3,5-tetrabromobenzene
(*k*) 2-bromo-4-nitrobenzoic acid

F. ALTERATION OF —COOH

The replacement of —COOH on an aromatic ring can be accomplished by the use of the Hofmann or Curtius reaction, in which —COOH is replaced by $—NH_2$. The latter can then be replaced through diazotization as already described.

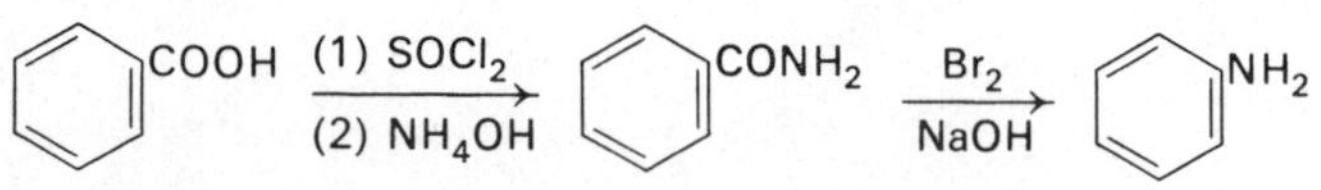

$$C_6H_5COOH \xrightarrow[(2)\ NH_4OH]{(1)\ SOCl_2} C_6H_5CONH_2 \xrightarrow[NaOH]{Br_2} C_6H_5NH_2 \qquad (28)$$

Reduction of —COOH to —CHO can be accomplished by converting it to the acid chloride (—COCl) and reducing the latter. An efficient method for this reduction is by the use of the complex lithium aluminum hydride, $LiAl(OtBu)_3H$.

$$C_6H_5COOH \xrightarrow{SOCl_2} C_6H_5COCl \xrightarrow{LiAl(OtBu)_3H} C_6H_5CHO \qquad (29)$$

Reduction of carboxylic acids or acid derivatives (e.g., esters, acid halides) with lithium aluminum hydride leads to the corresponding benzyl alcohols. The conversion of the benzyl alcohol into the benzyl chloride permits the formation of the Grignard reagent with consequent access to many further avenues of synthesis:*

$$C_6H_5CH_2OH \longrightarrow C_6H_5CH_2Cl \longrightarrow C_6H_5CH_2MgCl \qquad (30)$$

$$C_6H_5CH_2MgCl \xrightarrow{(i)\ RCN} C_6H_5CH_2C(=O)R \qquad C_6H_5CH_2MgCl \xrightarrow{(ii)\ RCHO} C_6H_5CH_2CH(OH)R \qquad C_6H_5CH_2MgCl \xrightarrow{CO_2\ (iii)} C_6H_5CH_2COOH$$

G. ALTERATION OF $-COCH_3$

Since the direct introduction of acyl groups into the aromatic nucleus is easily accomplished by the Friedel-Crafts reaction (Equation 3), a route is provided for the preparation of compounds derived by modification of an acyl group.

The acetyl group ($-COCH_3$) is the most versatile in this respect. It can be oxidized to the —COOH group by means of the haloform reaction.†

$$C_6H_5COCH_3 \xrightarrow[Br_2]{NaOH} C_6H_5COOH + CHBr_3 \qquad (31)$$

The Beckmann or Schmidt rearrangement provides a route to aromatic amines:

$$C_6H_5COCH_3 \xrightarrow{H_2NOH} C_6H_5C(=NOH)CH_3 \xrightarrow{H_2SO_4} C_6H_5NHCOCH_3 \qquad (32)$$

* It must be borne in mind that the Grignard reagent can be prepared only if any additional substituents that may be present are inert to the action of Grignard reagents (for example, halogen, alkyl, alkoxy, dialkylamino).

† The reaction is, of course, confined to those compounds in which the aromatic ring is not readily brominated.

Oxidation of aromatic ketones by means of peroxycarboxylic acids (the Baeyer-Villiger reaction) provides a convenient route to phenols:

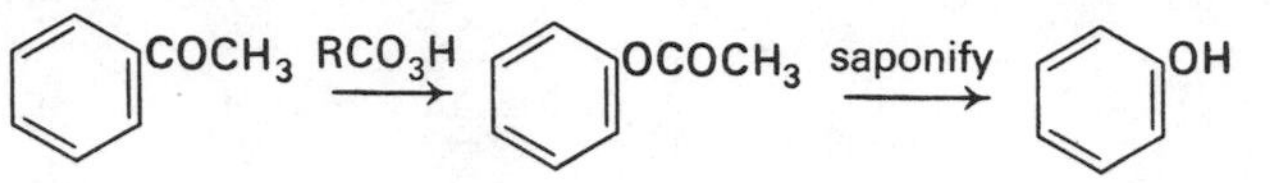

$$C_6H_5COCH_3 \xrightarrow{RCO_3H} C_6H_5OCOCH_3 \xrightarrow{\text{saponify}} C_6H_5OH \qquad (33)$$

Various other unexceptional reactions that need no further discussion (reduction, addition of Grignard reagents, etc.) are also possible.

Exercise 9.6

With reference to Sections F and G above and to earlier discussions, show how you could carry out the following transformations (more than one step may be needed):

(*a*) $C_6H_5CHO \rightarrow C_6H_5COCH_3$

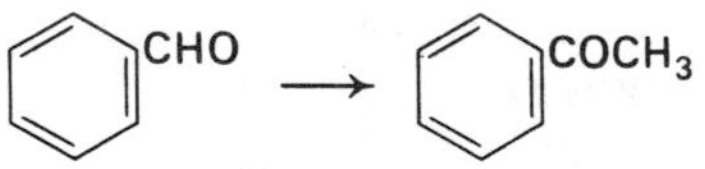

(*b*) $C_6H_5COOH \rightarrow C_6H_5CH_2N(CH_3)_2$

(*c*) Br, Br (para-dibromobenzene) $\rightarrow$ Br, OH, Br

(*d*) $C_6H_5COOH \rightarrow CH_3{-}C{=}CH_6H_5$ with $CH_2C_6H_3$ on the C

(*e*) $C_6H_5NH_2 \rightarrow (C_6H_5)_2CHOH$

(*f*) $C_6H_5CH_3 \rightarrow$ CH$_3$, Br, Br, OH

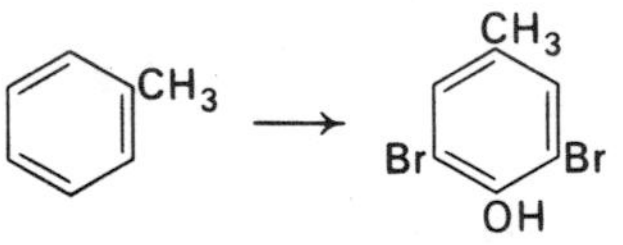

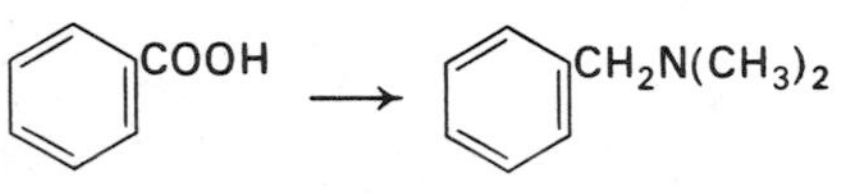

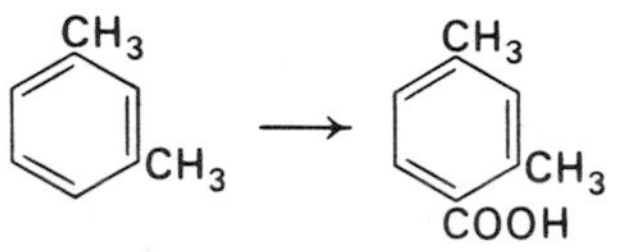

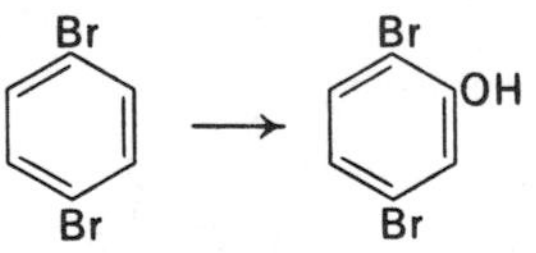

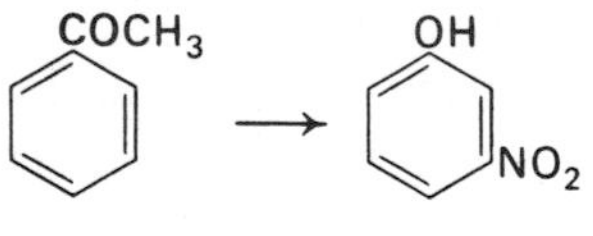

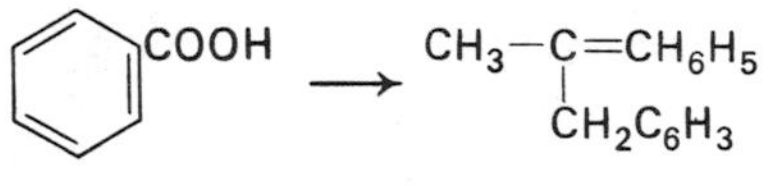

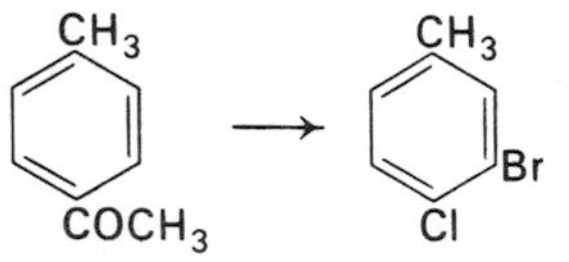

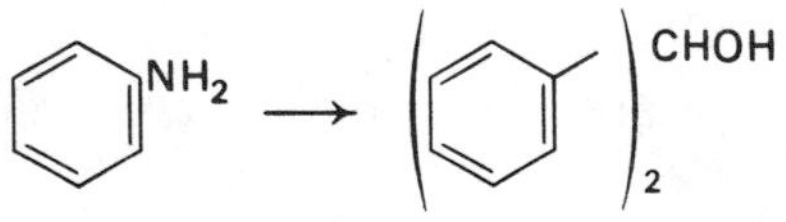

(*g*) CH_3, CH_3 $\rightarrow$ CH_3, CH_3, COOH

(*h*) $C_6H_5COCH_3 \rightarrow$ OH, NO_2

(*i*) CH_3, $COCH_3$ $\rightarrow$ CH_3, Br, Cl

(*j*) $C_6H_5COCH_3 \rightarrow$ Br, Br, Br

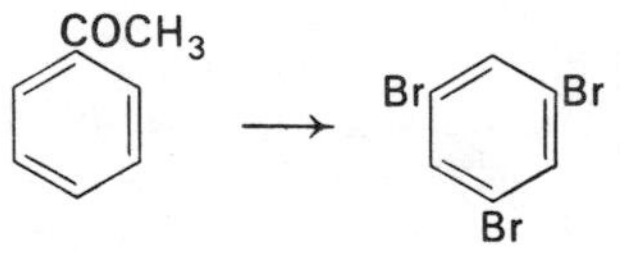

H. OXIDATION OF AROMATIC SIDE CHAINS

Groups attached to the benzene ring by a carbon-carbon linkage can be oxidized (usually by the use of KMnO or CrO_3) to the corresponding benzoic acid.

There are limitations to this general reaction.

(1) Readily oxidizable aromatic nuclei (such as with phenols, and aromatic amines) are attacked and destroyed under the conditions ordinarily used.

(2) Selective oxidation of one alkyl group when two or more are present is not usually possible.

(3) *t*-Alkyl groups are resistant to the oxidation.

(4) The Ar-CO-Ar grouping (that is, benzophenone) is stable to the oxidation.

The generality of the reaction is shown by the following examples:

$$H_3C,\ CH_3,\ CH_3 \xrightarrow{KMnO_4} HOOC,\ COOH,\ COOH \qquad (34)$$

$$CH_2\text{–}CH_2\text{–}CH_2 \xrightarrow{KMnO_4} COOH,\ COOH \qquad (35)$$

$$C_6H_5 \xrightarrow{CrO_3} COOH,\ COC_6H_5 \qquad (36)$$

$$OH \xrightarrow{KMnO_4} COOH,\ COOH \qquad (37)$$

$$\text{graphite} \xrightarrow{HNO_3} COOH,\ HOOC,\ COOH,\ HOOC,\ COOH,\ COOH \qquad (38)$$

$$CH_3O,\ CH_3O,\ OCH_3 \xrightarrow{KMnO_4} CH_3O,\ CH_3O,\ COOH,\ COOH \ + \ CH_3O,\ COOH \qquad (39)$$

$$CH_3,\ O_2N,\ NO_2,\ NO_2 \xrightarrow{CrO_3} COOH,\ O_2N,\ NO_2,\ NO_2 \qquad (40)$$

As is to be expected, partially oxidized side chains such as $-COCH_3$, $-CHO$, $-CH_2OH$, $-CH_2COOH$, $-CH_2COCH_3$, and so forth, are readily converted into $-COOH$ by oxidation, and can often be preferentially oxidized.

Exercise 9.7

Show the result of oxidizing the following compounds with $KMnO_4$ (in excess, under vigorous experimental conditions):

(*a*) O

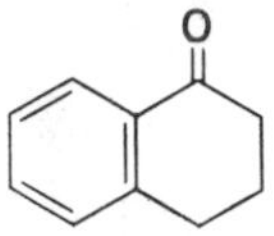

(*b*) CH_3O $CH{=}CH_2$ CH_3O

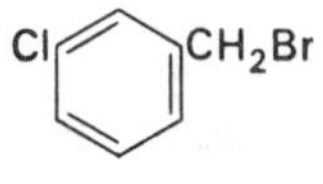

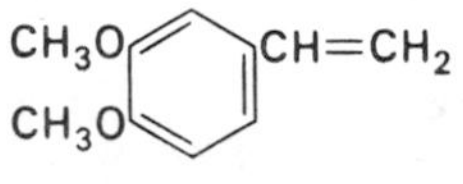

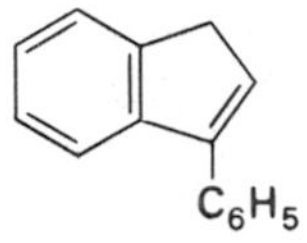

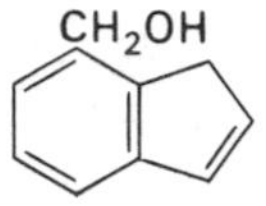

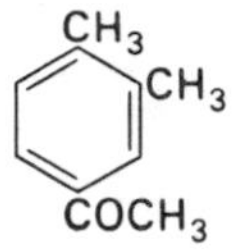

(*c*) CH_2OH

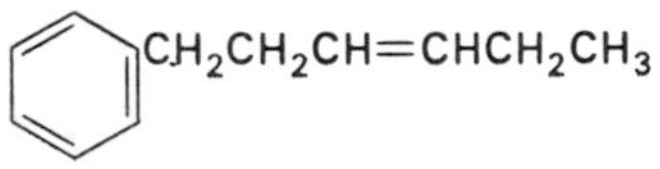

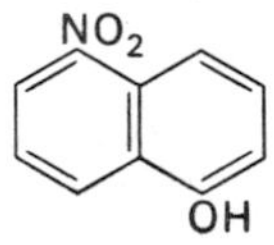

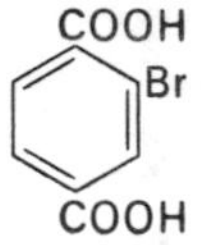

(*d*) $CH_2CH_2CH{=}CHCH_2CH_3$

(*e*) COOH Br COOH

(*f*) Cl CH_2Br

(*g*) C_6H_5

(*h*) CH_3 CH_3 $COCH_3$

(*i*) NO_2 OH

(*j*) COOH

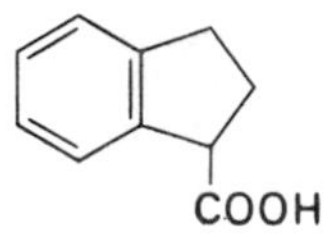

I. MISCELLANEOUS SUBSTITUTION REACTIONS. PHENOLS

Electrophilic substitution into the aromatic ring is enormously facilitated when the ring contains strongly electron-donating substituents. The most commonly encountered of such substituents is the hydroxyl group (and its ionized form, the $-\ddot{\underset{..}{O}}:^-$ grouping). Alkoxyl groups (for example, CH_3O-) are also activating toward nucleophilic attack, but somewhat less so than the hydroxyl group.

Phenol itself is the simplest example. It is very reactive to electrophilic attack. Nitration and bromination not only proceed with ease but tend to yield polysubstituted products. Controlled nitration to give mono-nitro phenols is carried out with dilute aqueous nitric acid, in marked contrast with the nitric-sulfuric acid mixtures needed for nitration of benzene itself. The presence of two or three hydroxyl groups, as in resorcinol and phloroglucinol, leads to greatly increased reactivity.

The highly nucleophilic character of the phenol anion is demonstrated by its reaction with carbon dioxide under pressure (the Kolbe reaction):

$$C_6H_5O^- + CO_2 \longrightarrow o\text{-}HOC_6H_4COO^- \xrightarrow{H^+} o\text{-}HOC_6H_4COOH \quad \text{(salicylic acid)} \tag{41}$$

Resorcinol and phloroglucinol can be carboxylated under even milder conditions.

The Hoesch reaction of phenols is shown in the acid-catalyzed acylation of resorcinol by a nitrile. This synthesis, catalyzed by zinc chloride —HCl (or, with phloroglucinol, by HCl alone) is related to Friedel-Crafts acylation and is a general procedure for the preparation of phenolic ketones and aldehydes:

$$m\text{-}C_6H_4(OH)_2 + CH_3CN \xrightarrow[HCl]{ZnCl_2} (HO)_2C_6H_3C(=\overset{+}{N}H_2)CH_3\ Cl^- \xrightarrow{H_2O} (HO)_2C_6H_3COCH_3 \tag{42}$$

$$m\text{-}C_6H_4(OH)_2 + Zn(CN)_2 \xrightarrow{HCl} (HO)_2C_6H_3CH{=}\overset{+}{N}H_2\ Cl^- \xrightarrow{H_2O} (HO)_2C_6H_3CHO \tag{43}$$

The use of chloroacetonitrile provides the chloroketone:

$$m\text{-}C_6H_4(OH)_2 \xrightarrow[ZnCl_2,\ HCl]{ClCH_2CN} (HO)_2C_6H_3COCH_2Cl \tag{44}$$

Treatment of the chloroketone with a mild base (for instance, sodium acetate is sufficient) leads to an internal nucleophilic displacement reaction and ring formation:

$$(HO)_2C_6H_3COCH_2Cl \underset{}{\overset{AcO^-}{\rightleftarrows}} HO(O^-)C_6H_3COCH_2Cl \longrightarrow HOC_6H_3(O\text{-}CH_2\text{-}CO) \tag{45}$$

Although diazonium salts are electrophiles, they are not sufficiently electrophilic to react with benzene or toluene; they do, however, couple with phenols to give azo compounds:

$$C_6H_6 + C_6H_5N_2^+Cl^- \longrightarrow \text{no reaction} \tag{46}$$

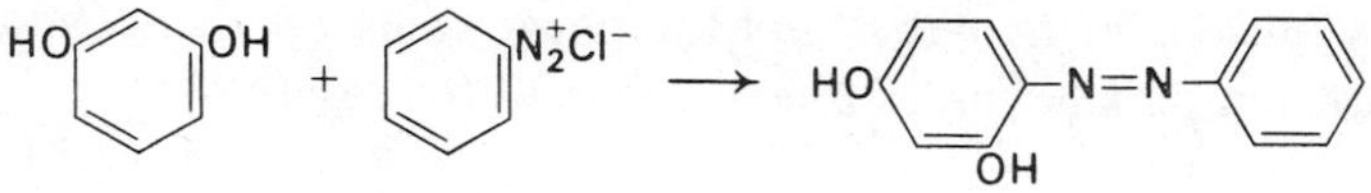

(47)

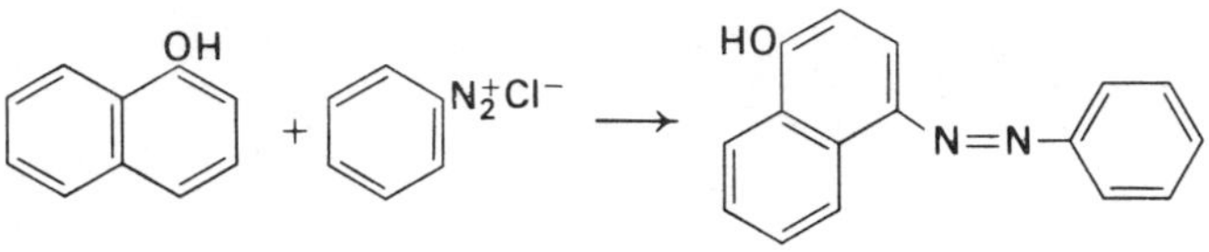

(48)

Azo compounds are important compounds in their own right, for they represent the large class of dyestuffs that contain the azo grouping. They are also useful in synthetic work, for the azo linkage is readily reduced to provide the amine:

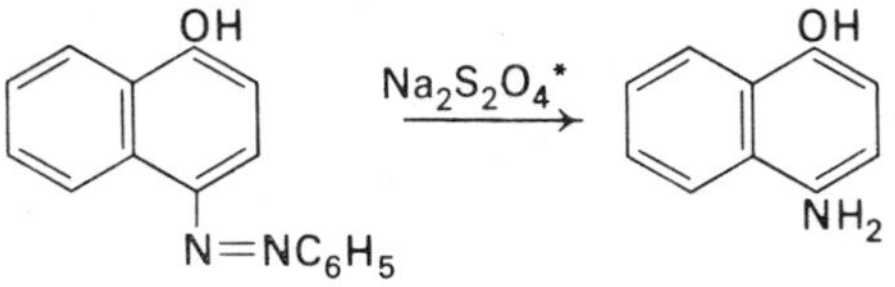

(49)

Phenolic aldehydes and ketones are useful in many ways. Oxidation with alkaline hydrogen peroxide (Dakin reaction) converts the —CHO or $—COCH_3$ group to the —OCHO and $—OCOCH_3$ groupings, which undergo alkaline hydrolysis to the corresponding phenol:

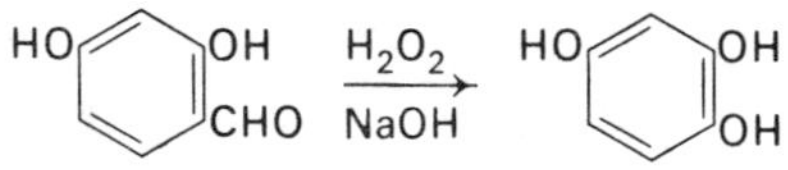

(50)

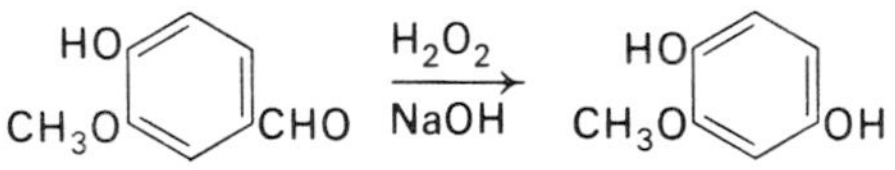

(51)

Condensation reactions lead to useful products of relatively complex structures:

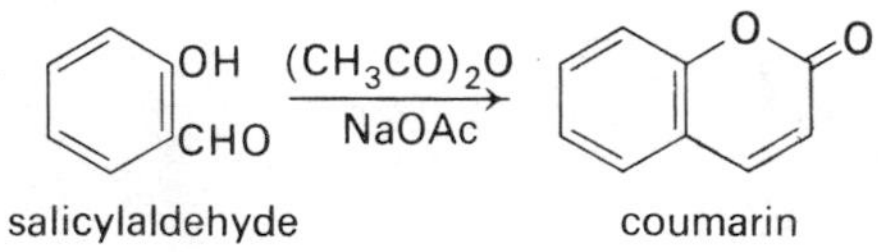

salicylaldehyde coumarin (52)

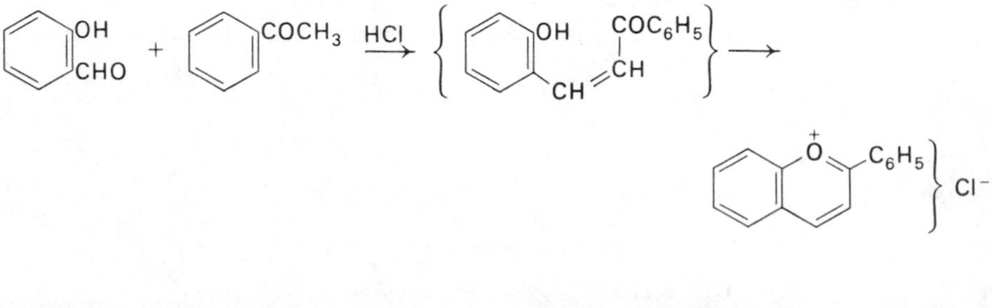

(53)

Although salicylaldehyde cannot be prepared by the Hoesch reaction (the reaction between phenol and HCN occurs principally at the hydroxyl group), it is formed by the reaction of phenol with chloroform in the presence of sodium hydroxide (the Reimer-Tiemann reaction):

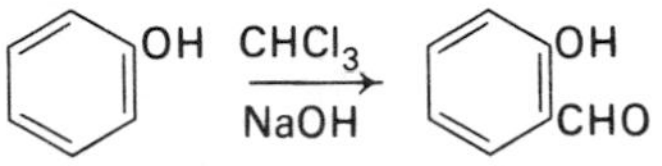
OH
CHCl3
NaOH
OH
CHO

(54)

Exercise 9.8

Starting with phenol, resorcinol or phloroglucinol (1,3,5-trihydroxybenzene), and other necessary reagents, show how the following compounds could be prepared:

(*a*) 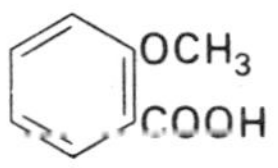
OCH3
COOH

(*f*) 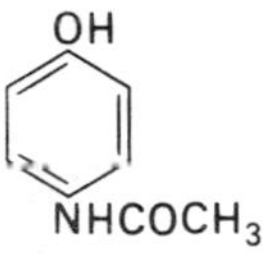
OH
NHCOCH3

(*b*) 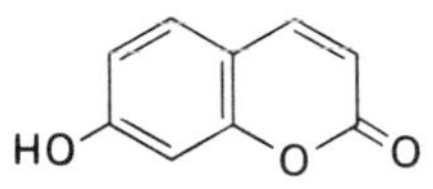
HO
O
O

(*g*) 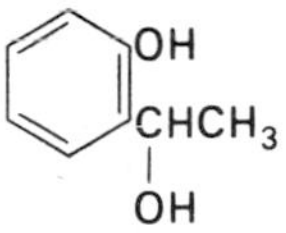
OH
CHCH3
OH

(*c*) 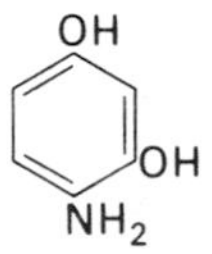
OH
OH
NH2

(*h*) 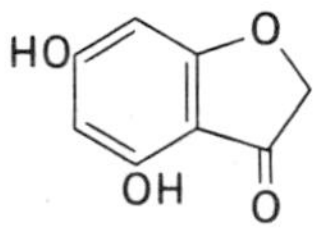
HO
O
OH
O

(*d*) 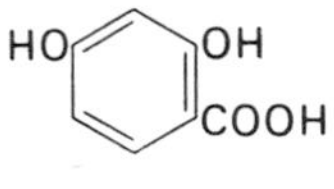
HO
OH
COOH

(*i*) 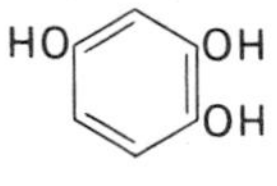
HO
OH
OH

(*e*) 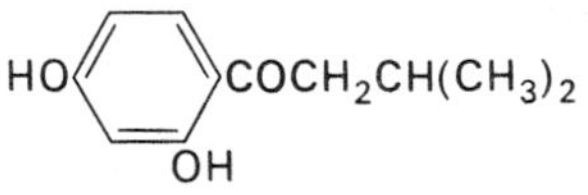
HO
COCH2CH(CH3)2
OH

(*j*) 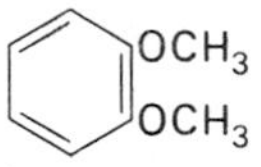
OCH3
OCH3

---

* A convenient reducing agent for this reaction is sodium dithionite, $Na_2S_2O_4$.

PART 10

Oxidation reactions

A. INTRODUCTION

Oxidation reactions are used in a great many ways in the synthesis, degradation and transformations of organic compounds. They cover so wide a variety of types of reactions and employ so varied a range of oxidizing reagents that an attempt to treat the subject exhaustively would require an encyclopedic account. Only a few will be dealt with here; these will include those of the most general application, principally the following:

$$R{-}CH(R')OH \xrightarrow{CrO_3} R{-}C(R'){=}O \quad (R' = H, \text{alkyl, aryl}) \tag{1}$$

$$R^1R^2C{=}CR^3R^4 \xrightarrow{O_3} R^1R^2C{=}O + O{=}CR^3R^4 \tag{2}$$

$$R{-}C(CH_3){=}O \xrightarrow{Br_2/NaOH} R{-}COOH \tag{3}$$

$$Ar{-}CH_2R \xrightarrow{KMnO_4} ArCOOH \quad (Ar = \text{aryl}) \tag{4}$$

B. BALANCING OXIDATION-REDUCTION EQUATIONS

The balancing of an equation for an oxidation reaction is of use principally in the calculation of the requisite amount of reagent needed for the reaction. Two items of fact are required: (1) the products of the oxidation of the organic compounds; and (2) the change in oxidation state of the oxidant.

Some of the most commonly used reagents are compounds of Cr^{VI} (chromium trioxide or potassium dichromate); compounds of Mn^{VII} (potassium permanganate); periodic acid (HIO_4); and compounds of Pb^{IV} (lead dioxide, lead tetraacetate). The use of these as oxidizing agents results in their change in valence state as in the following half-reactions:

$$CrO_3 + 6\,H^+ + 3\,e \longrightarrow Cr^{+++} + 3\,H_2O^* \tag{5}$$

$$MnO_4^- + 4\,H^+ + 3\,e \longrightarrow MnO_2 + 2\,H_2O \tag{6}$$

$$HIO_4 + 2\,H^+ + 2\,e \longrightarrow HIO_3 + H_2O \tag{7}$$

$$Pb(OAc)_4 + 2\,H^+ + 2\,e \longrightarrow Pb(OAc)_2 + 2\,HOAc \tag{8}$$

Calculation of the oxidation change in an organic oxidation reaction can be performed in a number of ways, most of them empirical and often involving artificial devices, such as the assignment of "valence" numbers to the atoms. The method to be described here is also empirical, but it depends upon the calculation of the number of electrons involved in the reaction. Coupled with the calculation of the electron change in the oxidant (Equations 5–8), the method consists in the writing and combination of two "half-reactions."

The charge and material balance in an organic oxidation-reduction half-reaction is mostly simply accomplished with the use of the "operators" H_2O (as a source of oxygen), H^+, and e (electrons). For example, the oxidation of CO to CO_2 can be written

$$CO + H_2O \longrightarrow CO_2 + 2\,H^+ + 2\,e \tag{9}$$

It is apparent that the material and charge balances are correct.

The oxidation of NH_3 to HNO_3 can be written:

$$\underset{9\,H}{NH_3 + 3\,H_2O} \longrightarrow \underset{9\,H}{HNO_3 + 8\,H^+} + 8\,e \tag{10}$$

Consider the following example:

$$RCH_2OH \longrightarrow RCOOH \tag{11}$$

* Note the balance of charges on the two sides of the equation: left, 6 +, 3 – ; right, 3 +. Note also the charge balances in Equations 6–8.

How many electrons are involved (that is, how many are provided to an oxidant) in this change? Using the charge-material-balance method, it is seen that one oxygen atom must be provided on the left-hand side of the equation. The number of protons (as H^+) required on the right-hand side is balanced by the number of electrons involved:

$$\underbrace{RCH_2OH + H_2O}_{5\,H,\ 2\,O} \longrightarrow \underbrace{RCOOH + 4\,H^+}_{\substack{5\,H,\ 2\,O \\ 4+,\ 4-}} + 4\,e \qquad \begin{matrix}\text{(material balance)} \\ \text{(charge balance)}\end{matrix}$$

Consider the extensive oxidation

$$\text{cyclohexanone} \xrightarrow{\text{oxidation}} HOOC(CH_2)_4COOH \tag{12}$$

It is seen that 3 oxygen atoms must be provided. Using H_2O as the source of these, and balancing the charges as was done above, we have

$$\text{cyclohexanone} + 3H_2O \longrightarrow HOOC(CH_2)_4COOH + 6\,H^+ + 6\,e \tag{13}$$

Consider the extensive oxidation described by the following *transformation* (not an "equation"!):

$$CH_3CH_2CH_2COOH \longrightarrow CH_3COOH + 2\,CO_2 \tag{14}$$

Since the right-hand side contains 4 more atoms of oxygen than the left-hand side, $4\,H_2O$ are provided, with the final charge and material balance as follows:

$$CH_3CH_2CH_2COOH + 4\,H_2O \longrightarrow CH_3COOH + 2\,CO_2 + 12\,H^+ + 12\,e \tag{15}$$

A final example, illustrating the general reaction in which an aromatic side chain is degraded to the aromatic carboxylic acid, is the following:

$$C_6H_3(CH_3)_3 + 6\,H_2O \longrightarrow C_6H_3(COOH)_3 + 18\,H^+ + 18\,e \tag{16}$$

material balance: (excluding aromatic ring) 9 H + 12 H, 6 O → 3 H; 6 O + 18 H

A completely analogous procedure can be used when oxidative cleavage of the molecule (as at a carbon-carbon double bond) occurs:

$$CH_3CH{=}CHCH_3 \xrightarrow{\text{oxidation}} 2\,CH_3COOH \tag{17}$$

Four oxygen atoms must be supplied:

$$CH_3CH{=}CHCH_3 + 4\,H_2O \longrightarrow CH_3COOH + CH_3COOH + 8\,H^+ + 8\,e \tag{18}$$

Exercise 10.1

Calculate the number of electrons provided by each of the following oxidations by writing a balanced half-reaction for each:

(*a*) $CH_3CH_2OH \longrightarrow CH_3CHO$

(*b*) $(CH_3)_2CHOH \longrightarrow (CH_3)_2C{=}O$

(*c*) $CH_3CH{=}CHCH_3 \longrightarrow CH_3CH(OH){-}CH(OH)CH_3$

(*d*) $CH_3COOH \longrightarrow 2\,CO_2$

(*e*) $CH_3CHOHCOCH_3 \longrightarrow CH_3CHO + CH_3COOH$

(*f*) cyclohexene ⟶ cyclohexane-1,2-dione (O, O)

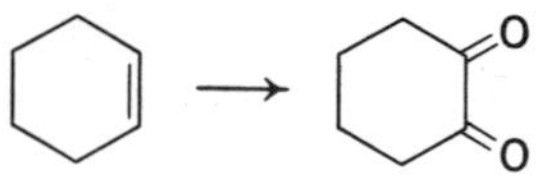

(*g*) cyclohexene ⟶ cyclohex-2-enone (O)

(*h*) $C_6H_5CH_2CH_2CH_2C_6H_5 \longrightarrow 2\,C_6H_5COOH + CO_2$

(*i*) phenol (OH) ⟶ p-benzoquinone (O, O)

(*j*) tetralin ⟶ 1-tetralone (O)

(*k*) p-$CH_3C_6H_4COCH_3$ ($COCH_3$, CH_3) ⟶ p-$C_6H_4(COOH)_2$ (COOH, COOH) + CO_2

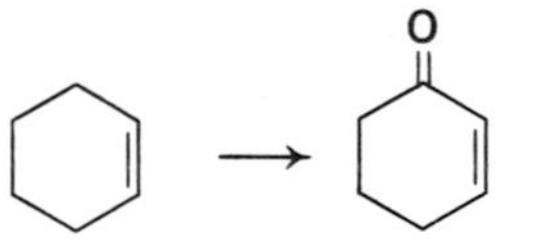

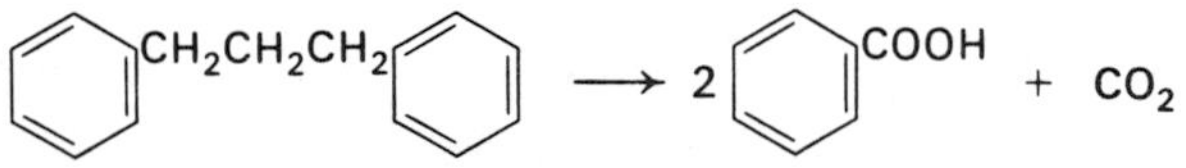

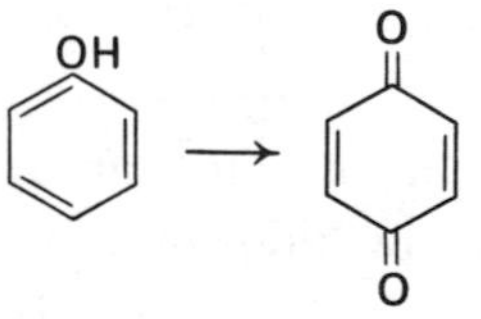

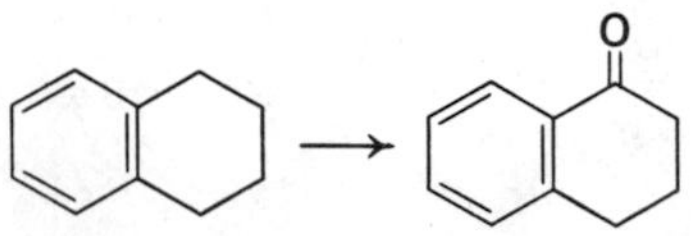

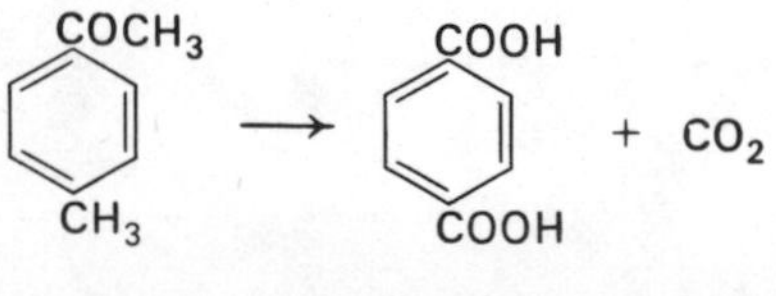

(*l*) [indane with CH_3 on the five-membered ring] $\longrightarrow$ [benzene-1,2-dicarboxylic acid: COOH, COOH] $+ 2CO_2$

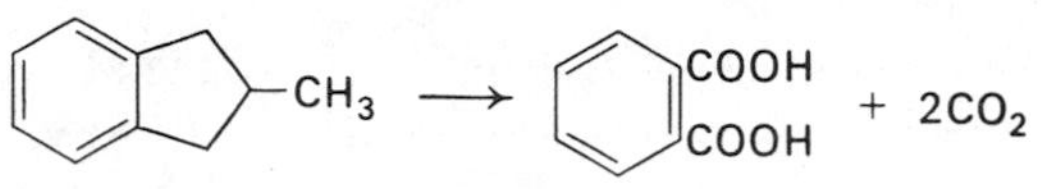

(*m*) [indene with C_6H_5] $\longrightarrow$ [benzene with COOH and COC_6H_5] $+ CO_2$

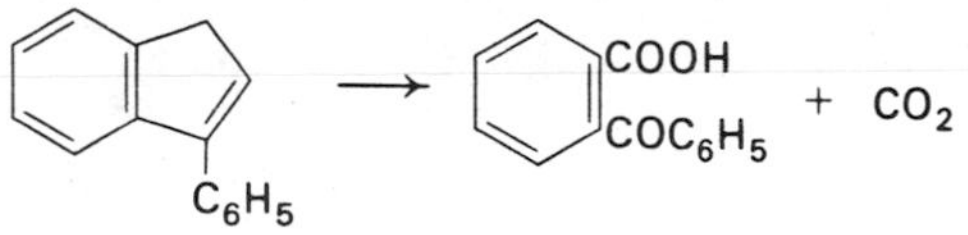

(*n*) $CH_3CH{=}CH_2 \longrightarrow CH_3COOH + CO_2$

(*o*) $CH_3CH{=}CHCH{=}CHCH{=}CHCH_3 \longrightarrow 2\,CH_3COOH + 4\,CO_2$

Notes to exercise 10.1

(*c*)

$$CH_3CH{=}CHCH_3 + 2\,H_2O \longrightarrow CH_3\underset{}{\overset{OH}{C}}H{-}\overset{OH}{C}HCH_3 + 2\,H\cdot$$

Since this half-reaction represents a source of electrons (eventually utilized by some unspecified oxidant), 2 H· can be expressed as 2 H^+ + 2 e. Thus, the total material- and charge-balanced equation is:

$$CH_3CH{=}CHCH_3 + 2\,H_2O \longrightarrow CH_3\overset{OH}{C}H{-}\overset{OH}{C}HCH_3 + 2\,H^+ + 2\,e$$

(*k*)

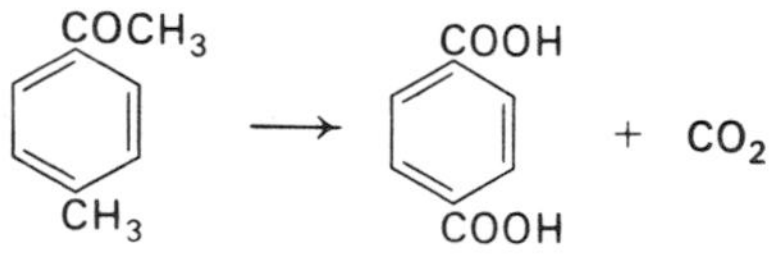

[benzene with $COCH_3$ and CH_3 (para)] $\longrightarrow$ [benzene with COOH and COOH (para)] $+ CO_2$

The starting compound contains one O; the products contain a total of 6 O; thus, 5 O are needed and are provided by 5 H_2O:

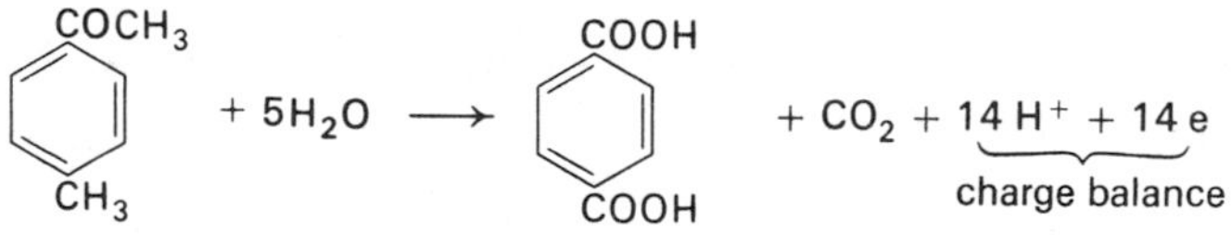

[benzene with $COCH_3$ and CH_3] $+ 5H_2O \longrightarrow$ [benzene with COOH and COOH] $+ CO_2 + \underbrace{14\,H^+ + 14\,e}_{\text{charge balance}}$

$$\begin{array}{rcl} 10 + 10 = 20\,H & \longrightarrow & 6\,H + 14\,H \\ 1 + 5 = 6\,O & & 4\,O + 2\,O \end{array} \Big\}\ \text{material balance}$$

(*o*)

$$CH_3CH{=}CH{-}CH{=}CH{-}CH{=}CH{-}CH_3 \longrightarrow 2\,CH_3COOH + 4\,CO_2$$

O present = none O needed = 12

Thus,

$$CH_3CH{=}CH{-}CH{=}CH{-}CH{=}CH{-}CH_3 + 12\,H_2O \longrightarrow$$

H balance = 12 + 24 = 36 $\longrightarrow$

$$2\,CH_3COOH + 4\,CO_2 + 28\,H^+ + 28\,e$$

8 H + 28 H

In order to calculate the amount of oxidizing reagent for any of the above reactions, it is necessary to know the change in oxidation state of the atom (the oxidant) which gains the electrons in the overall reaction. If Cr^{VI} (for example, as CrO_3) is used, the usual course of its reduction is that shown again in Equation 5, or the shorter form shown in (19):

$$Cr^{VI} + 3\,e \longrightarrow Cr^{III} \tag{19}$$

Thus, for the oxidation whose half-reaction is that of Equation 11:

$$RCH_2OH + H_2O \longrightarrow RCOOH + 4\,H^+ + 4\,e \tag{11}$$

$$CrO_3 + 6\,H^+ + 3\,e \longrightarrow Cr^{+++} + 3\,H_2O \tag{5}$$

In order to permit these to be added conveniently to give the balanced equation for the oxidation, they are converted to an equivalent electron change:

$$3\,RCH_2OH + 3\,H_2O \longrightarrow 3\,RCOOH + 12\,H^+ + 12\,e \tag{20}$$

$$4\,CrO_3 + 24\,H^+ + 12\,e \longrightarrow 4\,Cr^{+++} + 12\,H_2O \tag{21}$$

Sum:

$$3\,RCH_2OH + 4\,CrO_3 + 3\,H_2O + 24\,H^+ + 12\,e \longrightarrow \tag{22}$$

$$3\,RCOOH + 12\,H^+ + 12\,e + 4\,Cr^{+++} + 12\,H_2O$$

Cancel like terms on opposite sides:

$$3\,RCH_2OH + 4\,CrO_3 + 12\,H^+ \longrightarrow 3\,RCOOH + 4\,Cr^{+++} + 9\,H_2O \tag{23}$$

A complete balance, in nonionic form can now be written if the experimental conditions are stated. Suppose the oxidation is carried out with CrO_3 in aqueous H_2SO_4:

$12\,H^+$ is replaced by $6\,H_2SO_4$

$4\,Cr^{+++}$ is replaced by $2\,Cr_2(SO_4)_3$

and a completely balanced equation results.

Exercise 10.2

In each of the following are given the products of the oxidation of the organic compound and the change in oxidation state of the inorganic reagent. Write a completely balanced equation for each (assume formation or availability of H_2O if needed):

(*a*) $CH_3CHOHCH_2CH_3 \longrightarrow CH_3COCH_2CH_3$; $CrO_3 \longrightarrow Cr^{+++}$ (in H_2SO_4)

(*b*) cyclohexanone $\longrightarrow$ HOOC–(CH$_2$)$_4$–COOH ; $K_2Cr_2O_7 \longrightarrow Cr^{+++}$ (in H_2SO_4)

(*c*) $C_6H_5CH_2COCH_3 \longrightarrow C_6H_5COOH$; $KMnO_4 \longrightarrow MnO_2$

(*d*) cyclohexene $\longrightarrow$ cyclohexane-1,2-diol (OH, OH) ; $OsO_4 \longrightarrow H_2OsO_4(\equiv OsO_3)$

(*e*) cyclohexane-1,2-diol (OH, OH) $\longrightarrow$ cyclohexane-1,2-dione (O, O) ; $CrO_3 \longrightarrow Cr^{+++}$ (in H_2SO_4)

(*f*) $CH_3CH{=}CHCOOH \longrightarrow CH_3COOH + 2\,CO_2$; $KMnO_4 \longrightarrow MnO_2$

Answer to exercise 10.2(*c*)

$$C_6H_5CH_2COCH_3 + 5\,H_2O \longrightarrow C_6H_5COOH + 2\,CO_2 + 14\,H^+ + 14\,e \qquad \text{(C-1)}$$

$$MnO_4^- + 4\,H^+ + 3\,e \longrightarrow MnO_2 + 2\,H_2O \qquad \text{(C-2)}$$

Multiplying (C-1) by 3 and (C-2) by 14, adding and cancelling, one obtains:

$$3\,C_6H_5CH_2COCH_3 + 14\,MnO_4^- + 14\,H^+ \longrightarrow 3\,C_6H_5COOH + 6\,CO_2 + 14\,MnO_2 + 13\,H_2O \qquad \text{(C-3)}$$

To accommodate 14 K^+ (in $KMnO_4$), add 14 K^+ and 14 OH^- to each side and cancel 13 H_2O:

$$3\,C_6H_5CH_2COCH_3 + 14\,KMnO_4 + H_2O \longrightarrow 3\,C_6H_5COOH + 6\,CO_2 + 14\,MnO_2 + 14\,KOH \qquad \text{(C-4)}$$

Check:

H: 30 + 2 = 32, O: 3 + 56 + 1 = 60 } left

H: 18 + 14 = 32, O: 6 + 12 + 28 + 14 = 60 } right

C. OXIDATION OF ALCOHOLS AND ALDEHYDES

The oxidation of alcohols to aldehydes and ketones, and in particular of secondary alcohols to ketones, is expressed in the general equation

$$>CHOH \longrightarrow >C{=}O \qquad (24)$$

and in the following examples:

$$(CH_3)_2CHOH \xrightarrow{CrO_3} (CH_3)_2C{=}O$$

$$\text{cyclopentanol (OH)} \xrightarrow{CrO_3} \text{cyclopentanone (O)}$$

$$HC{\equiv}CCH_2OH \xrightarrow{CrO_3} HC{\equiv}CCHO$$

$$\text{RO–decalol (OH)} \xrightarrow{CrO_3} \text{RO–decalone (O)} \qquad (R = C_6H_5CO{-})$$

$$\text{CH}_3,\ C_6H_5\text{–cyclohexenol (OH)} \xrightarrow{CrO_3} \text{CH}_3,\ C_6H_5\text{–cyclohexenone (O)}$$

$$CH_3CH{=}CHCH(OH)CH_2CH_3 \xrightarrow{CrO_3} CH_3CH{=}CHCOCH_2CH_3$$

A common and convenient reagent for oxidations of this kind (often called the "Jones reagent") is a standardized solution of CrO_3 in dilute aqueous sulfuric acid. The reaction is usually carried out by adding the calculated amount of the reagent to an acetone solution of the alcohol to be oxidized, using the change from orange, Cr^{VI}, to green, Cr^{III}, as an indicator. Other organic solvents that may be used for Cr^{VI} oxidations are acetic acid, pyridine and *t*-butyl alcohol.

It should be noted that in some of these examples other oxidizable groupings ($C{=}C$, $C{\equiv}C$) are present, but are not attacked. This selectivity is, of course, dependent upon the use of controlled experimental conditions. Under more vigorous conditions, extensive degradation can occur.*

Tertiary alcohols are not oxidized under conditions that lead to ready oxidation of primary and secondary hydroxyl groups. Thus, a reaction such as the following is readily accomplished:

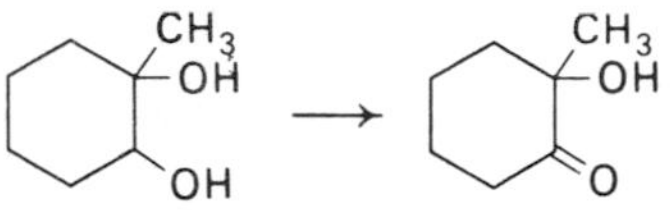

CH_3 OH OH $\longrightarrow$ CH_3 OH O

Exercise 10.3

Write the equations for the oxidation of the following compounds under the following conditions: (1) Jones reagent is used (2) at room temperature (3) in the proportion of one two-electron equivalent of reagent:

(*a*) $CH_3{-}\underset{\large OH}{\underset{|}{C}}HCH_2CH_3$

(*b*) $CH_3{-}\underset{\large OH}{\underset{|}{C}}HCOCH_3$

(*c*) $CH_3CH_2CH_2OH$

(*d*) OH CH_3

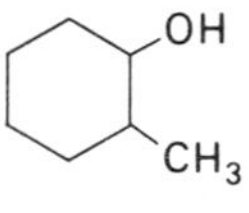

(*e*) OH

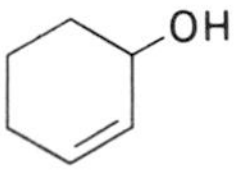

(*f*) $CH_3{-}\overset{\large CH_3}{\overset{|}{\underset{\large OH}{\underset{|}{C}}}}{-}CH_2\underset{\large OH}{\underset{|}{C}}HCH_3$

(*g*) HO

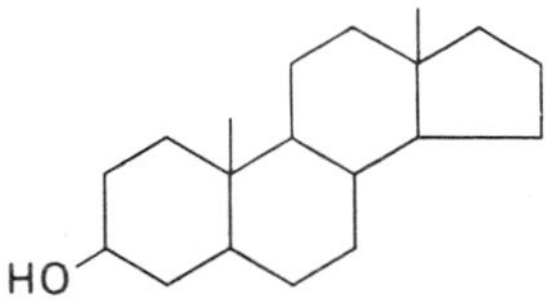

(*h*) $CH_3OCH_2CH_2CH_2\overset{\large OH}{\overset{|}{C}}HCH_3$

(*i*) $CH_3C{\equiv}C{-}\underset{\large OH}{\underset{|}{C}}H{-}C{\equiv}CCH_3$

(*j*) $CH_3C{\equiv}C{-}\underset{\large OH}{\underset{|}{C}}HCH_3$

---

* A valuable analytical procedure (called the Kuhn-Roth analysis) consists in the oxidation of organic compounds with a large excess of CrO_3 in aqueous H_2SO_4 at reflux temperatures. Acetic acid is resistant to oxidation; consequently compounds containing CH_3 groups, particularly at $>C{=}\overset{|}{C}{-}CH_3$ groupings, give acetic acid, often in nearly quantitative yield. Determination (by titration) of the acetic acid formed gives the number of $-CH_3$ groups in the compounds.

Oxidation of most primary alcohols to aldehydes is in general not easily accomplished because the aldehyde is susceptible to further oxidation to the corresponding carboxylic acid, or to the ester formed by oxidation of the hemiacetal:

$$\mathrm{RCH_2OH} \xrightarrow{\mathrm{CrO_3}} \mathrm{RCHO} \xrightarrow{\text{further}} \mathrm{RCOOH}$$

$$\mathrm{RCH_2OH + RCHO} \rightleftharpoons \mathrm{R\overset{\displaystyle OH}{\overset{|}{C}}H{-}O{-}CH_2R} \xrightarrow{\mathrm{CrO_3}} \mathrm{RCO{-}OCH_2R}$$

Satisfactory yields of aldehydes from primary alcohols can in some cases be obtained by removing the aldehyde from the reaction mixture by steam distillation or extraction into an immiscible solvent as it is formed.

Exercise 10.4

Calculate the amount of $KMnO_4$ (in grams) needed to carry out the following oxidations:

(*a*) $\mathrm{CH_3CHO}$ (11 g) $\longrightarrow$ $\mathrm{CH_3COOH}$

(*b*) $\mathrm{CH_2CHO{-}CH_2CHO}$ (17.2 g) $\longrightarrow$ $\mathrm{CH_2COOH{-}CH_2COOH}$

(*c*) 3,4-$(\mathrm{CH_3O})_2\mathrm{C_6H_3CHO}$ (83 g) $\longrightarrow$ 3,4-$(\mathrm{CH_3O})_2\mathrm{C_6H_3COOH}$

D. OXIDATION OF CARBON-CARBON DOUBLE BONDS

Carbon-carbon double bonds can be oxidized in discrete stages, described by the following expression

$$\mathrm{R^1R^2C{=}CR^3R^4} \longrightarrow \mathrm{R^1R^2C(OH){-}C(OH)R^3R^4} \longrightarrow \mathrm{R^1R^2C{=}O + O{=}CR^3R^4} \qquad (25)$$

If either R^1 or R^2, or R^3 or R^4, is hydrogen, the oxidation can proceed further to give the corresponding carboxylic acid(s). Thus, the reaction can be considered in terms of (a) hydroxylation of the double bond, and (b) oxidative cleavage of the 1,2-glycol.

The hydroxylation of carbon-carbon double bonds can lead to two stereochemically different results: the hydroxyl groups may be added in the *cis* manner, or in the *trans* manner. For example, the hydroxylation of *cis*-2-butene can be performed to lead to the *meso* glycol or to the resolvable (±)-2,3-glycol:

$$CH_3(H)C{=}C(H)CH_3 \xrightarrow[\text{hydroxylation}]{cis\text{-}} \underset{meso\text{-glycol}}{H(CH_3)(OH)C{-}C(CH_3)(OH)H} \quad (26)$$

$$\xrightarrow[\text{hydroxylation}]{trans\text{-}} \underset{(\pm)\text{-glycol}}{H(CH_3)(OH)C{-}C(OH)(CH_3)H + CH_3(HO)(H)C{-}C(CH_3)(OH)H} \quad (27)$$

Cis-hydroxylation can be accomplished with the use of dilute, alkaline potassium permanganate. Labeling experiments have shown that both oxygen atoms of the glycol are derived from the MnO_4^-. The course of the reaction can be described briefly by the following equations:

$$RCH{=}RCH + O_2\overset{VII}{Mn}(O)O^- \longrightarrow \left\{ \overset{+}{RCH}(RCH{-}O)\;\overset{-}{O}{-}Mn(O)(O^-) \right\} \longrightarrow (RCH{-}O)(RCH{-}O)Mn(O)O^- \xrightarrow{H_2O} \quad (28)$$

$$RCHOH{-}RCH{-}OMnO_3H^- \xrightarrow{H_2O} RCHOH{-}RCHOH + H_2MnO_4^-(\equiv MnO_3^- = Mn^V)$$

Since the glycol is the product of the hydrolysis of the cyclic ester, and since both C—O bonds remain intact in the hydrolytic step, the overall hydroxylation is necessarily *cis*.

Cis-hydroxylation is also the result of hydroxylation with osmium tetroxide, OsO_4.

$$RCH{=}RCH + O_2\overset{VIII}{Os}O_2 \longrightarrow (RCH{-}O)(RCH{-}O)OsO_2 \xrightarrow{H_2O} RCHOH{-}RCHOH + H_2OsO_4\ (\equiv OsO_3 = Os^{VI}) \quad (29)$$

Both of these hydroxylations are two-electron changes. This can be seen by the fact that the oxidants change from Mn^{VII} to Mn^V, or from Os^{VIII} to Os^{VI}; and also by the half-reaction:

$$RCH{=}CHR + 2\,H_2O \longrightarrow RCH(OH){-}CH(OH)R + 2\,H^+ + 2\,e \quad (30)$$

Exercise 10.5

Show by appropriate perspective formulas the stereochemical result of the following *cis*-hydroxylation reactions (with $KMnO_4$ or OsO_4):

(*a*) *cis*- $CH_3CH{=}CHCH_3$

(*b*) *trans*- $CH_3CH{=}CHCH_2CH_3$

(*c*) oleic acid

(*d*) 3-methylcyclohexene

(*e*) cholesterol

Trans-hydroxylation is the final result achieved in two stages: (1) formation of an epoxide, and (2) opening of the epoxide by nucleophilic attack of water (30*a*) or by a carboxylate anion (30*b*).

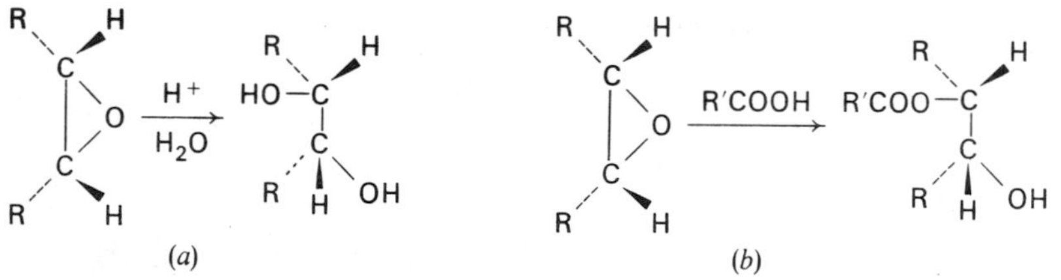

$\xrightarrow[H_2O]{H^+}$ (*a*) $\xrightarrow{R'COOH}$ (*b*) (30)

Exercise 10.6

In the ring opening steps of Reaction 30, nucleophilic attack is shown to occur only at one (the upper) terminus of the oxide ring. Write the corresponding equations for ring opening by attack in the opposite way (that is, at the other terminus of the oxide) and state whether (first) the two glycols, and (second) the two glycol monesters so formed are, respectively, *meso* or enantiomeric forms.

Trans-hydroxylation by way of the intermediate epoxide is most conveniently carried out with the use of peroxyformic or peroxyacetic acids. These are the effective reagents in mixtures of hydrogen peroxide and the corresponding acids:

$$HCOOH + H_2O_2 \rightleftharpoons HCO_3H + H_2O$$

$$CH_3COOH + H_2O_2 \rightleftharpoons CH_3CO_3H + H_2O$$

E. OZONOLYSIS IN STRUCTURE PROOF AND DEGRADATION

Ozone, generated by the action of a high-voltage silent discharge upon dry oxygen, reacts with a carbon-carbon double bond in the following manner:

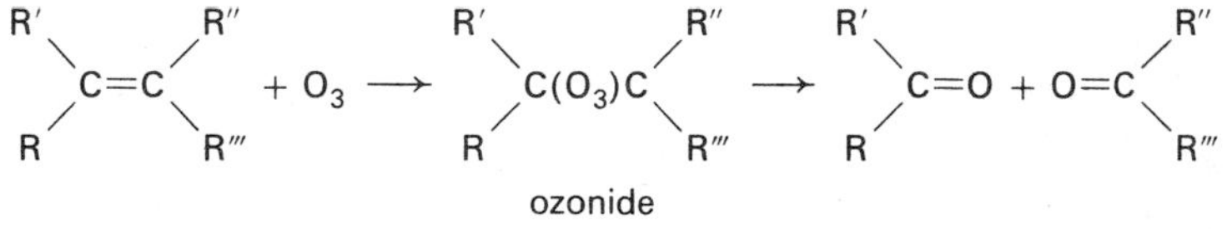

$$\begin{matrix} R' & & R'' \\ & C{=}C & \\ R & & R''' \end{matrix} + O_3 \longrightarrow \underset{\text{ozonide}}{\begin{matrix} R' & & R'' \\ & C(O_3)C & \\ R & & R''' \end{matrix}} \longrightarrow \begin{matrix} R' & \\ & C{=}O \\ R & \end{matrix} + \begin{matrix} & R'' \\ O{=}C & \\ & R''' \end{matrix} \qquad (31)$$

The intermediate ozonide is written here in a noncommittal representation; it is a compound, usually quite unstable but often isolable, in which the carbon-carbon bond has been broken and replaced by a C—O—C bond. Decomposition of this ozonide in one of several ways (which are not detailed here) results in the reductive removal of one of the three oxygen atoms and the formation of two carbonyl groups. It will be seen that the formulation of the overall reaction is readily comprehensible if it is represented as follows:

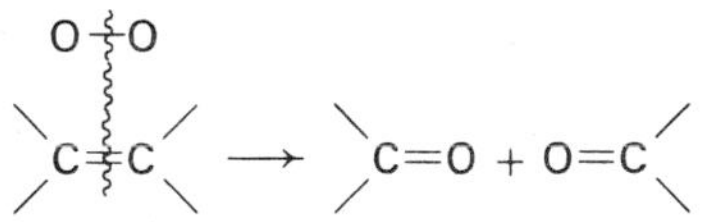

$$\begin{matrix} O \wr O \\ \rangle C \wr C \langle \end{matrix} \longrightarrow \rangle C{=}O + O{=}C\langle \qquad (32)$$

Ozonolysis is applicable to double bonds of all kinds (in other words, with the R, R′, R″ and R‴ in Equation 31 any of a variety of groups: alkyl, aryl, alkoxyl, and so forth). Some examples of ozonolytic degradations are shown in Equations 33–42, which are selected to show the scope of the reaction. In these equations, the intermediate ozonide is omitted:

$$CH_3CH_2CH{=}CH_2 + O_3 \longrightarrow CH_3CH_2CHO + CH_2O \qquad (33)$$

$$CH_3{-}\underset{CH_3}{\underset{|}{C}}{=}CHCH(CH_3)_2 + O_3 \longrightarrow CH_3{-}\underset{CH_3}{\underset{|}{C}}{=}O + (CH_3)_2CHCHO \qquad (34)$$

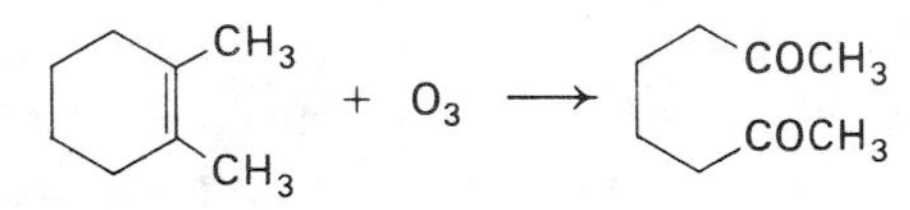

$$\text{1,2-dimethylcyclohexene} + O_3 \longrightarrow \text{(ring-opened chain with two } COCH_3 \text{ ends)} \qquad (35)$$

$$\text{(bicyclic alkene with angular } CH_3) + O_3 \longrightarrow \text{(cyclohexanone bearing } CH_3 \text{ and } CH_2CH_2CH_2CHO \text{ side chain)} \qquad (36)$$

$$\text{1-methoxycyclohexene } (OCH_3) + O_3 \longrightarrow \text{(chain with } COOCH_3 \text{ and } CHO \text{ ends)} \qquad (37)$$

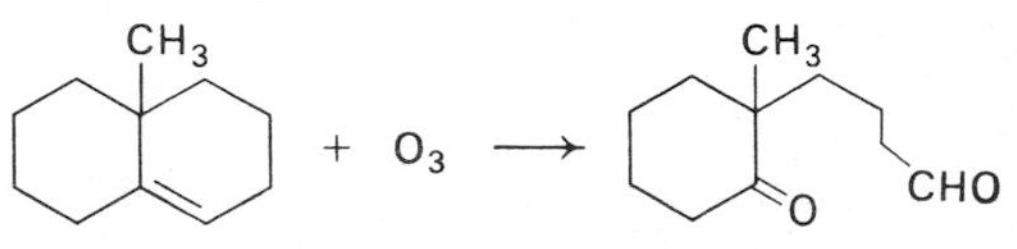

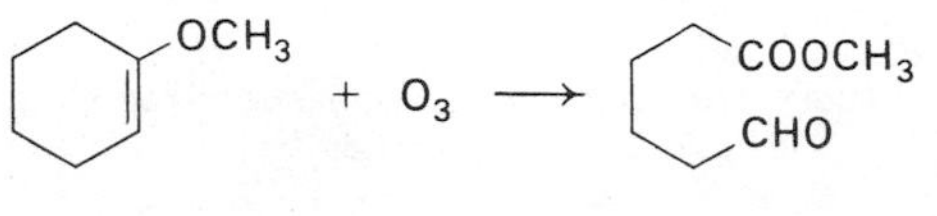

$$\text{3,4,5-triphenylisoxazole} + O_3 \longrightarrow R{-}CO{-}C(R){=}N{-}O{-}CO{-}R \quad (R = \text{phenyl}) \qquad (38)$$

$$\text{o-xylene} + O_3 \longrightarrow CH_3COCOCH_3 + CH_3COCHO + O{=}CHCH{=}O^* \qquad (39)$$

$$\left(\cdots CH_2\overset{CH_3}{\overset{|}{C}}{=}CHCH_2CH_2\overset{CH_3}{\overset{|}{C}}{=}CHCH_2\cdots\right)_x + O_3 \longrightarrow CH_3COCH_2CH_2CHO \qquad (40)$$

rubber

Ozonolysis is used for

(a) Proof of structure; that is, locating the double bonds in a compound.

(b) Degradation of a complex compound to a simpler compound, or to a compound of known structure.

(c) Preparative work. Although it is not often practicable to prepare large quantities of material by ozonolysis, the reaction can be used in special cases for the preparation of compounds difficult to obtain in other ways.

Compounds containing multiple double bonds are broken down into the expected fragments; for example,

$$CH_3CH{=}\overset{CH_3}{\overset{|}{C}}CH_2CH_2CH{=}CH_2 \xrightarrow{O_3} CH_3CHO + CH_3COCH_2CH_2CHO + H_2CO$$

$$CH_3CH{=}CH{-}\overset{CH_3}{\overset{|}{C}}{=}CHCH_3 + O_3 \longrightarrow CH_3CHO + CH_3COCHO + CH_3CHO \qquad (41)$$

$$\text{(methoxy-vinyl-methoxystyryl benzene, } CH_3O\text{-substituted)} + O_3 \longrightarrow CH_3O\text{-}C_6H_3(CHO)_2 + p\text{-}OCH_3C_6H_4CHO + CH_2O \qquad (42)$$

When ozonolysis is used in structure determination or degradation, the aldehydes that may be formed from —CH= units are often oxidized in a separate step to give carboxylic acids, which are usually easier to handle and identify. Ozonolysis followed by oxidation may also be used to carry degradation further.

* Note that the products do not correspond with the formal (arbitrary) Kekulé structure that is used for the benzene ring.

Exercise 10.7

Write the structure of compounds the ozonolysis of which would give the products shown:

(*a*) $CH_2O + (CH_3)_2CO$

(*b*) $CH_3COCH_2CH_3 + CH_3CHO$

(*c*) =O + CH_3CHO

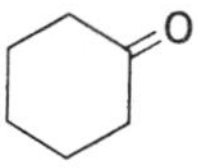

(*d*) O + $(CH_3)_2CO$

(*e*) COC_6H_5 CH_2COCH_3

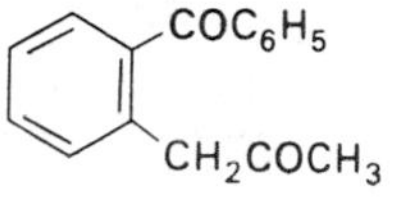

(*f*) 2 $CH_2(CHO)_2$

(*g*) $CH_3COCH_2COCH_3 + CH_3COCHO + O{=}CH{-}CH{=}O$

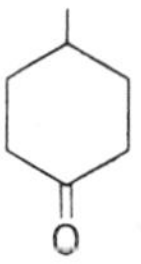

(*h*) $CH_3COCHO + CH_3COCH_2CH_2CHO$

(*i*) O O + CH_2O

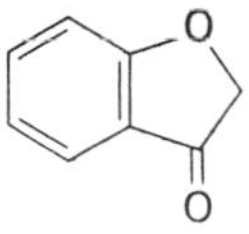

(*j*) =O + C_6H_5CHO

(*k*) $CH_2CH_2CH_2CH_2CH_2CHO + CH_3OOCCH_2CH_2CH_2CHO$ OH OH

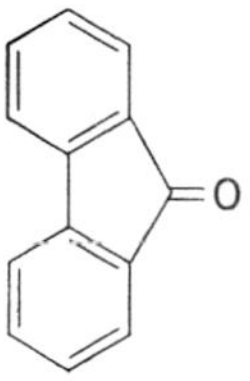

(*l*) CH_2 $CH_2CH_2CH_2COOCH_3$ $CH_2CH_2CH_2CHO$

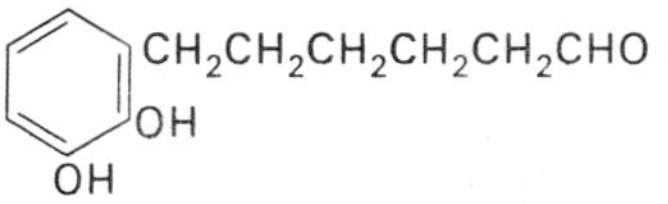

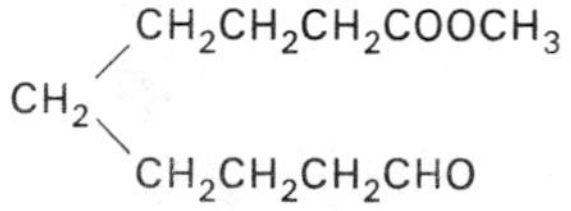

PART **11**

Planning the synthesis of an organic compound

A. INTRODUCTION

The synthetic problems found in the Exercises of the foregoing Parts have for the most part been related directly to a single class or type of reaction: Grignard, Claisen, carbon alkylation, oxidation of single functions, etc. Thus, such Exercises represent only fragments of practical synthetic problems, most of which involve a series of several or even numerous different types of synthetic operations, carried out in a sequence of distinct steps. Each step represents a specific transformation, and most steps include the isolation and purification (and occasionally characterization) of the intermediate product before it is submitted to the next operation.

As an example, consider the series

$$CH_3COCH_3 \xrightarrow{A} (CH_3)_2C{=}CHCOCH_3 \xrightarrow{B} (CH_3)_2\underset{\displaystyle CH(COOEt)_2}{\underset{|}{C}}CH_2COCH_3 \xrightarrow{C} \quad (1)$$

COOEt $\xrightarrow{D}$ COOH $\xrightarrow{E}$

This synthesis proceeds by way of an aldol condensation (*A*), a Michael reaction (*B*), a Claisen reaction (*C*), a saponification (*D*), and a decarboxylation (*E*).

The sequence

$$p\text{-}BrC_6H_4CH_3 \xrightarrow{A} p\text{-}BrC_6H_4COOH \xrightarrow{B} p\text{-}BrC_6H_4NH_2 \xrightarrow{C} p\text{-}BrC_6H_4Cl \qquad (2)$$

includes an oxidation (*A*), a Hofmann degradation (*B*), a diazotization-replacement (*C*).

It is often difficult to visualize in complete detail the whole series of steps that may be required for the preparation of a desired compound. Consequently, it is usually convenient to view the synthetic sequence in the reverse direction; that is, *to work backwards step by step from the final product* until a point is reached at which the early stages become obvious.

In designing a synthesis, the organic chemist has at his command an extensive array of chemicals that can be purchased from commercial sources. In setting up the Exercises in this workbook, it is assumed that the student will have at his command organic chemicals that are available in quantity, but compounds of elaborate and unusual structures should not be chosen as starting materials. It may be assumed that most monosubstituted benzenes, most aliphatic alcohols, ketones and acids of six or less carbon atoms are readily available. When the phrase "readily accessible starting materials" is used the student may choose compounds of the above classes, including such simple bifunctional compounds as the alkanedioic acids, unsaturated alcohols, olefins, and so on.

B. EXAMPLES OF MULTISTEP SYNTHESES

Example 1

Devise a practical synthesis for *A*, which is (2-hydroxy-2-methylcyclohexanone: ring with CH_3, OH, and =O), using reasonably accessible reagents.

(a) The immediate precursor for *A* can be the glycol *B*:

$$\underset{B}{\text{1-methylcyclohexane-1,2-diol}} \xrightarrow{CrO_3} \underset{A}{\text{2-hydroxy-2-methylcyclohexanone}}$$

The oxidation of the secondary alcohol to the ketone can be accomplished without affecting the tertiary alcohol. The stereochemistry of the glycol *B*—that is, whether it is the *cis*- or *trans*-glycol—is unimportant, for the oxidation of either will give the same product.

(b) The glycol *B* can be formed by the hydroxylation of the olefin *C*:

$$\text{1-methylcyclohexene } (C) \xrightarrow[\text{or } KMnO_4]{OsO_4} \text{1-methylcyclohexane-1,2-diol } (B)$$

(c) The olefin *C* is readily prepared by the dehydration of the alcohol *D*:

$$\text{1-methylcyclohexanol } (D) \xrightarrow{H_2SO_4} \text{1-methylcyclohexene } (C)$$

and the alcohol *D* is the product of the addition of a methyl Grignard reagent to cyclohexanone; these are readily available, inexpensive commercial chemicals.

Example 2

Devise a synthesis for hexane-2,5-dione, *A*. Working backwards, we can see that there are several ways of carrying out a final step leading to *A*:

B $CH_3CH(OH)CH_2CH_2CH(OH)CH_3 \xrightarrow{CrO_3} CH_3COCH_2CH_2COCH_3$ *A*

C $CH_3COCH{=}CHCOCH_3 \xrightarrow{H_2/Pt} CH_3COCH_2CH_2COCH_3$ *A*

D $CH_3COCH_2{-}CH(COOH)COCH_3 \xrightarrow{-CO_2} CH_3COCH_2CH_2COCH_3$ *A*

Which of the precursors *B*, *C* or *D* would be the most practical to prepare? The diol *B* could be prepared by the addition of methylmagnesium iodide to succinic dialdehyde; but the dialdehyde is not readily prepared. The unsaturated diketone *C* is in fact most conveniently made *from* the desired compound *A*, and so would not be a practical precursor to the latter. Compound *D* is readily made from unexceptional starting materials:

$$CH_3COCH_2Br + CH_3COCH_2COOEt \xrightarrow{NaOEt} CH_3COCH_2{-}CH(COOEt)COCH_3$$

Saponification of the keto ester would give *D*.

Example 3

Devise a practical synthesis for *A*, which is

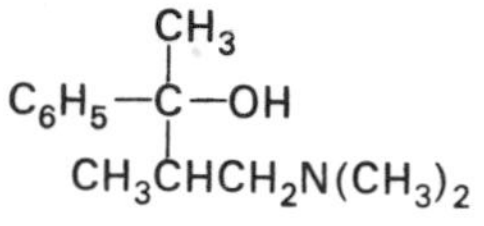

$$C_6H_5-\underset{CH_3\overset{|}{C}HCH_2N(CH_3)_2}{\overset{CH_3}{\overset{|}{\underset{|}{C}}}}-OH$$

Since the desired compound is a tertiary alcohol, a possible final step is a Grignard reaction; for example:

$$B \qquad C_6H_5-CO-\underset{CH_3}{\underset{|}{C}}HCH_2N(CH_3)_2 + CH_3MgI \longrightarrow A$$

$$C \qquad CH_3-CO-\underset{CH_3}{\underset{|}{C}}HCH_2N(CH_3)_2 + C_6H_5MgBr \longrightarrow A$$

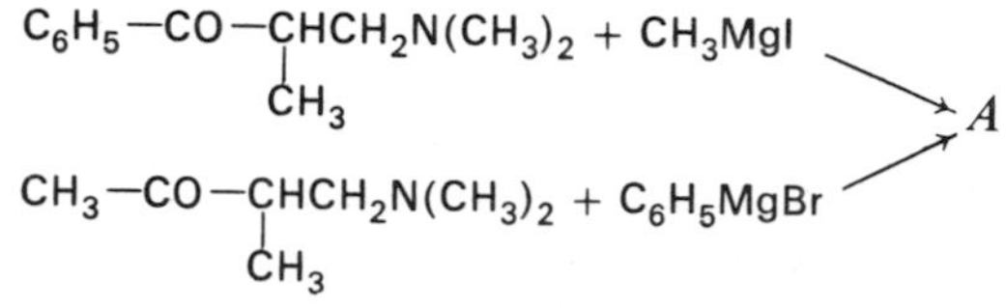

Which of these is preferable? It will be noted that the ketone in each case is a Mannich base. Compound *B* can be prepared by the reaction between propiophenone ($C_6H_5COCH_2CH_3$), formaldehyde, and dimethylamine. It would appear that *C* could be prepared in an analogous way, using 2-butanone. But it will be seen that while propiophenone will give only *B*, the Mannich reaction with 2-butanone could proceed in two ways, for both the CH_3- group and the $-CH_2-$ group are susceptible to the Mannich condensation. Consequently, it would be prudent to use the route leading unambiguously to the desired product, and *B* would be preferred.

Example 4

Devise a synthesis for *A*, 2,5-dibromobenzoic acid.

This compound cannot be prepared satisfactorily by the bromination of benzoic acid. (Why?) Several possible routes, starting with *p*-dibromobenzene, are conceivable; two of them are the following:

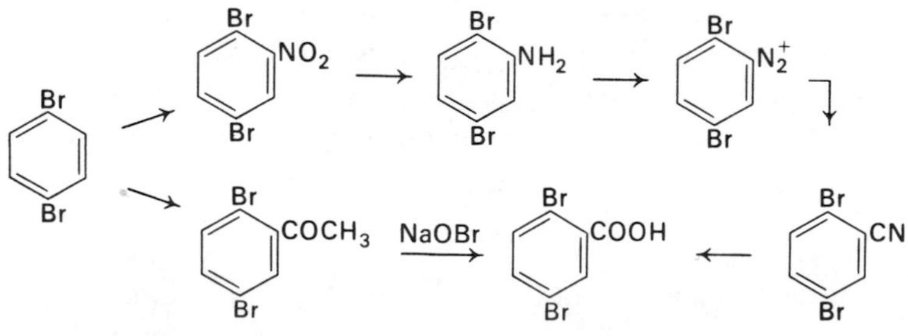

Br, Br (p-dibromobenzene) → Br, NO_2, Br → Br, NH_2, Br → Br, N_2^+, Br → Br, CN, Br → Br, COOH, Br

Br, Br (p-dibromobenzene) → Br, $COCH_3$, Br $\xrightarrow{NaOBr}$ Br, COOH, Br

Since the route starting with the Friedel-Crafts acetylation reaction is direct and requires fewer steps, it would be chosen.

Example 5

Show how 2,2,5,5-tetramethyltetrahydrofuran,

$$\text{(2,2,5,5-tetramethyltetrahydrofuran ring: } (H_3C)_2C \text{ and } C(CH_3)_2 \text{ joined through O)}$$

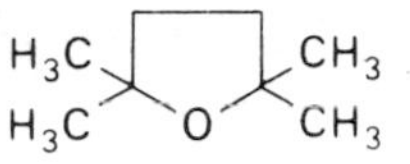

A

can be prepared.

The synthesis of this cyclic ether can be carried out by the acid-catalyzed dehydration shown in the following equation

$$HOC(CH_3)_2CH_2CH_2C(CH_3)_2OH \xrightarrow{H_2SO_4} A$$

B

The diol *B* could be prepared in either of two ways:

$$\underset{C}{CH_3COCH_2CH_2COCH_3} + 2\,CH_3MgI \longrightarrow (CH_3)_2C(HO)CH_2CH_2C(HO)(CH_3)_2$$

or

$$\underset{D}{\begin{matrix}CH_2COOEt\\ |\\ CH_2COOEt\end{matrix}} + 4\,CH_3MgI \longrightarrow (CH_3)_2C(HO)CH_2CH_2C(HO)(CH_3)_2$$

Clearly, the synthesis involving the ethyl ester of the readily available succinic acid would be a convenient and practicable one.

Exercise 11.1

Devise the practical syntheses of the following compounds, using the indicated starting material or other readily available starting materials, and other necessary reagents:

(*a*) $CH_3CH_2OCH_2CH_2OH$
(from $BrCH_2COOEt$)

(*b*) $HOCH_2(CH_2)_4CH_2OH$
(from cyclohexanol)

(*c*) 2-allylcyclohexanol (cyclohexane ring bearing $CH_2CH{=}CH_2$ and OH on adjacent carbons)
(from cyclohexanone)

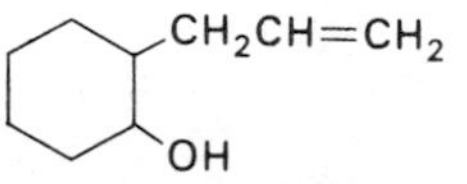

(*d*) $(C_6H_5)_2C{=}CH_2$
(from bromobenzene)

(*e*) 5,6-dimethoxy-1-indanone (CH_3O, CH_3O on the benzene ring; C=O on the five-membered ring)
(from vanillin)

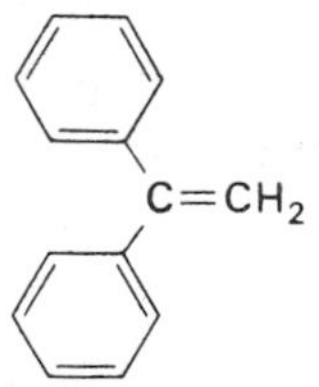

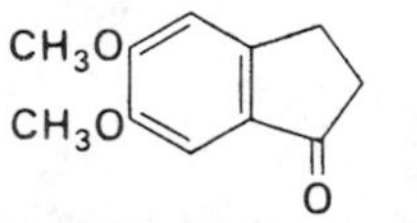

(*f*) $C_6H_5CH_2-\underset{\displaystyle CH_3}{\underset{|}{C}}HCOOH$

(from benzaldehyde)

(*g*) $CH_3COCH_2-\underset{\displaystyle CH_3}{\underset{|}{C}}HCOOH$

(from ethyl malonate)

(*h*) (cyclopentyl)$-CH_2OH$

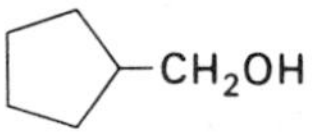

(from succinic acid)

(*i*) (cyclohexenyl)$CH_2CH(CH_3)_2$

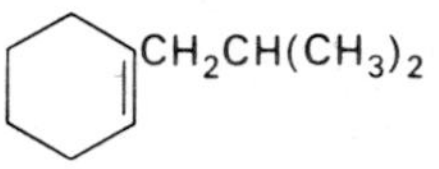

(from cyclohexanol)

(*j*) $C_6H_5-\underset{\displaystyle COOH}{\underset{|}{C}}HCH_2COC_6H_5$

(from benzalacetophenone)

(*k*) CH_3O CH_3O $COCH_3$

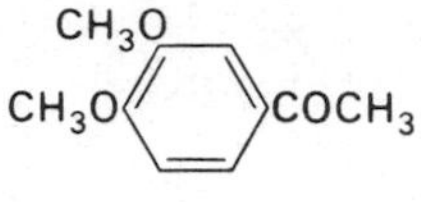

(from vanillin)

(*l*) $(CH_3)_2CHCH_2CH_2CH_3$

(*m*) OH CH_2OH

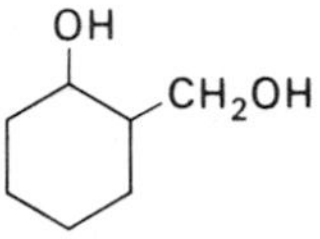

Example 6

Suppose it is desired to prepare

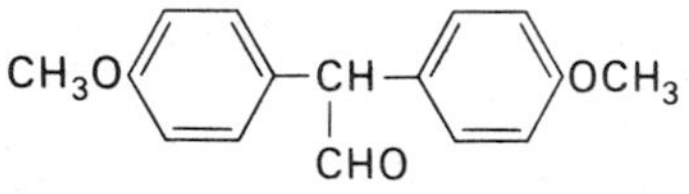

$$CH_3O-C_6H_4-\underset{\displaystyle CHO}{\underset{|}{CH}}-C_6H_4-OCH_3$$

In the first place, whatever the route chosen, the *p*-methoxyphenyl groups would be present in the starting materials, for it would be highly impractical to attempt to introduce the methoxyl substituents at a late stage in the synthesis.

There are several general ways to produce the formyl (—CHO) group. For example:

$$R_2CHCH{=}CH_2 \xrightarrow{O_3} R_2CHCHO \qquad (a)$$

$$R_2CHCH_2OH \xrightarrow{CrO_3} R_2CHCHO \qquad (b)$$

$$R_2CH\overset{\displaystyle OH}{\overset{|}{C}}HCH_2OH \xrightarrow[Pb(OAc)_4]{HIO_4,\text{ or}} R_2CHCHO \qquad (c)$$

$$R_2C(OH)CH_2OR' \xrightarrow[\text{(pinacol rearrangement)}]{H_2O/H^+} R_2CHCHO \quad (R' = H \text{ or } CH_3) \qquad (d)$$

$$R_2CHMgBr + HC(OEt)_3 \longrightarrow R_2CHCH(OEt)_2 \xrightarrow{H^+} R_2CHCHO \qquad (e)$$

Method *a* suffers from the fact that the synthesis of $(CH_3OC_6H_4)_2CHCH{=}CH_2$ is in itself not without difficulty; and ozonolysis as a *preparative* procedure is far from ideal. Method *b* is feasible insofar as the preparation of the alcohol is concerned. This compound could be prepared by the route:

$$CH_3OC_6H_4CHO \xrightarrow{KCN} CH_3OC_6H_4CH(OH)COC_6H_4OCH_3 \longrightarrow \qquad (f)$$

$$CH_3OC_6H_4COCOC_6H_4OCH_3 \xrightarrow{KOH} CH_3OC_6H_4C(OH)(COOH)C_6H_4OCH_3 \xrightarrow{HI}$$

$$CH_3OC_6H_4CH(COOH)C_6H_4OCH_3 \xrightarrow{LiAlH_4} CH_3OC_6H_4CH(CH_2OH)C_6H_4OCH_3$$

All of these steps are practicable, but the number of manipulations makes this route very unattractive. If each step were capable of a 90% yield, the overall yield would be $0.9 \times 0.9 \times 0.9 \times 0.9 \times 0.9 =$ about 59%; and if any one step were, say, 70%, the yield would be 45%. But this is not the principal disadvantage: the final step, the oxidation of the alcohol to the aldehyde (method *b*) would be very difficult to carry out in high yield. Even if this step gave a 75% yield the overall yield of final compound would be somewhere in the region 35–45%. Thus, the time required, coupled with the questionable yield of final product, makes this overall method unappealing.

Method *c* requires a starting material (the glycol) that could be prepared from the olefin of method *a*, but there is no simple and reliable method of preparing this olefin. An attempt to prepare it by dehydration of $R_2CHCH_2CH_2OH$ or $R_2CHCHOHCH_3$ (where R is *p*-anisyl) would probably be difficult to accomplish satisfactorily, for the formation of $R_2C{=}CHCH_3$ can be expected to predominate.

Method *d* is practical for several reasons. The glycol can be prepared by hydroxylation of the olefin $R_2C{=}CH_2$, or by reduction ($LiAlH_4$) of the hydroxy acid, $R_2C(OH)COOH$, shown in the scheme above. Thus, the scheme in which R is *p*-anisyl is feasible.

$$RCHO \longrightarrow RCHOHCOR \longrightarrow RCOCOR \longrightarrow R_2C(OH)COOH \longrightarrow$$

$$R_2C(OH)CH_2OH \longrightarrow R_2CHCHO$$

The starting material, anisaldehyde (or *p*-hydroxybenzaldehyde, from which the methoxy compound is readily prepared), can be regarded as an available chemical.

An alternative method, essentially the same as (*d*) is the following:

$$R_2\overset{\overset{\displaystyle OH}{|}}{C}CH_2OCH_3 \xrightarrow{H^+} R_2CHCHO \qquad (g)$$

The pinacol rearrangement proceeds just as well with the ether, in (*g*), as with the alcohol, in (*d*), for the intermediate steps

$$R_2\overset{\overset{\displaystyle OH}{|}}{C}-CH_2OH \xrightarrow{H^+} R_2\overset{+}{C}CH_2OH \longrightarrow R_2CH\overset{+}{C}HOH \qquad (h)$$

and

$$R_2\overset{\overset{\displaystyle OH}{|}}{C}CH_2OCH_3 \xrightarrow{H^+} R_2\overset{+}{C}CH_2OCH_3 \longrightarrow R_2CH\overset{+}{C}HOCH_3 \qquad (i)$$

lead to intermediate oxonium compounds that are essentially equivalent to the aldehyde in the presence of an excess of water. It should be noted that the corresponding epoxide can also undergo rearrangement in the same way.

How could $R_2\overset{\overset{\displaystyle OH}{|}}{C}CH_2OCH_3$ be prepared? It is to be noticed that this compound is a tertiary alcohol in which the R groups are the same. Tertiary alcohols of this type are prepared by the reaction of a Grignard reagent with an ester.

$$RMgBr + CH_3OCH_2COOMe \longrightarrow CH_3OCH_2C(R)_2-OH \qquad (j)$$

Thus, the necessary starting materials are $CH_3O-C_6H_4-Br$ and CH_3OCH_2COOMe. The former may be regarded as available. The ester can be prepared by the straightforward displacement reaction

$$CH_3O^- + BrCH_2COOMe \longrightarrow CH_3OCH_2COOMe + Br^- \qquad (k)$$

Exercise 11.2

Suppose you wished to prepare the unsymmetrical aldehyde,

$$CH_3O-C_6H_4-\underset{\underset{\displaystyle CHO}{|}}{CH}-C_6H_5$$

How could you modify the procedures above for this purpose? [*Hint:* The problem resolves itself to the preparation of

$$\begin{matrix} C_6H_5 \\ & \diagdown \\ & & CCH_2OR \\ & \diagup & | \\ CH_3O{\cdot}C_6H_4 & & OH \end{matrix}$$

where R = OH or OCH_3].

Example 7

Starting with available materials, prepare compounds of the structure [3-R-cyclohex-2-enone],

designing the procedure so that various R groups can be introduced.

Various synthetic routes could be devised to effect this result. One, for example, would be the following:

$$RCOCH_2CH_2CH_2COCH_3 \xrightarrow[\text{condensation}]{\text{aldol}} \text{3-R-cyclohex-2-enone} \qquad (a)$$

The problem in this case would be the preparation of the required diketone. It could be made in the following way

$$\text{1-R-2-methylcyclopentane-1,2-diol} \xrightarrow{Pb(OAc)_4} RCO(CH_2)_3COCH_3 \qquad (b)$$

or by ozonolysis of 2-R-1-methylcyclopentene. However, the problem of designing a *general* synthesis of the cyclopentene or the required diol would present problems comparable to that originally posed.

The diketones needed for Reaction *a* could be prepared by a Michael reaction:

$$RCOCH{=}CH_2 + \underset{}{CH_2(COCH_3)COOEt} \xrightarrow{NaOEt} RCOCH_2CH_2CH(COOEt)COCH_3 \xrightarrow[(2)\ -CO_2]{(1)\ \text{saponify}}$$

$$RCOCH_2CH_2CH_2COCH_3 \qquad (c)$$

Although this sequence seems practicable, there is a difficulty in the aldol ring closure in (*a*) that is not at once apparent as it is formulated. When R is CH_3 the ring closure condensation proceeds as shown in (*a*). If R is CH_3CH_2—, for example, the ring closure could (and probably would) take a different course:

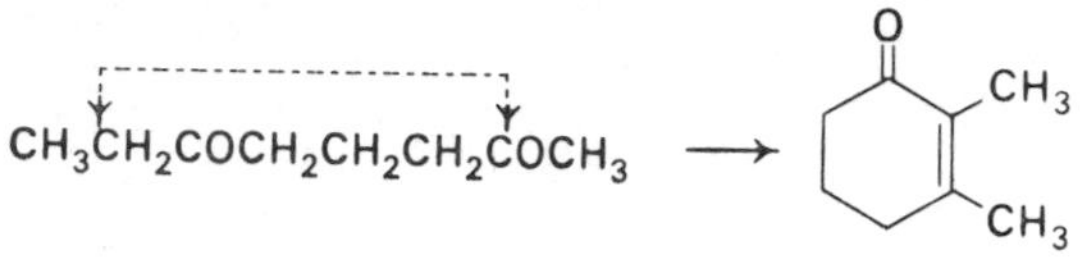

$$CH_3CH_2COCH_2CH_2CH_2COCH_3 \longrightarrow \qquad (d)$$

Thus, this first plan of a *general* synthesis is faulty.

An alternative procedure is one in which the six-membered ring is present in the starting material. One might suppose that the 1,3-diketone could be used in a Grignard reaction:

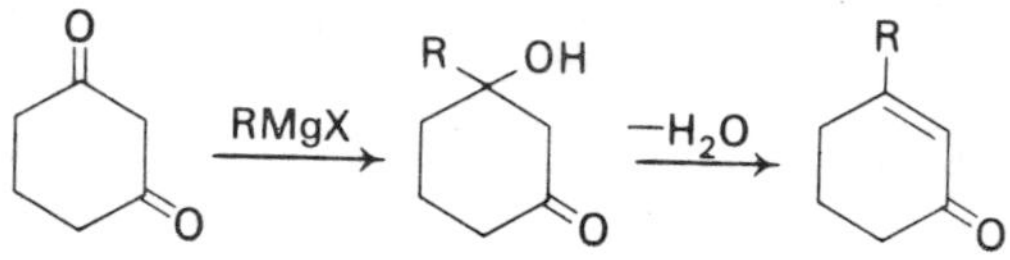

$$\xrightarrow{RMgX} \quad \xrightarrow{-H_2O} \qquad (e)$$

but this might be impractical for two reasons: (1) even if the Grignard reaction were feasible, it would be necessary to control the conditions so that only monoaddition occurred, and (2), the Grignard reaction would be impractical because the diketone would react as the enol:

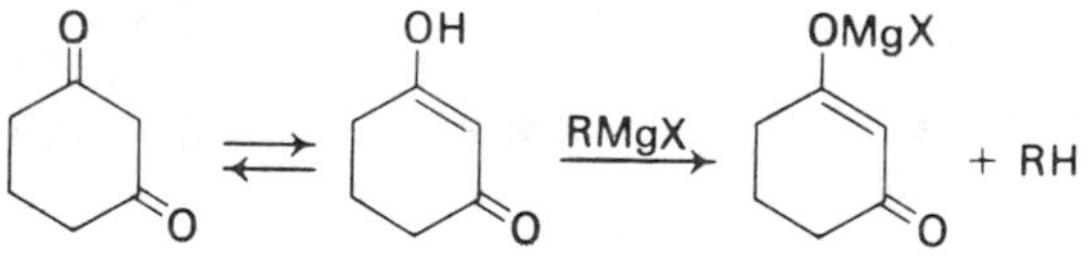

$$\rightleftarrows \quad \xrightarrow{RMgX} \quad + \text{RH} \qquad (f)$$

Although addition to the enolate carbonyl group could occur, the reaction might be expected to be a poor one (for example, the enolate-MgX might be insoluble in the reaction medium).

It is apparent, however, that if an enol ether could be made, the reaction would be expected to lead in good yield to the desired product:

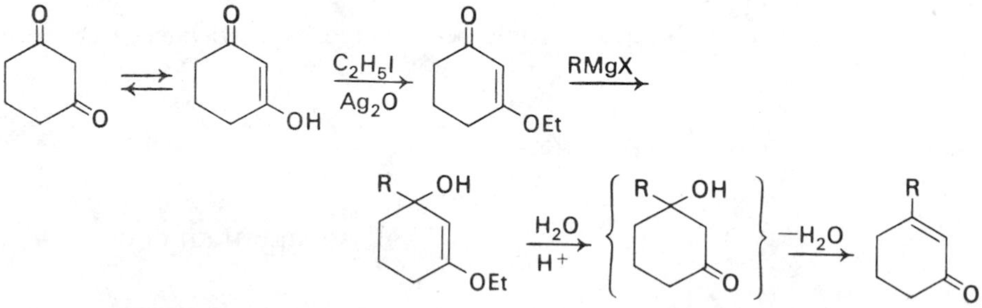

$$\rightleftarrows \quad \xrightarrow[Ag_2O]{C_2H_5I} \quad \xrightarrow{RMgX}$$

$$\xrightarrow[H^+]{H_2O} \left\{ \quad \right\} \xrightarrow{-H_2O} \qquad (g)$$

The problem of the synthesis is now reduced to the preparation of 1,3-cyclohexanedione, or dihydroresorcinol:

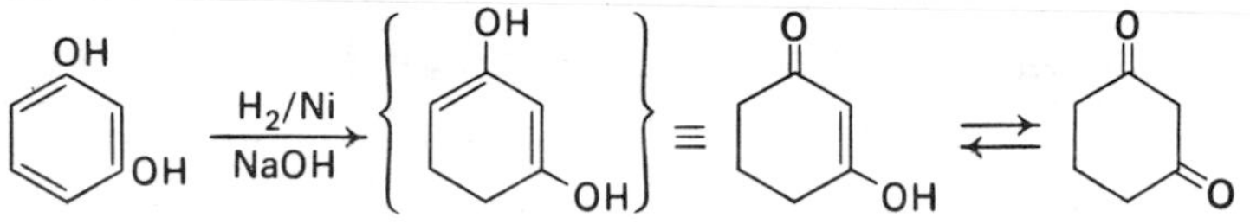

(h)

This compound is readily prepared. Catalytic reduction of resorcinol gives the dihydro compound, O-alkylation of which is readily accomplished. Thus, the overall scheme of synthesis is:

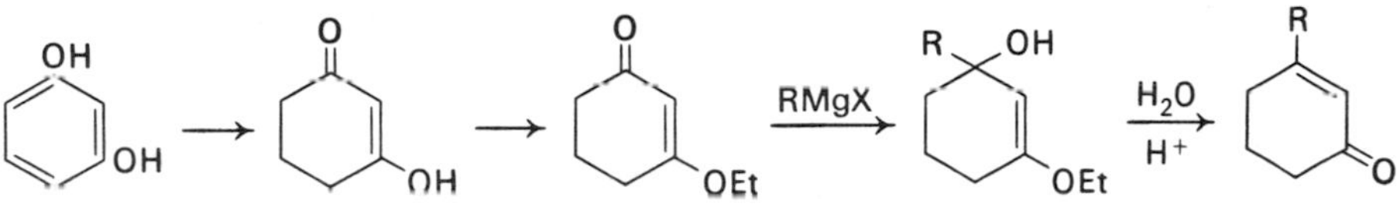

(i)

It is clear that this is a versatile synthesis; its limitations are few.

Exercise 11.3

Devise a procedure for the preparation of: (a) *m*-diphenylbenzene, (b) 3-*p*-anisyl-4′-chlorobiphenyl, (c) 1-methyl-1,3-cyclohexadiene. *Hint:* Dihydrobenzenes (cyclohexadienes) are readily dehydrogenated to the aromatic compounds by heating with sulfur or selenium; for example,

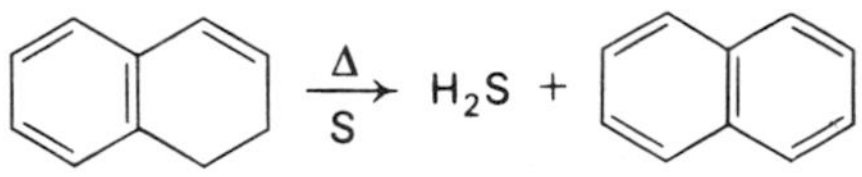

$$\xrightarrow[S]{\Delta} H_2S +$$

Example 8

Prepare $C_6H_5CH_2COCH_2CH_3$

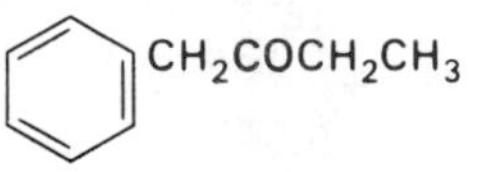

The most general route for the preparation of ketones is by the oxidation of the corresponding secondary alcohol:

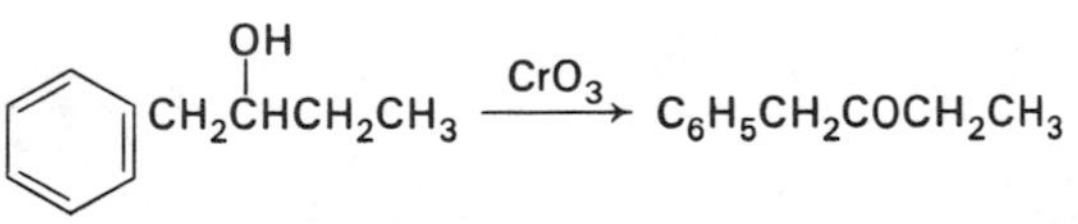

$$C_6H_5CH_2CH(OH)CH_2CH_3 \xrightarrow{CrO_3} C_6H_5CH_2COCH_2CH_3 \qquad (a)$$

Thus, the problem could be solved by the preparation of the alcohol of (*a*). This can be accomplished by either of the Grignard reactions *b* or *c*:

$$C_6H_5CH_2CHO + CH_3CH_2MgBr \longrightarrow C_6H_5CH_2\underset{}{\overset{OH}{\overset{|}{C}}}HCH_2CH_3 \qquad (b)$$

$$C_6H_5CH_2MgCl + CH_3CH_2CHO \longrightarrow C_6H_5CH_2\overset{OH}{\overset{|}{C}}HCH_2CH_3 \qquad (c)$$

A third route leading to the ketone itself is the addition of a Grignard reagent to a nitrile:

$$C_6H_5CH_2CN + CH_3CH_2MgBr \longrightarrow C_6H_5CH_2COCH_2CH_3 \qquad (d)$$

$$C_6H_5CH_2MgCl + CH_3CH_2CN \longrightarrow C_6H_5CH_2COCH_2CH_3 \qquad (e)$$

Method *d* is found to give very poor yields. The reason for this is that the active methylene group of phenylacetonitrile reacts with the Grignard reagent faster than addition to the —CN group occurs:

$$C_6H_5CH_2CN + RMgX \longrightarrow (C_6H_5CHCN)MgX + RH \qquad (f)$$

Method *e*, however, is quite feasible.

Still a third method of preparation of the desired ketone takes advantage of the reactive methylene group of phenylacetonitrile. A Claisen condensation of phenylacetonitrile with ethyl propionate occurs as follows:

$$C_6H_5CH_2CN + CH_3CH_2COOEt \xrightarrow{NaOEt} C_6H_5\overset{COCH_2CH_3}{\overset{|}{C}}HCN \xrightarrow[\text{(2) decarboxylation}]{\text{(1) hydrolysis (CN to COOH)}} C_6H_5CH_2COCH_2CH_3 \qquad (g)$$

It may now be asked: if several routes, all feasible, are available for the preparation of a compound, which is to be preferred? The choice will usually depend upon (1) the ready availability or ease of preparation of one or another of the reagents; (2) the relative yields of the various (overall) routes; (3) the possibility of the intervention of side reactions, with consequent problems of final purification of the product.

Example 9

Synthesize $CH_3OCH_2\overset{OH}{\overset{|}{C}}\underset{\underset{CH_2CH_3}{|}}{}CH_2CH_3$

A tertiary alcohol in which two of the groups are the same is conveniently synthesized by the addition of a Grignard reagent to an ester.

$$\underset{\text{ethyl methoxyacetate}}{CH_3OCH_2COOEt} + C_2H_5MgBr \longrightarrow \underset{\text{3-methoxymethyl-3-pentanol}}{CH_3OCH_2C(OH)(CH_2CH_3)CH_2CH_3} \qquad (a)$$

When the three groups are all different, addition of a Grignard reagent to a ketone is the method used (Example 3).

Example 10

Synthesize $C_6H_5C(OH)(CH_3)C_6H_4Br$

What method should be used to prepare this?

$$C_6H_5\text{—CO—}C_6H_4Br + CH_3MgBr \qquad (a)$$

$$BrC_6H_4COCH_3 + C_6H_5MgBr \qquad (b)$$

$$C_6H_5COCH_3 + BrC_6H_4MgBr \qquad (c)$$

The answer to a question of this kind depends upon a number of quite independent factors. When each reaction is in itself practicable, the choice often depends upon the (1) *immediate* availability of the necessary reagents; (2) relative ease of accessibility (by synthesis) of the necessary reagents; (3) relative cost of the use of one reagent over another; (4) relative yields of the product prepared by the three reactions.

Thus, it is clear that no *general* rule can be used as a guide. In the example presented, the ready availability of CH_3MgI and C_6H_5MgBr and the generally excellent yields in their reactions would lead to the choice of methods *a* or *b*. The mono-Grignard reagent of *p*-dibromobenzene can be prepared, but this would be an unnecessarily involved method. Each method requires the preparation of one reagent that is not likely to be readily available in any but a well-stocked laboratory. Methyl iodide, bromobenzene and acetophenone are, of course, quite unexceptional.

The other compounds could be prepared as follows:

$$p\text{-}BrC_6H_4COC_6H_5 \text{ from } p\text{-}BrC_6H_4COCl + C_6H_6\ (AlCl_3) \qquad (d)$$

$$p\text{-}BrC_6H_4COCH_3 \text{ from } C_6H_5Br + CH_3COCl\ (AlCl_3) \qquad (e)$$

$$p\text{-}BrC_6H_4Br \text{ from } C_6H_5Br + Br_2 \qquad (f)$$

If *p*-bromobenzoic acid were available, the most practicable route to the desired product would be by a combination of Reactions *a* and *d*.

Example 11

Synthesize $CH_2{=}CHCH_2CH_2OH$

There are several routes to this compound that include the use of a Grignard reagent, two of which involve the preparation of an allylmagnesium halide:

$$CH_2{=}CHCH_2Cl \xrightarrow[\text{ether}]{Mg} CH_2{=}CHCH_2MgCl \qquad (a)$$

$$CH_2{=}CHCH_2MgCl + HCHO \longrightarrow CH_2{=}CHCH_2CH_2OH \qquad (b)$$

or

$$CH_2{=}CHCH_2MgCl + CO_2 \longrightarrow CH_2{=}CHCH_2COOH \qquad (c')$$

$$CH_2{=}CHCH_2COOH + LiAlH_4 \longrightarrow CH_2{=}CHCH_2CH_2OH \qquad (c'')$$

Another related method makes use of vinyllithium:

$$CH_2{=}CHCl + Li \longrightarrow CH_2{=}CHLi \qquad (d')$$

$$CH_2{=}CHLi + \underset{\text{O}}{CH_2{-}CH_2} \longrightarrow CH_2{=}CHCH_2CH_2OH \qquad (d'')$$

Example 12

Synthesize 1-(2-methylpropyl)cyclohexanol: $C_6H_{10}(OH)CH_2CH(CH_3)_2$

The most practicable route to the compound is the following:

$$\text{cyclohexanone} + (CH_3)_2CHCH_2MgBr \longrightarrow \text{1-}(CH_2CH(CH_3)_2)\text{-cyclohexan-1-ol (HO)}$$

This is simple and straightforward and would certainly be the method of choice.

Example 13

Synthesize 2-phenyl-1-benzopyrylium ($\overset{+}{O}$, C_6H_5) Cl^-

This cyclic oxonium salt is related to the neutral compound, called the "pseudobase," by the acid-base equilibrium

$$\text{cation } (\overset{+}{O}, C_6H_5) \underset{H^+}{\overset{OH^-}{\rightleftarrows}} \text{pseudobase } (O, HO, C_6H_5) \qquad (a)$$

cation pseudobase

The pseudobase is the cyclic hemiketal of the phenolic ketone:

$$\text{(OH, } COC_6H_5) \rightleftarrows \text{(O, HO, } C_6H_5) \qquad (b)$$

The phenolic ketone of (*b*) is the product of the following aldol condensation:

$$\text{(OH, CHO)} + \underset{CH_3}{COC_6H_5} \xrightarrow{HCl} \text{(OH, } COC_6H_5) \qquad (c)$$

When the condensation Reaction *c* is carried out with HCl catalysis, the oxonium salt is formed directly.

Example 14

Synthesize 1-methyl-7-isopropylnaphthalene.

It is clear that an attempt to prepare this compound by somehow introducing the isopropyl and methyl groups into naphthalene, and in the desired positions, could not succeed. Controlled mono-alkylation is seldom practicable, and it is not to be expected that substitution would occur uniquely and exclusively in the desired positions.

The most attractive plan of attack would be to construct one of the two rings, starting from a substituted benzene in which one of the two substituents is already present. This would include the necessity for providing a functional group at such a position as to make possible the introduction of the second alkyl group.

Since the final aromatic system can be produced by dehydrogenation (S or Se), most of the early stages can be performed with a single aromatic (benzene) ring present. Thus, one can visualize such routes as the following:

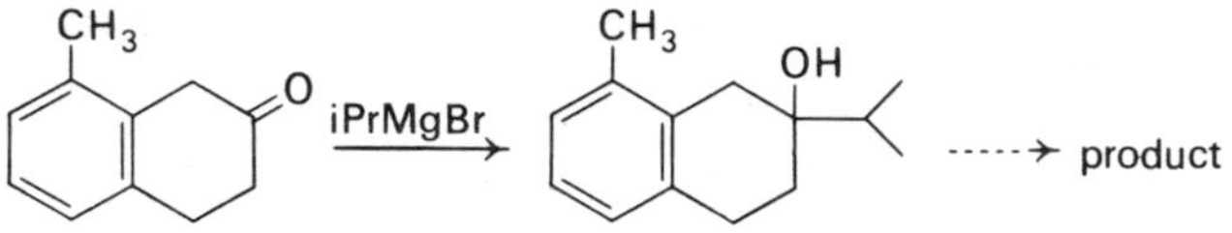

CH_3 ... O —iPrMgBr→ CH_3 ... OH ----→ product (*a*)

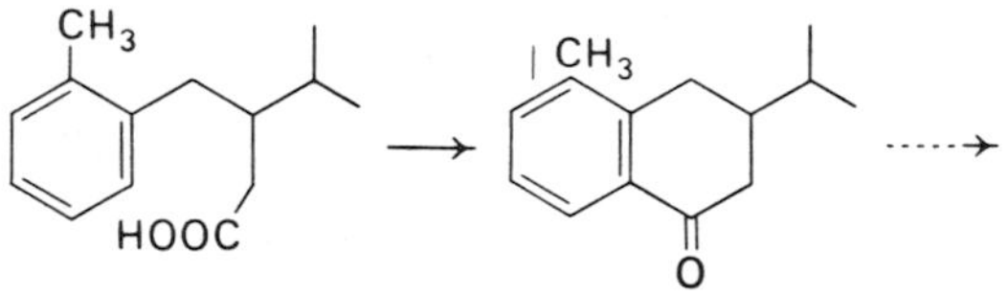

CH_3 ... HOOC → CH_3 ... O ----→ (*b*)

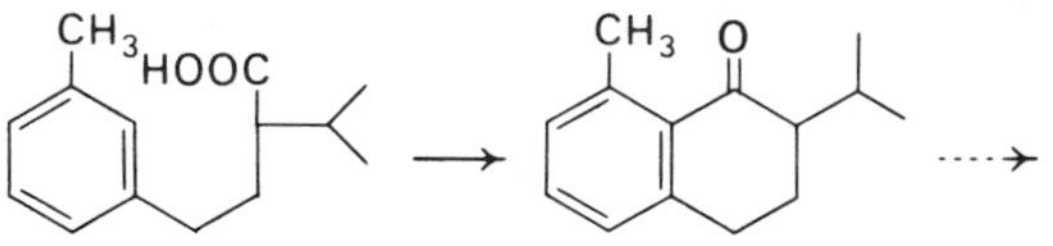

CH_3 HOOC ... → CH_3 O ... ----→ (*c*)

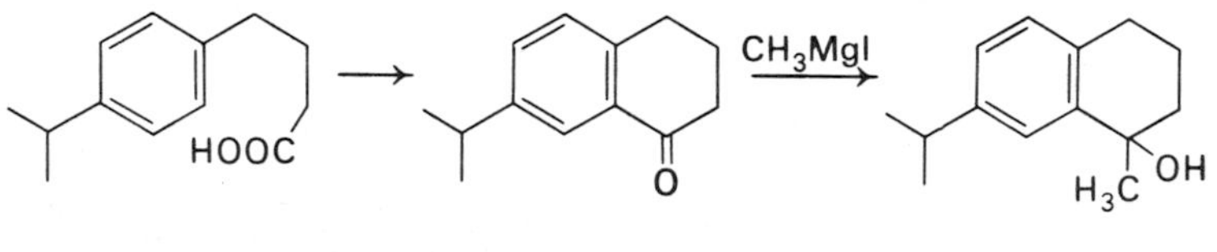

HOOC ... → ... O —CH_3MgI→ ... H_3C OH (*d*)

Other, more elaborate, routes could be devised, but the above will serve for discussion.

Route a. This requires the preparation of 8-methyl-2-tetralone. To construct this starting with toluene would be a long and arduous process.

Route b. Starting with *o*-tolualdehyde, the Stobbe condensation would provide the following compound

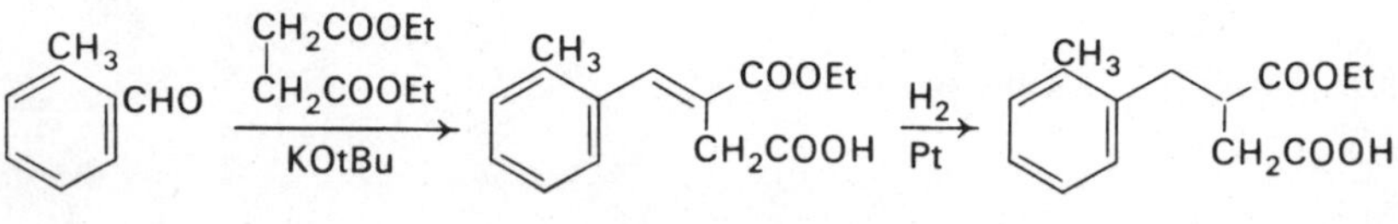

CH_3 CHO —(CH_2COOEt / CH_2COOEt, KOtBu)→ CH_3 ... COOEt / CH_2COOH —H_2, Pt→ CH_3 ... COOEt / CH_2COOH

Conversion of —COOEt into —C(OH)(CH$_3$)CH$_3$ and then into —CH(CH$_3$)$_2$ would lead to the desired carboxylic acid, which could be cyclized to 3-isopropyl-5-methyl-1-tetralone.

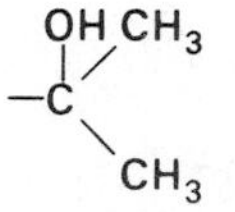

Route c. The construction of the desired acid could be accomplished as follows:

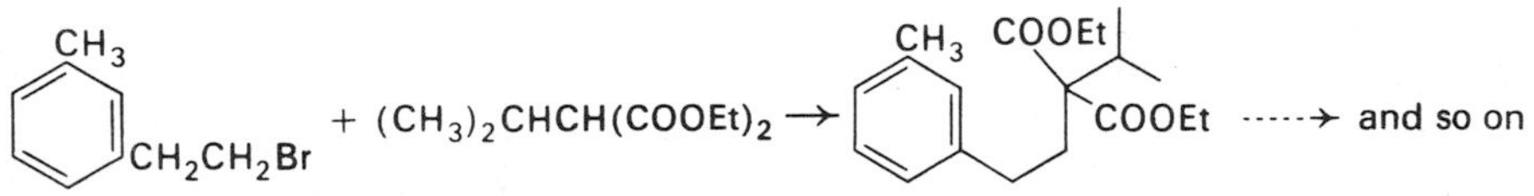

$$\text{CH}_3\text{C}_6\text{H}_4\text{CH}_2\text{CH}_2\text{Br} + (\text{CH}_3)_2\text{CHCH(COOEt)}_2 \rightarrow \text{product (CH}_3, \text{COOEt, COOEt)} \dashrightarrow \text{and so on}$$

but it is apparent that ring closure could proceed in two ways:

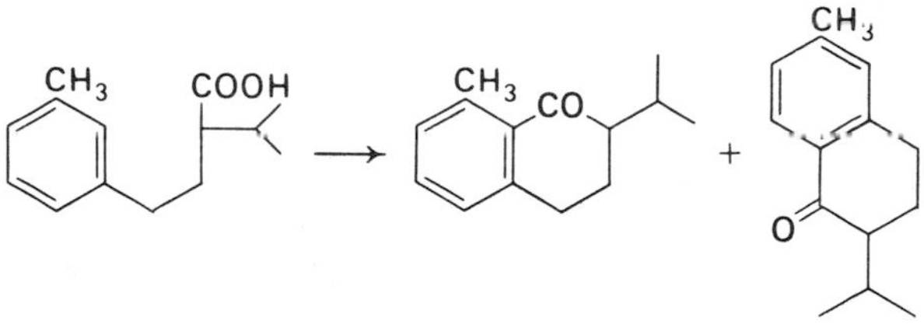

$$\text{CH}_3, \text{COOH} \longrightarrow \text{CH}_3, \text{CO} + \text{CH}_3$$

This would pose problems in respect to (1) yield of the desired compound, and (2) separation and purification.

Route d. This is practical and uncomplicated:

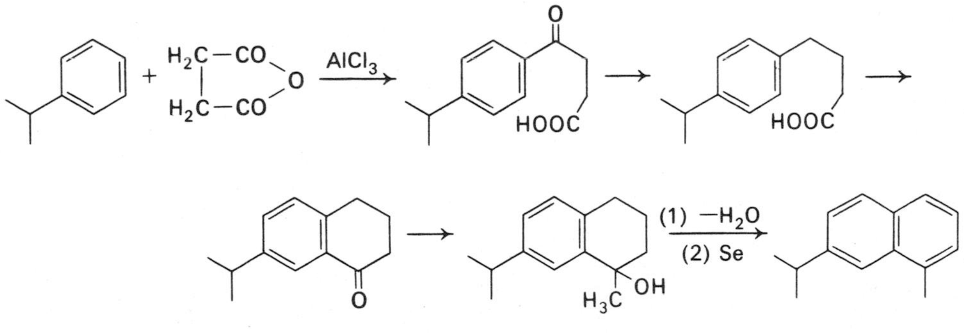

$$+\ \text{H}_2\text{C—CO, O, H}_2\text{C—CO} \xrightarrow{\text{AlCl}_3} \text{O, HOOC} \longrightarrow \text{HOOC} \longrightarrow$$

$$\longrightarrow \text{O} \longrightarrow \text{H}_3\text{C OH} \xrightarrow[\text{(2) Se}]{\text{(1) }-\text{H}_2\text{O}}$$

Note that the ring closure of the γ-(p-isopropylphenyl) butyric acid can lead to a single isomer only.

Exercise 11.4

Devise practical syntheses for the following compounds. Show all steps and reagents to be used.

(*a*) CH_2CH_2COOH, OH

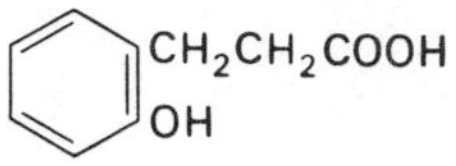

(*b*) $CH_3COCH_2CH_2COOH$

(*c*) $(C_6H_5COCH_2CH_2)_2NCH_3$

(*d*)

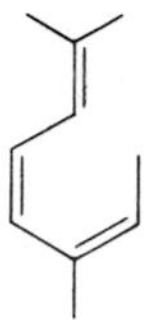

(*e*) 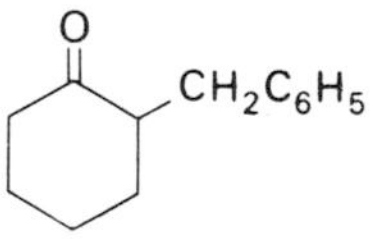 O, $CH_2C_6H_5$

(*f*) C_2H_5, N, CH_3

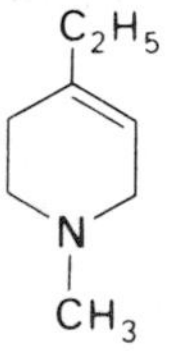

(*g*) $CH_3CH_2C{=}CH$, CH_3CH_2

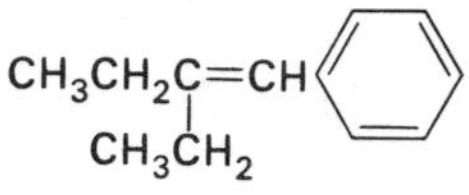

(*h*) $CH_2{=}CHCH_2CH_2OCOCH_3$

(*i*) $C_6H_5CH_2CH_2CH_2CH_2CH_2OH$

(*j*) CH_3O, OH, $CHCH_2NH_2$

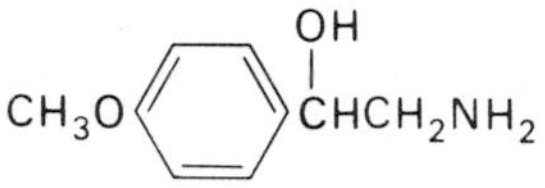

(*k*) CH_3, $COCH_2$

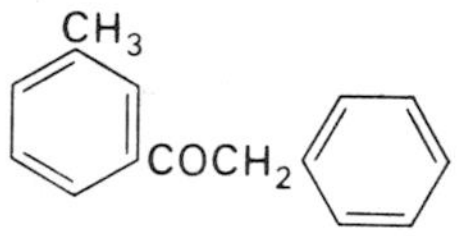

(*l*) C_6H_5, COOH, N, CH_3

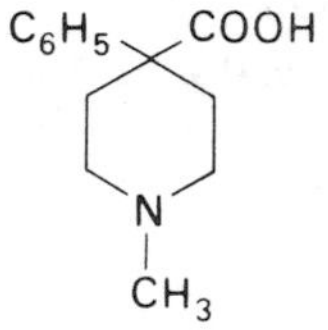

Suggested principal starting materials are

(*a*) benzaldehyde
(*b*) ethyl acetoacetate
(*c*) acetophenone
(*d*) methyl ethyl ketone, acetylene, isobutyraldehyde
(*e*) pimelic acid
(*f*) methyl acrylate
(*g*) phenylacetic acid
(*h*) allyl chloride
(*i*) benzaldehyde
(*j*) anisaldehyde
(*k*) toluene
(*l*) phenylacetonitrile

PART **12**

Problems in interpretation of experimental data

The following exercises involve the use of various kinds of experimental evidence —physical constants, composition, chemical behavior—in arriving at the structures of organic compounds. Most of the information provided concerns chemical behavior, both chemical degradation into fragments that are readily recognized and chemical transformations that define the nature of functional groups. Exercises of this kind are valuable in that they require the student to apply, in typical laboratory situations, the information that he has gained by the study of individual reaction types and of typical functional groups. They are the "putting together" of information that has earlier been dealt with in a somewhat detached and abstract way into practical solutions of practical problems.

It should be noted that the modern organic chemist makes use of many physical methods that shortcut the often laborious procedures of degradation and functional group analysis. For pedagogical purposes at the introductory level, however, it is far more instructive to distinguish between, say, phenylacetaldehyde and acetophenone by the experimental behavior of an aldehyde *versus* a ketone, by the use of the iodoform reaction, and so on, rather than (as in fact most organic chemists would) through the differences in the carbonyl stretching frequency in the infrared or the presence of the signal for the methyl group in the nuclear magnetic resonance spectrum. Thus, the emphasis here is on chemical behavior rather than spectral properties; this is not to minimize the utility and importance of the latter; rather it is because an understanding of spectral properties is not a substitute for a knowledge of the chemical behavior of organic compounds and of organic reactions.

The limits of this workbook have not permitted a treatment of the use and interpretation of ultraviolet, infrared, nuclear magnetic resonance, and mass spectroscopy, subjects adequately dealt with in numerous textbooks and monographs. It is expected that a substantial modern course in organic chemistry will include instruction in these physical methods. Consequently, in the exercises to follow, occasional allusion is made to UV, IR, and n.m.r. spectra to define certain details of structure. It is assumed that the average student will be competent to cope with these, all of which are chosen to fall within the limits of instruction of a typical one-year course.

A. PROBLEMS WITH INTERPRETATIONS

Example 1

A liquid compound *A*, b.p. 98°, gave the following result on elemental analysis: C, 87.4%; H, 12.5%. The compound decolorized a solution of bromine in CCl_4 and decolorized a $KMnO_4$ solution in ethanol (with formation of MnO_2).

Treatment of *A* with ozone and reductive decomposition of the ozonide yielded a solution which, upon steam distillation and extraction of the distillate with ether, yielded a product that could be fractionally distilled to give two compounds *B* and *C*.

Compound *B* reacted at once with 2,4-dinitrophenylhydrazine to give a derivative, $C_{11}H_{12}N_4O_4$. Compound *B* itself did not decolorize cold ethanolic $KMnO_4$.

What is the structure of *A*?

Solution. The composition of *A* calculated from the analytical figures is C_7H_{12}. The alternative formulas, $C_{14}C_{24}$, $C_{21}H_{36}$, and so on, can be discarded, for the b.p. of a typical C_7 hydrocarbon, methylcyclohexane, is 101°. The compound is olefinic (Br/CCl_4, $KMnO_4$ tests), and thus is a diene or a cyclic mono-olefin.*

Ozonolysis gives a carbonyl compound, C_5H_8O (*B*):

$$C_5H_8O + (NO_2)_2C_6H_3NHNH_2 \longrightarrow \underbrace{(C_5H_8){=}NNHC_6H_3(NO_2)_2}_{C_{11}H_{12}N_4O_4}$$

Since *B* is stable to cold $KMnO_4$ it is not olefinic, and thus is a cyclic ketone. (An aldehyde would react with $KMnO_4$.)

The other ozonolysis product is necessarily a C_2 compound, C_2H_4O, which is *acetaldehyde*.

From these data, *A* could be

CHCH$_3$ (cyclopentylidene) CHCH$_3$ / CH$_3$ (cyclobutylidene) H$_3$C / CHCH$_3$ (cyclobutylidene) CHCH$_3$ / H$_3$C, H$_3$C (cyclopropylidene) H$_3$C, H$_3$C / =CHCH$_3$ (cyclopropylidene)

* *Query:* What simple quantitative experiment can be performed to show whether *A* is a diene or a mono-olefin?

That A cannot be a cyclopropane derivative can be assumed from the properties of B. If B were a cyclopropanone it would exist as the steam-involatile hydrate. The clearest evidence for the structure of B would be the infrared spectrum. Cyclopentanones show a C=O stretching absorption at about 1740 cm^{-1}, cyclobutanones at about 1780 cm^{-1}.

Example 2

A colorless liquid A decolorized bromine in CCl_4. Upon catalytic hydrogenation, 200 mg of A consumed 94 ml of hydrogen (at STP).

When A was subjected to ozonolysis, the only compound that could be isolated was acetone. A quantitative estimation showed that one mole of A gave two moles of acetone.

Solution. The hydrogenation showed that 200 mg of compound A corresponds to $94/22.4 = 4.2$ mM of hydrogen. Thus, if the compound were a mono-olefin, the mol wt would be $200/4.2 = 48$. If it were a diene, mol wt = 96; a triene, 144; and so forth.

(*a*) Since acetone is formed on ozonolysis, the unit $(CH_3)_2C{=}$ is present. Thus, of a mol wt of 48, 42 are represented by $(CH_3)_2C{=}$. There is no unit X that could be attached as in $(CH_3)_2C{=}X$, where X is 6. Therefore, A must be a diene, triene, etc.

(*b*) If A were a diene, mol wt 96, many combinations containing $(CH_3)_2{=}$ are possible; for example, $(CH_3)_2C{=}C$ [+42]; then B, $(CH_3)_2C{=}CHCH_2CH{=}CH_2$; C, $(CH_3)_2C{=}CH{-}CH{=}CHCH_3$, and so on.

(*c*) If A were a triene, many combinations are also possible. However, the mol wt of 144 would be reflected in the physical properties of A—for example, in the boiling point. Let it be assumed the b.p. of A is consistent with a mol wt in the range of 96 (or, that A is a diene).

Note that *only* acetone was found in the ozonolysis. Both B and C (of (*b*) above) would give other ozonolysis products. There is, however, one combination, D, that would be included under (*b*) that would satisfy the above conditions. This is tetramethylallene

$$(CH_3)_2C{=}C{=}C(CH_3)_2$$

which upon ozonolysis would give two molecules of acetone and one of CO_2, which would ordinarily escape detection. Repetition of the ozonolysis in order to show the formation of CO_2 could be carried out for confirmation.

Example 3

A hydrocarbon A, C_8H_{12}, decolorized $KMnO_4$ in ethanol and bromine in CCl_4. When it was oxidized with chromic acid it yielded only one product, an acid $C_4H_6O_4$

(*B*). The yield of *B* was high: 5.4 g of *A* yielded 9.8 g of pure *B*. Compound *B* was stable to moderate heating, but when heated until distillation occurred, the distillate —obtained in good yield—was a crystalline, neutral compound $C_4H_4O_3$ (*C*). Solution of *C* in aqueous alkali, followed by acidification, gave *B*. Write a structure for *A*.

Solution. Compound *B*, an acid, yields a neutral compound *C* that differs from *B* by the elements of water. It is a reasonable assumption that *C* is the anhydride of the dibasic acid *B*. Therefore *B* must be succinic acid, for the isomeric methylmalonic acid would decarboxylate on heating.

Since a yield of one mole of succinic acid ($C_4H_6O_4$, mol wt 118) from one mole of C_8H_{12} (108) would be a maximum of 5.9 g of *B* from 5.4 g of *A*, the actual yield of 9.8 g of *B* indicates that *A* contains *two* of the units $=C-CH_2CH_2-C=$. Since this constitutes all 8 carbon atoms of C_8H_{12}, the structure of *A* is 1,5-cyclooctadiene. A noncyclic triene C_8H_{12} could not contain two pairs of adjacent $-CH_2-$ groups in the required arrangement.

Example 4

A compound *A*, $C_7H_{16}O$, is readily oxidized by chromic acid to yield *B*, $C_7H_{14}O$. Compound *B* is not readily oxidized further; it does not form a bisulfite addition compound and gives no iodoform when treated with iodine and alkali; and it *does* form a crystalline 2,4-dinitrophenylhydrazone.

(*a*) With this information alone, write all of the structures that could represent *A*.

(*b*) If *A* were optically inactive and nonresolvable into enantiomeric forms, which of the structures written for *A* would be ruled out?

(*c*) Dehydration of *A* and ozonolysis of the resulting product (without purification) gives acetone as one of the ozonolysis products. Does this, along with the other evidence, establish the structure of *A*?

Solution. Compound *A* is a saturated alcohol or an ether. Its oxidation to the ketone, *B*, shows that it is a secondary alcohol (for if it were a primary alcohol, $C_7H_{14}O$ would be an aldehyde and should form a bisulfite addition compound). Since the ketone does not contain the $-COCH_3$ group (shown by iodoform test), *A* could be only one of the following:

(*i*) $CH_3CH_2CH_2CH(OH)CH_2CH_2CH_3$

(*ii*) $CH_3CH_2CH(OH)$butyl(-*n*, -*sec*, -iso, or -*tert*)

(*iii*) $(CH_3)_2CHCH(OH)CH_2CH_2CH_3$

(*iv*) $(CH_3)_2CHCH(OH)CH(CH_3)_2$

Because the alcohol cannot exist as an optically active compound, (*ii*) and (*iii*) are eliminated, leaving (*i*) and (*iv*) as the remaining possibilities.

Dehydration of (*i*) or (*iv*), followed by ozonolysis of the olefin would lead to the formation of acetone from (*iv*) but not from (*i*). Thus, *A* is alcohol (*iv*).

Example 5

A compound *A*, $C_8H_{10}O$, was inert to bromine in CCl_4 and to cold alcoholic $KMnO_4$. It did not react with carbonyl derivative-forming reagents and gave no iodoform with iodine and alkali. It reacted with acetyl chloride with evolution of HCl and the formation of a fragrant, oily product.

Vigorous oxidation with $KMnO_4$ converted *A* into an acid *B*, $C_7H_6O_2$, which was stable to further oxidation.

Write a structure for *A*.

Solution. The composition of *A* shows that it has four ring or double bond equivalents, an indication that it may contain a benzene ring. The oxidation to *B*, which is stable to further oxidation, supports this, for *B* is evidently benzoic acid, C_6H_5COOH. Thus,

$$\underset{A}{C_6H_5(C_2H_5O)} \longrightarrow \underset{B}{C_6H_5(COOH)}$$

The fragment C_2H_5O is clearly saturated. It contains an OH group (shown by acetyl chloride reaction), and is not $-\underset{\text{OH}}{\underset{|}{\text{CH}}}\text{CH}_3$ (shown by negative iodoform test).

Thus *A*, is $C_6H_5CH_2CH_2OH$ (2-phenylethanol; phenethyl alcohol).

Example 6

A compound *A*, $C_{10}H_7NO_2$, is quite inert toward halogenation or oxidation under mild (room temperature) conditions. It gives a positive test for the presence of a nitro group. When oxidized vigorously with $KMnO_4$, it is oxidized slowly, with the eventual formation of *B*, $C_8H_5NO_6$, an acid with a neutralization equivalent of 105 ± 1. Compound *B* also contains a nitro group. When *B* is strongly heated it is transformed into a neutral compound *C*, $C_8H_3NO_5$.

When *A* is reduced with tin and HCl it is converted into *D*, $C_{10}H_9N$. Oxidation of *D* with $KMnO_4$ transforms it into an acid *E*, $C_8H_6O_4$. Heating *E* converts it into the neutral compound *F*. Nitration of *F* yields *C*. Write the equations for these reactions, starting with the structure of *A*.

Solution. Compound *A* is 2-nitronaphthalene, and *D* is 2-aminonaphthalene. *B* and *E* are, respectively, 4-nitrophthalic acid and phthalic acid. It is to be noted that in the oxidation of *A*, the nitro group deactivates the ring toward attack of the oxidant, while in *D* the amino group activates the ring toward oxidation.

Example 7

A compound *A*, $C_{10}H_{12}O_3$, is an acid with a neutralization equivalent (N.E.) = 179 ± 1. Vigorous oxidation of *A* with $KMnO_4$ yields *B*, also an acid, $C_8H_8O_3$. When *A* is allowed to react with $SOCl_2$ and the product is treated with anhydrous aluminum chloride, a neutral compound *C*, $C_{10}H_{10}O_2$ is formed. In the reaction in which *C* is formed, *C* is the only product; no isomeric compound can be detected. Compound *C* forms a 2,4-dinitrophenylhydrazone, and when treated with benzaldehyde and alcoholic sodium ethoxide, it gives *D*, $C_{17}H_{14}O_2$. Vigorous $KMnO_4$ oxidation of *C* or *D* yields an acid *E*, $C_9H_8O_5$, N.E. = 98 ± 1.

Write two structures for *A* that accommodate these observations. How could you distinguish between them?

Solution. Since *B* is the product of vigorous oxidation, it can be assumed to be an aromatic acid, $C_6H_4COOH(CH_3O)$. The fragment CH_3O can be assumed to be a methoxyl group, since a substituent $—CH_2OH$ would not have survived the oxidation, nor can the oxygen be present as a phenolic hydroxyl group for the same reason. Therefore, *A* must be $CH_3O—C_6H_4—C_3H_5O_2$, of which $—C_3H_5O_2$ must be C_2H_4COOH. Let it be assumed that *A* and the other compounds are represented as follows:

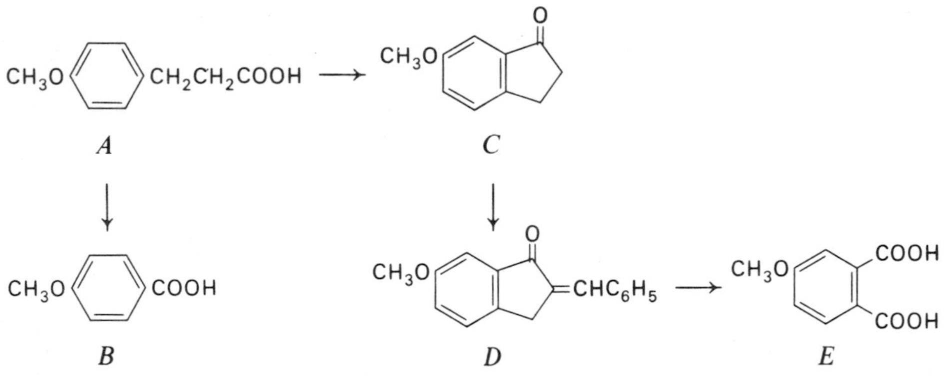

CH_3O–C_6H_4–CH_2CH_2COOH (*A*) → *C* (methoxy-indanone, CH_3O, O)

A ↓ CH_3O–C_6H_4–COOH (*B*)

C ↓ *D* (CH_3O, O, $=CHC_6H_5$) → *E* (CH_3O, COOH, COOH)

Compound *A* could also be (*ortho*: CH_2CH_2COOH, OCH_3), but not (*meta*: CH_2CH_2COOH, OCH_3) Why?

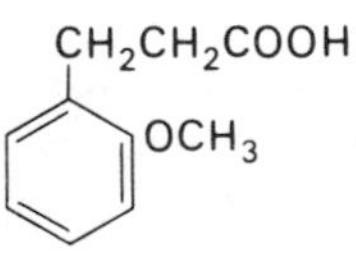

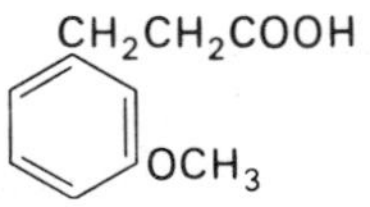

How could one distinguish between *p*-methoxyhydrocinnamic acid the the *ortho* isomer? This could be accomplished most readily by nuclear magnetic resonance (n.m.r.) spectrometry, for the *para* compound would show the aromatic protons as a symmetrical A_2B_2 quartet, and the *ortho* compound would show these protons as a more complex set of lines. The distinction could be made chemically in the following way. Compound *B* is a methoxybenzoic acid, and could be demethylated to a hydroxybenzoic acid. Salicylic acid gives an intense violet color when reacted with ferric chloride; *p*-hydroxybenzoic acid does not give a ferric chloride color in aqueous or alcoholic solution.

Example 8

A neutral compound *A*, $C_{10}H_{16}O$, reacts readily with Tollens' reagent and with Fehling's solution. Catalytic hydrogenation readily yields *B*, $C_{10}H_{20}O$, and can be carried further to yield *C*, $C_{10}H_{22}O$.

Ozonolysis of *A* gives acetone and a compound *D*, which upon gentle oxidation gives an acid *E*, $C_5H_8O_3$. The acid *E* gives a positive iodoform test.

When *A* is heated with sulfuric acid it is converted into a hydrocarbon which proves to be *p*-cymene (*p*-isopropyltoluene). What is the structure of *A*?

Solution. *A* is an unsaturated aldehyde. The presence of two double bonds is shown by the formation of *B* on hydrogenation; the last step of the complete reduction, leading to C, converts the aldehyde into a primary alcohol.

The ozonolysis shows that the structural unit $(CH_3)_2C{=}$ is present. The acid *E* contains the $CH_3CO{-}$ grouping (iodoform test), showing that *A* has the partial structure $CH_3{-}\overset{|}{C}{=}/(CH_3)_2C{=}/CHO/C_2H_4$.

The formation of *p*-cymene shows that the acid-catalyzed reaction is a cyclization (*A* must be an *acyclic* dienal; why?), and thus that the carbon skeleton of *A* is

```
       C—C      C
      /        /
  C—C      C—C
      \    /   \
       C—C      C
```

Since the elements $CH_3{-}\overset{|}{C}{=}C$ and $(CH_3)_2C{=}$ are present, and a five-carbon acid containing $CH_3CO{-}$ is formed on ozonolysis, *A* is formulated as follows:

$$A \xrightarrow{O_3} \left\{ CH_3-CO-CH_2-CH_2-COOH \right\} + H_3C-CO-CH_3 \left(+ \begin{array}{c} CHO \\ | \\ CHO \end{array} \right)$$

A *E*

Problem: Show how *A* is converted into *p*-cymene by the action of sulfuric acid.

Example 9

A neutral, colorless compound *A*, $C_9H_6O_2$, was insoluble in cold alkali but dissolved on heating. When dimethyl sulfate was added to the alkaline solution, an insoluble oil separated; this again dissolved when more alkali was added and the solution heated. Acidification of the final clear alkaline solution caused the separation of a colorless, crystalline compound *B*, $C_{10}H_{10}O_3$. Oxidation of *B* with $KMnO_4$ gave an acid *C*, $C_8H_8O_3$.

Solution. Compound *A* has the properties of an ester or a lactone (why not an acid anhydride?). From its composition it has seven double bond or ring equivalents ($C_9H_{20} - C_9H_6 = H_{14}$); and since the final oxidation leads to a C_8 compound, the conclusion is that it is probably aromatic:

$$\underset{A}{C_9H_6O_2} + H_2O \longrightarrow C_9H_8O_3{=}C_9H_7O_2(OH) \longrightarrow \underset{B}{C_9H_7O_2(OCH_3)}$$

Since *B* is an acid (it is soluble in alkali and is precipitated by acidification), its formula can be expanded to $C_8H_6(OCH_3)(COOH)$; and if it is aromatic, $C_8H_6 = C_6H_4(C_2H_2)$. A structure that accommodates these data is

$$\left\{ C_6H_4 \begin{matrix} CH{=}CHCOOH \end{matrix} \right\} OCH_3$$

The position of $-OCH_3$ must be *ortho*, since it arose by opening of a lactone. Thus, the compounds are

A (coumarin) → *B* ($C_6H_4(CH{=}CHCOOH)(OCH_3)$) → *C* ($C_6H_4(COOH)(OCH_3)$)

A — *via* → $C_6H_4(OH)(CH{=}CHCOOH)$ (as Na salt) → $C_6H_4(OCH_3)(CH{=}CHCOOCH_3)$ — saponify → *B*

Example 10

Treatment of 5-methyl-3-hexen-2-one with diethyl malonate, in the presence of a catalytic amount of sodium ethoxide in ethanol, gives *A*, $C_{14}H_{24}O_5$. When *A* is treated further with a molar equivalent of NaOEt it is converted into *B*, $C_{12}H_{18}O_4$. Saponification of *B*, followed by heating, leads to the loss of CO_2 and formation of *C*, $C_9H_{14}O_2$. Compound *C* is a weakly acidic compound (soluble in aqueous alkali) but is not a carboxylic acid. Treatment of *C* with methyl iodide and sodium ethoxide gives two products, *D* and *E*, both $C_{10}H_{12}O_2$. One of them (*D*) is reconverted to *C* upon acid hydrolysis; E does not undergo this change. When E is treated with PCl_5 it is transformed into F, $C_{10}H_{15}ClO$, a neutral compound that gives a crystalline derivative with 2,4-dinitrophenylhydrazine.

Reduction of F with zinc leads to the formation of G, $C_{10}H_{16}O$, a neutral compound whose UV spectrum shows a maximum at 245 nm ($\varepsilon > 10000$).

Solution

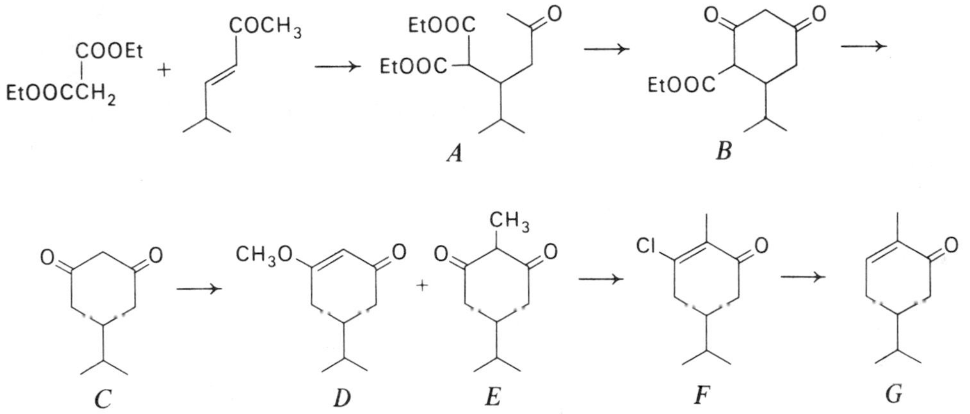

COOEt | EtOOCCH$_2$ + COCH$_3$ → EtOOC, EtOOC → EtOOC →

A *B*

CH$_3$O, CH$_3$, Cl

C *D* *E* *F* *G*

Example 11

Reduction of resorcinol (using a nickel catalyst and hydrogen, in alkali) leads smoothly to a dihydrocompound A, $C_6H_8O_2$. Compound A is moderately acidic (pK_a 5.2) and has UV absorption maxima at 255 nm and 280 nm in alcoholic solution (282 nm in dilute alkali). Treatment of A with ethyl iodide and Ag_2O leads to the formation of B, $C_8H_{12}O_2$, which is no longer acidic, and which is readily hydrolyzed to A with aqueous acid.

Reaction of B with CH_3MgI yields C, $C_9H_{16}O_2$. Treatment of C with aqueous sulfuric acid gives D, $C_7H_{10}O$. Compound D is neutral and gives a red-orange 2,4-dinitrophenylhydrazone. Reduction of D (with $LiAlH_4$) followed by heating of the product with Se gives toluene.

Solution

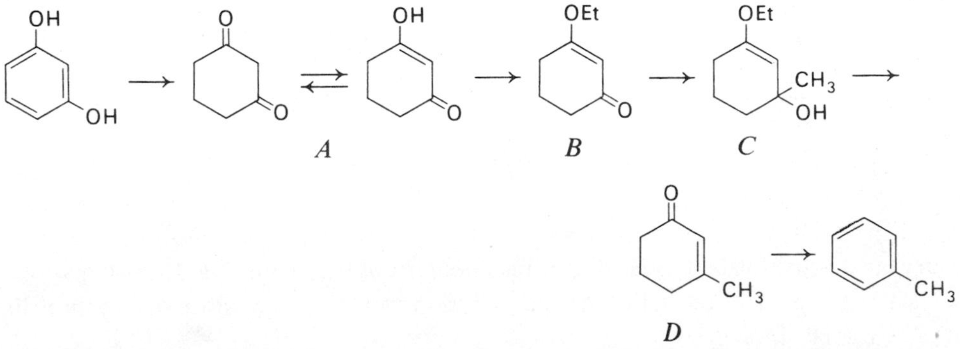

OH, OH → O, O ⇄ OH, O → OEt, O → OEt, CH$_3$, OH →

A *B* *C*

O, CH$_3$ → CH$_3$

D

Note: the possible dihydro compounds derived from resorcinol are the following:

OH OH OH

OH OH OH

1 2 3

Since 1 and 3 would exist in their keto forms (O, OH) and (OH, O), neither would have the properties of the moderately strong acid *A*. Thus dihydroresorcinol 2 must be *A*, as the keto form.

Example 12

Phthalimide, $C_8H_5NO_2$, is reduced by a zinc-copper couple in alkali to yield *A*, $C_8H_6O_2$. Compound *A* is a neutral compound, but when heated with aqueous sodium hydroxide it dissolves. Acidification of the alkaline solution causes the precipitation of *A*.

When *A* is heated with alcoholic KCN it dissolves, and acidification yields a compound *B*, $C_9H_7NO_2$; *B* is an acid, soluble in sodium bicarbonate; it has N.E. = 161.

Vigorous hydrolysis of *B* (in aqueous H_2SO_4) converts it into *C*, $C_9H_8O_4$, N.E. = 90. When *C* is reduced with $LiAlH_4$, it yields *D*, $C_9H_{12}O_2$, a neutral compound. Treatment of *D* with *p*-toluenesulfonyl chloride in alkali leads to the formation of *E*, $C_9H_{10}O$, a neutral compound. Vigorous oxidation of *A*, *B*, *C*, *D*, or *E* (with $KMnO_4$) leads to *F*, $C_8H_6O_4$, an acid with N.E. = 83.

Solution

CO, NH, CO → (CH_2, O, CO) *A* → (CH_2CN, COOH) *B* → (CH_2COOH, COOH) *C* →

(CH_2CH_2OH, CH_2OH) *D* → (O) *E*

A, B, C, D, E $\xrightarrow{KMnO_4}$ (COOH, COOH)

Example 13

A neutral compound *A*, $C_{15}H_{10}O_2$, is insoluble in cold, aqueous NaOH but dissolves on heating. Acidification of the aqueous solution causes precipitation of a carboxylic acid *B*, $C_{15}H_{12}O_3$.

When *A* was treated with sodium methoxide in methanol it dissolved, after which acidification yielded *C*, $C_{15}H_{10}O_2$ (isomeric with *A*). An alkali solution of *C* yielded unchanged *C* on acidification. Vigorous $KMnO_4$ oxidation of *A*, *B*, or *C* yielded two acids: *D*, $C_8H_6O_4$ (N.E. = 83) and *E*, $C_7H_6O_2$ (N.E. = 122), in excellent yields.

Solution

$$A \xrightarrow[(2)\ H^+]{(1)\ OH^-} B\ (o\text{-}C_6H_4(COCH_2C_6H_5)(COOH)) \xrightarrow{KMnO_4} o\text{-}C_6H_4(COOH)_2 + C_6H_5COOH$$

$$A \xrightarrow{NaOCH_3} C\ (C_6H_4(CO)_2CHC_6H_5)$$

soluble in alkali as the enolate

C

Notes: Since *A* is neutral and is converted by alkali into an acid with the addition of the elements of water, *A* can be assumed to be a lactone. Since *A* (as well as *B* and *C*) is oxidized to a mixture of benzoic and phthalic acids, it can be represented by the partial formula $C_6H_4\{C, C\}C{-}C_6H_5$.

Exercise: Formulate the course of the rearrangement of *A* into *C*.

Example 14

A neutral, optically active compound *A*, $C_{11}H_{14}O_2$, is unaffected by treatment with alkali, but when it is added to 5% sulfuric acid and the solution is steam-distilled, the distillate contains a water-insoluble liquid compound *B*, C_7H_6O, which gives a positive test with Tollens' reagent and is readily oxidized with $KMnO_4$ to an acid *C*, $C_7H_6O_2$, N.E. = 122. The acid *C* is stable to further treatment with $KMnO_4$.

The aqueous solution D remaining after removal of *B* by steam distillation is still optically active. When benzoyl chloride and excess alkali are added, a neutral compound *E* is formed, $C_{18}H_{18}O_4$; this is optically active.

Treatment of the aqueous solution *D* with sodium periodate destroys the optical activity, and subsequent treatment with 2,4-dinitrophenylhydrazine yields one compound only; *F*, $C_8H_8N_4O_4$.

Solution

$$\underset{A\ (trans)}{C_6H_5CH\begin{matrix}O-CHCH_3\\ |\\ O-CHCH_3\end{matrix}} \xrightarrow{H^+/H_2O} \underset{B}{\begin{matrix}C_6H_5COOH\\ \uparrow\\ C_6H_5CHO\end{matrix}} + \underset{D}{\begin{bmatrix}CH_3CHOH\\ |\\ CH_3CHOH\end{bmatrix}}$$

$$\underset{F}{CH_3CH{=}NNH-C_6H_3(NO_2)_2} \leftarrow [CH_3CHO] \qquad \underset{E}{\begin{matrix}CH_3CHOCOC_6H_5\\ |\\ CH_3CHOCOC_6H_5\end{matrix}}$$

Example 15

A neutral compound *A*, $C_{14}H_{12}O_2$, did not react with bromine in CCl_4 and gave positive tests with 2,4-dinitrophenylhydrazine and with Tollens' reagent. Its n.m.r. spectrum showed a total of nine protons in the region centering at about 7.5 ppm, among which could be discerned a symmetrical four-line pattern of an A_2B_2 system (comprising 4 protons) with a coupling constant of 7 Hz (cycles per second). A sharp two-proton singlet appeared at 5.2 ppm.

Gentle oxidation of *A* with $KMnO_4$ yielded an acid *B*, with an n.m.r. spectrum very similar to that of *A*, and with a N.E. = 227 ± 2. Vigorous and prolonged oxidation of *A* with $KMnO_4$ did not give *B*, but afforded instead an acid with N.E. = 121 ± 1, and with a m.p. of 121°. No other significant products could be found.

Solution. Compound *A* is an aldehyde and, from its composition and the fact that it was unaffected by bromine in CCl_4, it may be assumed to be an aromatic aldehyde. Its n.m.r. spectrum suggests that it can be represented as *p*-$(C_7H_7O)-C_6H_4CHO$. The acid *B* is then *p*-$(C_7H_7O)-C_6H_4COOH$. The acid produced by vigorous oxidation of *A* is evidently benzoic acid, and can come only from the fragment (C_7H_7O). This can be expanded to $C_6H_5CH_2O-$ (n.m.r.), and thus the structure of *A* is established as *p*-benzyloxybenzaldehyde.

B. PROBLEMS WITHOUT ACCOMPANYING INTERPRETATIONS

(Answers in answer section)

Example 16

A compound *A*, $C_7H_{12}O_3$, is soluble in aqueous $NaHCO_3$ and has a N.E. = 142. It does not react with carbonyl reagents. When *A* is oxidized with chromic acid or lead tetraacetate it is converted into a neutral compound *B*, $C_6H_{10}O$, which readily forms a bisulfite addition compound. Compound *B* is stable to cold, alcoholic $KMnO_4$, but when oxidized with hot CrO_3 solution it is converted into an acid *C*, $C_6H_{10}O_4$, N.E. = 73.

Note: An *isomer D* of *A*, also an acid, can also be oxidized (with care) to an acid *E*, $C_7H_{10}O_3$, which when heated is readily decarboxylated to give *B*. No conditions can be found to oxidize *A* to an isomer of *E*.

Example 17

A neutral compound *A*, $C_5H_{12}O_2$, shows no reactions of a carbonyl compound and has no IR absorption in the 1700 cm^{-1} region. It reacts with acetic anhydride to give a neutral compound *B*, $C_7H_{14}O_3$, which is converted back to *A* by saponification. Compound *B* shows no IR absorption for a hydroxyl group.

When *A* is treated with *p*-bromobenzenesulfonyl chloride in pyridine it is converted into *C*, $C_{11}H_{15}SO_4Br$. When a solution of *C* in water (containing dioxan to aid solution) is heated, *C* is converted into *D*, C_4H_8O. Compound *D* reacts at once with 2,4-dinitrophenylhydrazine, and reduces Tollens' and Fehling's reagents. Among the products formed by chromic acid oxidation of *D* is an acid *E*, $C_4H_8O_2$ (and some minor amounts of products of further oxidation, including acetone). Explain the conversion of *C* into *D*. (Note that in the conversion of *C* into *D*, *p*-bromobenzenesulfonic acid is also formed; and examination of the solution discloses the presence of methanol.)

Example 18

The reaction of acetoacetic ester first with NaOEt, then with ethyl bromoacetate, gave *A*, $C_{10}H_{16}O_5$. Treatment of *A* with one mole-equivalent of CH_3MgI gave *B*, $C_{11}H_{20}O_5$. Hydrolysis of *B* (NaOH), followed by acidification yielded *C*, $C_7H_{10}O_4$. Compound *C* was an acid with N.E. = 158 by direct titration. When *C* was heated with an excess of standard NaOH and the solution back-titrated, an N.E. of 79 was found.

Example 19

A neutral compound *A*, $C_6H_{12}O_2$, did not react with 2,4-dinitrophenylhydrazine but dissolved slowly in hot aqueous alkali. Addition of iodine (in KI solution) to the alkaline saponification solution gave a precipitate of iodoform.

(*a*) Write the structures of all compounds *A* that accommodate these observations.

(*b*) What experimental approach would you adopt to choose between these possibilities?

Example 20

A yellow, crystalline compound *A*, $C_{16}H_{14}O_2$, gives a positive test with 2,4-dinitrophenylhydrazine. Vigorous oxidation of *A* with $KMnO_4$ gives only one product, an acid *B*, $C_8H_6O_4$. When heated, *B* is transformed into *C*, $C_8H_4O_3$, a neutral compound.

When *A* is treated with strong, hot aqueous alkali it gradually dissolves, and acidification yields a colorless crystalline compound *D*, $C_{16}H_{16}O_3$. Oxidation of *D* with chromic acid converts it into a colorless, neutral compound *E*, $C_{15}H_{14}O$. *E* is quite stable to further oxidation, but when subjected to prolonged treatment with hot, alkaline $KMnO_4$ a poor yield of *B* can be isolated.

Example 21

The sodium ethoxide-catalyzed reaction of ethyl β-iodopropionate with ethyl cyanoacetate gave *A*, $C_{15}H_{23}NO_6$. Saponification of *C* gave an acid that lost CO_2 when heated to give an acid *B*, $C_8H_{12}O_6$, (which had N.E. = 68).

When *B* was strongly heated in the presence of acetic anhydride and the product distilled, an acid *C* formed, $C_7H_{10}O_3$, with N.E. = 142. Compound *C* reacted readily with 2,4-dinitrophenylhydrazine to give a 2,4-DNPH derivative.

Reaction of the ethyl ester of *C* with an excess of CH_3MgI gives a neutral compound *D*, $C_{10}H_{20}O_2$. Treatment of *D* with hot, dilute sulfuric acid converts it into *E*, $C_{10}H_{16}$. When *E* is heated with sulfur (or selenium) it is converted into an aromatic hydrocarbon *F*, $C_{10}H_{14}$.

Example 22

An optically active, neutral compound *A*, $C_5H_8O_2$—assume it to be dextrorotatory, or (+)-*A*—gives a positive test with 2,4-dinitrophenylhydrazine and reacts with acetyl chloride to give *B*, $C_7H_{10}O_3$.

When *A* is reduced catalytically (H_2, PtO_2) it yields two products, both $C_5H_{10}O_2$. The principal product *C* is optically inactive; the other, *D*, is optically active.

Careful oxidation of *C* with CrO_3 (use 1 two-electron equivalent) yields *E*, $C_5H_8O_2$, which is (±)-*A*.

Careful oxidation of *D* gives *F*, $C_5H_8O_2$, which is (+)-*A*.

Neither *C* nor *D* reacts with HIO_4. Vigorous oxidation of *A*, *C*, or *D* with CrO_3 yields an acid *G*, $C_4H_6O_4$, having N.E. = 59, which melts without decomposition.

Show the stereochemistry involved in these transformations.

Example 23

Treatment of ethylene dibromide with KCN in ethanol yields *A*, $C_4H_4N_2$. When *A* is refluxed in ethanol in the presence of a small amount of concentrated H_2SO_4 it is converted into *B*, $C_8H_{14}O_4$. Addition of excess ethylmagnesium bromide to *B*, followed by hydrolysis of the reaction mixture with aqueous sulfuric acid, yields a neutral compound *C*, $C_{12}H_{24}O$. Compound *C* is inert toward metallic sodium and acetyl chloride and shows no IR absorption in the 1700 cm^{-1} region.

Example 24

Dihydropyran, C_5H_8O, is reduced with platinum and hydrogen to *A*, $C_5H_{10}O_2$. Treatment of *A* with hot, 48% HBr converts it into *B*, $C_5H_{10}Br_2$. When a solution of *B* in ethanol is allowed to react with one molar equivalent of potassium acetate, the principal product is *C*, $C_7H_{13}BrO_2$. Reaction of *C* with one molar equivalent of NaOH converts it to *D*, $C_5H_{11}BrO$. Treatment of *D* with methanesulfonyl chloride in pyridine yields *E*, $C_6H_{13}BrO_3S$. When *E* is allowed to react with excess methylamine it is converted to *F*, $C_6H_{13}N$. Compound *F* is a strongly basic amine, which is recovered unchanged after treatment with acetyl chloride.

Example 25

A neutral compound *A*, $C_{12}H_{14}O_3$, gives a positive test with 2,4-dinitrophenylhydrazine. When 0.412 g of *A* is refluxed in ethanol containing 5.0 ml of 1.00 *N* NaOH, and the resulting solution titrated with 0.100 *N* HCl to give a phenolphthalein endpoint, 30.0 ml of the standard acid is required.

Acidification of the saponification mixture causes the separation of a compound *B*, $C_{11}H_{12}O_3$. When *B* is heated with acetic anhydride it is transformed into *C*, $C_{11}H_{10}O_2$. Compound *C* is insoluble in cold NaOH solution but dissolves on heating and after acidification it regenerates compound *B*.

Compounds *A*, *B*, and *C* are all oxidized by $KMnO_4$ to give an acid *D*, $C_7H_6O_2$, N.E. = 122. Ozonolysis of *C*, followed by reductive decomposition of the ozonide, gives two compounds, *E* and *F*. Compound *E*, C_7H_6O, is readily oxidized to *D*. Compound *F*, $C_4H_4O_3$, is neutral but dissolves in NaOH on warming; acidification of the alkaline solution gives *G*, $C_4H_6O_4$.

Example 26

A weakly acidic compound *A*, C_8H_9NO, soluble in NaOH but not in $NaHCO_3$, reacts with benzoyl chloride in pyridine to give *B*, $C_{15}H_{13}NO_2$.

When *A* is dissolved in concentrated H_2SO_4 and the solution poured into water, there results a colorless, neutral compound *C*, C_8H_9NO (the compound is isomeric with *A*). Hydrolysis of *C* with H_2SO_4 converts it into *D*, C_5H_7N; *D* is reconverted to *C* by acetic anhydride.

Example 27

A compound *A*, $C_{20}H_{25}NO_3$ is not reduced by hydrogen-platinum oxide and is soluble in dilute HCl. It does not react with acetyl chloride, but it does react with methyl iodide to give *B*, $C_{21}H_{28}NO_3{}^+I^-$. When *B* is treated with silver oxide and, after removal of the silver iodide, the solution is concentrated and heated, there results *C*, $C_{21}H_{27}NO_3$, which is soluble in dilute HCl. Reaction of *C* with methyl iodide yields *D*, $C_{22}H_{30}NO_3{}^+I^-$, and treatment of *D* with silver oxide (as was done with *B*) results in the liberation of trimethylamine and the formation of *E*, $C_{19}H_{20}O_3$. Compound *E* absorbs two moles of hydrogen in the presence of a platinum catalyst to give *F*, $C_{19}H_{24}O_3$.

Oxidation of *E* with hot potassium permanganate solution yields an equimolal mixture of *p*-methoxybenzoic acid and 4,5-dimethoxyphthalic acid.

Example 28

Treatment of styrene (vinylbenzene) with formaldehyde and aqueous sulfuric acid leads to the formation of *A*, $C_9H_{12}O_2$. Permanganate oxidation of *A* gives benzoic acid. Treatment of *A* with one molar equivalent of *p*-toluenesulfonyl chloride in pyridine gives *B*, $C_{16}H_{18}SO_4$. Careful oxidation of *B* with chromic acid gives *C*, $C_{16}H_{16}SO_4$. Compound *C* is readily converted into *D*, C_9H_8O by treatment with dilute alkali. Addition to *D* of diethyl malonate in the presence of a catalytic amount of sodium ethoxide gives *E*, $C_{16}H_{20}O_5$.

Example 29

Cyclohexanone and piperidine, heated together with slow distillation of water, form *A*, $C_{11}H_{19}N$. Hydrolysis of *A* with aqueous HCl reconverts it to cyclohexanone and piperidine.

Treatment of *A* with acetyl chloride, followed by addition of water to the reaction mixture, results in the formation of *B*, $C_{10}H_{14}O_3$. Treatment of *B* with one equivalent of alkali under mild conditions converts it to *C*, $C_8H_{12}O_2$, but when excess alkali is used, with heating, the product is *D*, $C_8H_{14}O_3$. Compound *D* is a carboxylic acid (N.E. = 158), and gives a positive iodoform test. The products formed by the action of NaOI on *D* are iodoform and a compound *E*, $C_7H_{12}O_4$, which is an acid with N.E. = 80.

Example 30

The reaction between cyclohexanone, dimethylamine hydrochloride and formaldehyde yields *A*, $C_9H_{17}NO$. Treatment of *A* with ethyl acetoacetate and sodium ethoxide yields *B*, $C_{13}H_{18}O_3$. Hydrolysis of *B* with hot, aqueous HCl results in the formation of ethanol, carbon dioxide, and *C*, $C_{10}H_{14}O$. Compound *C* shows an IR absorption peak (in CS_2) at 1675 cm^{-1} and an UV λ_{max} of 240 nm (log ε about 4). In the n.m.r. spectrum, only one proton signal is present at low field; this is a sharp one-proton singlet at about δ 5.9 ppm (4.1 τ). All other protons are at fields higher than 5 ppm.

Example 31

Addition of bromine to pulegone (2-isopropylidene-5-methylcyclohexanone) gives *A*, $C_{10}H_{16}Br_2O$. The addition of NaOH, and then HCl, to *A* converts *A* to *B*, $C_{10}H_{16}O_2$. Ozonolysis of *B*, gives *C*, $C_7H_{10}O_3$, plus acetone. Heating *C* gives *D*, $C_6H_{10}O$. Oxidation of *D* with CrO_3 gives *E* and *F*, isomeric acids, $C_6H_{10}O_4$. *E* can be resolved into (+) and (−) forms. *F* is nonresolvable. The n.m.r. spectra of both *E* and *F* show a three-proton doublet (J = 7 Hz) at about 1 ppm (9 τ).

Example 32

Addition of phenylmagnesium bromide to ethyl methoxyacetate gives, after decomposition of the addition product with ammonium chloride solution, a compound *A*, $C_{15}H_{16}O_2$. Treatment of *A* with 5% sulfuric acid (in an inert solvent such as dioxan) converts it into *B*, $C_{14}H_{12}O$. Compound *B* gives a positive test with phenylhydrazine and reduces Tollens' reagent. Oxidation of *B* gives *C*, $C_{14}H_{12}O_2$, a carboxylic acid.

PART 13

Special problems in synthesis and structure analysis

Most of the problems in this Part are somewhat more advanced than those presented earlier. Some of them involve rather intricate syntheses, occasionally including rearrangements. Physical data are given in a number of cases to aid in the structural analysis: ultraviolet (UV), infrared (IR) and nuclear magnetic resonance (n.m.r.) data are given when their use gives important clues to structural details. Only those spectral data relevant to particular structural features are given.

1

Acetylene and methyl ethyl ketone react in the presence of potassium *t*-butoxide to give *A*, $C_6H_{10}O$.
A + EtMgBr ⟶ *B*.
B + $(CH_3)_2CHCHO$, followed by dilute HCL ⟶ *C*, $C_{10}H_{18}O_2$.
C + 1 mole H_2, with Pd catalyst ⟶ *D*, $C_{10}H_{20}O_2$.
D + hot, dilute H_2SO_4 ⟶ *E*, $C_{10}H_{16}$.
E has UV $\lambda_{max} = 275\ m\mu\ (\log \varepsilon > 4)$.

2

Ethyl acetoacetate and ethyl bromoacetate react in the presence of NaOEt to yield *A*, $C_{10}H_{16}O_5$. Treatment of *A* with one mole equivalent of each of NaOEt and methyl iodide gives *B*, $C_{11}H_{18}O_5$.
B + CH_3MgI, then HCl ⟶ *C*, $C_{12}H_{22}O_5$.
C + NaOH ⟶ *D*, $C_{10}H_{18}O_5$, and further, but much more slowly, *E*, $C_8H_{14}O_5$.
D + H_2O/H^+ ⟶ *F*, $C_{10}H_{16}O_4$.
F is a neutral compound. When it is treated with an excess of standard alkali and the solution is back-titrated with HCl it is found that 0.200 g of *F* consumes 2.0 ml of 1 *N* NaOH.

3

m-Isopropylanisole + Na(liquid NH_3) ⟶ *A*, $C_{10}H_{16}O$.
A + dilute HCl ⟶ *B*, $C_9H_{14}O$ (UV, λ_{max} 235 (log ε, 4.1)).
B + H_2(Pt) ⟶ *C*, $C_9H_{16}O$.
C + HCOOEt(NaOEt) ⟶ *D*, $C_{10}H_{16}O_2$ (UV, λ_{max} 264 (log ε, 4.1)).
D + $(CH_3)_2CHCH_2OH$ (*p*-toluenesulfonic acid) ⟶ *E*, $C_{14}H_{24}O_2$.
E + $LiAlH_4$ ⟶ *F*, $C_{14}H_{26}O_2$.
F + dilute H_2SO_4 ⟶ *G*, $C_{10}H_{16}O$ (UV, λ_{max} 228 (log ε, 4.3)).
G reacts with Tollens' reagent.

4

m-Cresol + H_2(Ni) ⟶ *A*, $C_7H_{14}O$.
A + CrO_3 ⟶ *B*, $C_7H_{12}O$.
B + ethyl oxalate (+ NaOEt) ⟶ *C*, $C_{11}H_{16}O_4$.
C + heated and distilled ⟶ *D*, $C_{10}H_{16}O_3$.
D + Na, then isopropyl iodide ⟶ *E*, $C_{13}H_{22}O_3$.
E hydrolyzed with hot, aqueous HCl ⟶ *F*, $C_{10}H_{18}O$.
F forms a 2,4-dinitrophenylhydrazone but has no high-intensity UV absorption.
F + $LiAlH_4$, then heated with selenium ⟶ *p*-cymene.

5

Salicylaldehyde + acetophenone ⟶ *A*, $(C_{15}H_{11}O)^+Cl^-$.
A + EtOH ⟶ *B*, $C_{17}H_{16}O_2$.
B has UV λ_{max} of about 250 nm (ε about 10,000).
B + cold, aqueous HCl ⟶ *A* + EtOH.

6

Phthalimide + Zn(NaOH)⟶phthalide, *A*, $C_8H_6O_2$.
A + KCN⟶*B*, $C_9H_7NO_2$.
B + $LiAlH_4$⟶*C*, $C_9H_{13}NO$.
C + trace HCl, heat ⟶ *D*, $C_9H_{11}N$.
D + CH_3I⟶*E*, $(C_{11}H_{16}N)^+I^-$.

7

Acetophenone, formaldehyde, and methylamine hydrochloride react to give *A*, $C_{19}H_{21}NO_2$.
A + NaOH⟶*B*, isomeric with *A*.
B + HCl⟶*C*, $(C_{19}H_{20}NO)^+Cl^-$.
C shows a UV absorption maximum at about 250 nm, as was expected. However, its n.m.r. spectrum shows, instead of the expected ten protons (for the two phenyl groups) at fields below 5 ppm, eleven protons in this region. Show the structure of *C* and explain why it is formed rather than the "expected" compound.

8

Treatment of 3,4-epoxy-1-butene with phenylmagnesium bromide gave three isomeric products, *A*, *B*, and *C*, two of them in minor amount. The principal product was *A*, $C_{10}H_{12}O$. When *A* was carefully oxidized with the Jones reagent (CrO_3, H_2SO_4) and the course of the reaction followed by observing the UV spectrum of the products, no high intensity absorption in the 220–240 nm range was observed.

What are the three products, and which is *A*?

9

When linalool (*A*) is brominated with N-bromosuccinimide (NBS) in CCl_4, the product is the bromotetrahydrofuran derivative *B*.

OH —NBS→ O Br

A *B*

When *B* is heated in boiling collidine (a high-boiling substituted pyridine), it dehydrobrominates and rearranges to give *C*, $C_{10}H_{18}O$, which can be catalytically reduced to *D*, $C_{10}H_{20}O$. Both *C* and *D* readily form 2,4-dinitrophenylhydrazones.

Compound *C* shows no UV absorption in the 200–280 nm range. Its n.m.r. spectrum shows a three-proton singlet (slightly split) at δ 1.70 (8.30 τ), and a six-proton singlet at δ 1.05 (8.95 τ). The n.m.r. spectrum of *D* showed a six-proton singlet at δ 1.05, and three-proton doublet (J = 6 Hz) at δ 0.95.
Treatment of *C* with D_2O/K_2CO_3 resulted in the formation of deuterated *C* with mol wt = 156.

10

Tryptamine (*A*) reacts with the acid chloride prepared from β,N-succinimidopropionic acid (*B*) to give *C*, $C_{17}H_{19}N_3O_3$. Treatment of *C* with $POCl_3$ gave *D*, $C_{17}H_{17}N_3O_2$. Vigorous reduction of *D* with $LiAlH_4$ gave two products, *E*, $C_{17}H_{21}N_3$, and *F*, $C_{17}H_{23}N_3$. Compound *F* reacted with acetic anhydride under mild conditions to give $C_{17}H_{22}N_3(COCH_3)$; compound *E* did not react. Under more vigorous conditions of acetylation, *E* gave $C_{17}H_{20}N_3(COCH_3)$, and *F* gave $C_{17}N_{21}N_3(COCH_3)_2$.

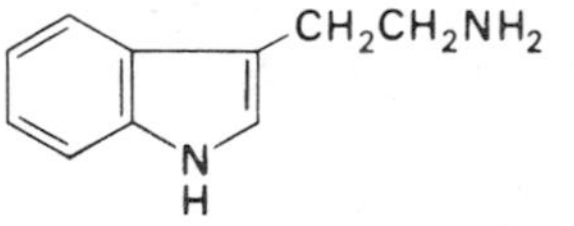

$CH_2CH_2NH_2$ N H

A

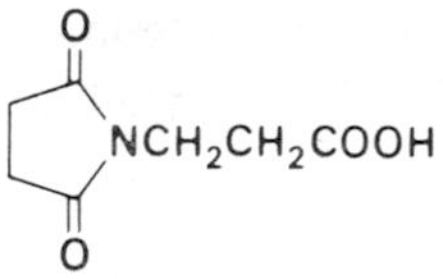

O O NCH_2CH_2COOH

B

11

Oxidation of phenanthrene, $C_{14}H_{10}$, with chromic acid gives a crystalline, orange compound *A*, $C_{14}H_8O_2$. When *A* is heated with strong aqueous KOH it dissolves to give a colorless solution, acidification of which yields *B*, $C_{14}H_{10}O_3$.
B + concentrated HI $\longrightarrow$ *C*, $C_{14}H_{10}O_2$
C + $LiAlH_4$ $\longrightarrow$ *D*, $C_{14}H_{12}O$
D + $SOCl_2$, then NH_4OH $\longrightarrow$ *E*, $C_{14}H_{11}NO$
E + $LiAlH_4$ $\longrightarrow$ *F*, $C_{14}H_{13}N$
F HCl/$NaNO_2$ $\longrightarrow$ phenanthrene.

12

Hydrocinnamic acid (β-phenylpropionic acid), treated with PCl_5 in benzene, followed by addition of anhydrous $AlCl_3$, gives *A*, C_9H_8O.
A + $C_6H_5CO_3H$ $\longrightarrow$ *B*, $C_9H_8O_2$.
B + aqueous NaOH $\longrightarrow$ *C*, sodium salt of $C_9H_{10}O_3$.
C + HCl $\longrightarrow$ *B*.
B + $LiAlH_4$ $\longrightarrow$ *D*, $C_9H_{12}O_2$.

D + CH_3SO_2Cl (pyridine) ⟶ E, $C_9H_{10}O$.
E is a neutral compound, insoluble in hot, alcoholic alkali, inert to cold $KMnO_4$ and to bromine in CCl_4. Vigorous oxidation of E with CrO_3 gives only succinic acid.

13

When limonene, A, $C_{10}H_{16}$, is treated with nitrosyl chloride (produced *in situ* from an alkyl nitrite and concentrated HCl) there is formed limonene nitrosochloride, B, $C_{10}H_{16}NOCl$. Treatment of B with hot dimethylformamide results in the loss of the elements of HCl and the formation of C, $C_{10}H_{15}NO$. Depending upon the conditions used, C is converted into D, $C_{10}H_{14}O$, by heating with 5*N* HCl; or into E, $C_{10}H_{14}O$, by heating with 5% aqueous oxalic acid.

Compound D is soluble in aqueous NaOH but not in aqueous $NaHCO_3$. Its n.m.r. spectrum shows that it contains three $—CH_3$ groups.

Compound E has UV maxima at 235 nm (log ε 3.93) and 318 (log ε 1.62). It is a neutral compound. Its n.m.r. spectrum shows two $—CH_3$ groups, both at 1.75 ppm, (8.25 τ).

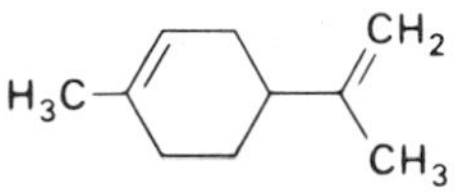

H_3C CH_2 CH_3

A

14

The N-methylation of amines by the Clarke-Eschweiler procedure is carried out as follows:

$$RNH_2 + 2\,H_2CO + 2\,HCOOH \longrightarrow RN(CH_3)_2 + 2\,CO_2 + 2\,H_2O$$

When this reaction was applied to α-allylbenzylamine, benzaldehyde was formed, along with CO_2 and a basic compound A, $C_6H_{13}N$. Catalytic hydrogenation of A gave B, $C_6H_{15}N$.

When α-allylbenzylamine was treated with formaldehyde in the absence of formic acid (but with a trace of mineral acid), benzaldehyde was again formed, along with a base C, C_4H_9N. When C was subjected to the Clarke-Eschweiler procedure, A was formed.

Synthesis of A:
1-Butanol + H_2CO + $(CH_3)_2NH$ ⟶ D, $C_7H_{17}NO$.
D + allylmagnesium bromide ⟶ A.
Formulate these transformations, including the synthesis of A.

15

Addition of methylmagnesium iodide to cycloheptanone gives *A*, $C_8H_{16}O$.
A + $KHSO_4$ (heated) ⟶ *B*, C_8H_{14}.
B, ozonolyzed, reductive decomposition ⟶ *C*, $C_8H_{14}O_2$.
C + $CH_2(COOH)_2$ (pyridine) ⟶ *D*, $C_{10}H_{16}O_3$.
D + H_2 (Pt) ⟶ *E*, $C_{10}H_{18}O_3$.
E + NaOI ⟶ CHI_3 + *F*, $C_9H_{16}O_4$.
D is the "queen substance" of the honeybee.

16

A naturally occurring compound *A*, $C_{10}H_{16}O$, is decomposed by the action of aqueous potassium carbonate into two compounds, *B*, $C_8H_{14}O$, and *C*, C_2H_4O. Both *B* and *C* give iodoform when treated with I_2/NaOH.

The iodoform reaction with *B* yields, along with iodoform, an acid, *D*, $C_7H_{12}O_2$. Ozonolysis of *A*, *B*, or *D* gives, along with other products, acetone. Oxidation of *A*, *B* or *D* with chromic acid gives succinic acid as one product.

17

Account for the following oxidative transformations:

(*a*)

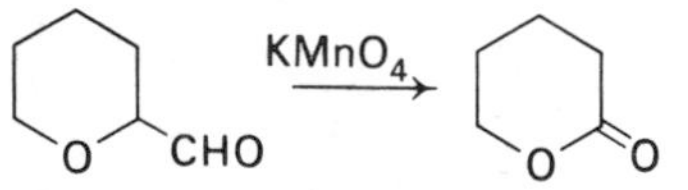

(*b*)

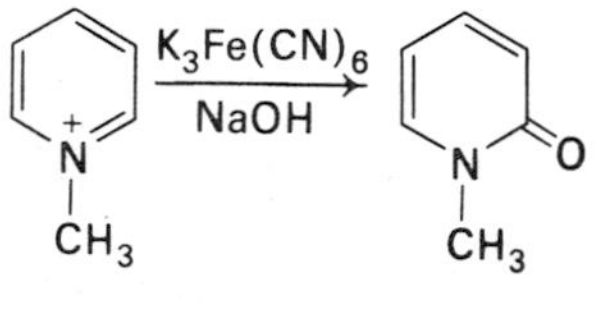

(*c*)

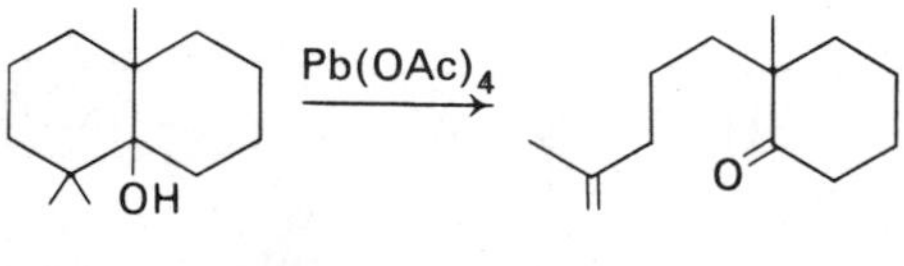

(*d*)

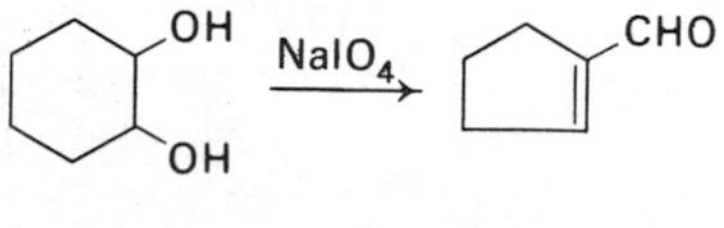

(*e*)

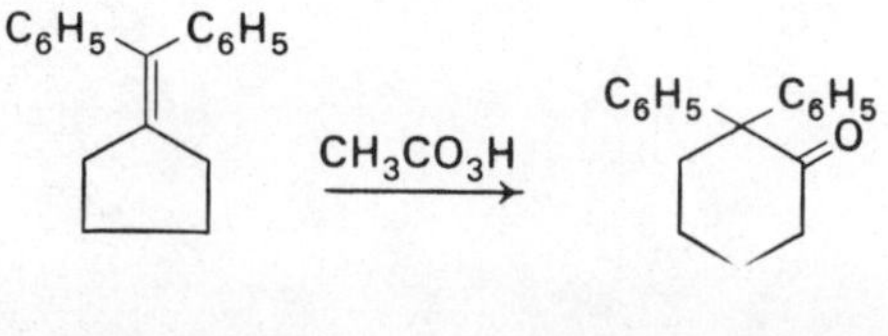

18

The "queen substance" (Example 15, Part 13) has also been synthesized as follows; formulate the steps shown.
Dihydropyran + H_2O (H^+) $\longrightarrow$ *A*, $C_5H_{10}O_2$.
A + $CH_2(COOH)_2$ (pyridine) $\longrightarrow$ *B*, $C_7H_{12}O_3$, along with an isomer *C*.
B + PBr_3; or C + HBr $\longrightarrow$ *D*, $C_7H_{11}BrO_2$.
D + CH_3COCH_2COOEt(NaOEt) $\longrightarrow$ *E*, $C_{13}H_{20}O_5$.
E + NaOH, then H_2O/H^+ $\longrightarrow$ *F*, $C_{11}H_{16}O_5$.
F, heated $\longrightarrow$ "queen substance" *G*, $C_{10}H_{16}O_3$.

19

A synthesis of the racemic form of the naturally occurring ketone, nootkatone, was carried out as follows; formulate the reactions.
Diethyl malonate + acrylonitrile $\longrightarrow$ *A*, $C_{13}H_{18}N_2O_4$.
A, saponified (NaOH) $\longrightarrow$ *B*, $C_9H_{10}N_2O_4$.
B, heated $\longrightarrow$ *C*, $C_8H_{10}N_2O_2$.
C + MeOH(HCl) $\longrightarrow$ *D*, $C_{11}H_{18}O_6$.
D + NaOMe $\longrightarrow$ *E*, $C_{10}H_{14}O_5$.
E + NaOMe + CH_3I $\longrightarrow$ *F*, $C_{11}H_{16}O_5$.
F, saponified (NaOH) $\longrightarrow$ *G*, $C_9H_{12}O_5$.
G, heated $\longrightarrow$ *H*, $C_8H_{12}O_3$.
H + $HSCH_2CH_2SH$ (+trace of acid; or BF_3) $\longrightarrow$ *I*, $C_{10}H_{16}O_2S_2$.
I + CH_3Li $\longrightarrow$ *J*, $C_{11}H_{18}OS_2$.
J + $(C_6H_5)_3P{=}CH_2$ $\longrightarrow$ *K*, $C_{12}H_{20}S_2$.
K + $HgCl_2$(MeOH) $\longrightarrow$ *L*, $C_{10}H_{16}O$
L + 3-penten-2-one (NaH) $\longrightarrow$ ($\pm$)-nootkatone, $C_{15}H_{22}O$, (*M*).
The n.m.r. spectrum of *M* showed three 3-proton signals for methyl groups: one singlet and one doublet (J about 7 Hz), both in the region of 1 ppm; and a singlet, slightly broadened, at about 1.7 ppm.

20

Treatment of methyl 2-methylcycloheptanone-5-carboxylate (*A*) with sodium methoxide causes it to be transformed into an isomer *B*, $C_{10}H_{16}O_3$. The isomer, *B* is also an ester, and shows two absorption peaks in the carbonyl region of the IR spectrum: one at 1735 cm^{-1}, the other at 1745 cm^{-1}.

Show the structure of *B* and the manner of its formation from *A*.

21

Condensation of pulegone (*A*) with ethyl acetoacetate in acetic acid solution, catalyzed by zinc chloride, gives rise to two principal products, *B* and *C*. The structure of *B* is correct as shown. The structure of *C* was at first regarded to be *C′*, but this has been corrected on the basis of the evidence presented below:

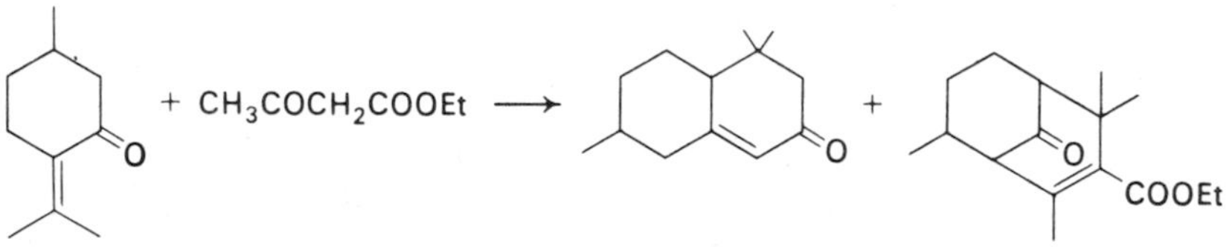

+ CH_3COCH_2COOEt ⟶ +

COOEt

A *B* *C′*?

Compound *C* can be transformed into *B* by the action of $ZnCl_2$ in acetic acid, showing that its formation is reversible.

Compound *C* is reduced by lithium aluminum hydride (LAH), but only the —COOEt group is reduced. *C* does not form a 2,4-dinitrophenylhydrazone. Neither of these observations is in accord with structure *C′*. The IR spectrum of *C* showed peaks at 1635 cm^{-1} and 1720 cm^{-1}, and the UV spectrum has maxima at 212 (ε 2140) and 272 (ε 1480) nm. (*Query:* Would *C′* show a maximum at 272 nm?)

Catalytic hydrogenation of *C* gave a tetrahydro derivative, indicating two carbon-carbon double bonds. The LAH reduction of *C* gives *D*, $C_{14}H_{22}O_2$, which could be oxidized to an acid *E*, $C_{14}H_{20}O_3$. *E* is the acid corresponding to the ester *C*. Decarboxylation of *E* gave an unsaturated compound *F*, $C_{13}H_{20}O$. The oxygen atom in *C* is not present in either a carbonyl nor a hydroxyl group. Treatment of *C* with 2,4-dinitrophenylhydrazine in acid solution gives a product that contains *two* $=NNHC_6H_3(NO_2)_2$ units, and no additional oxygen atoms; in other words, the derivative is a bis-dinitrophenylhydrazone of a dicarbonyl compound *G*, $C_{13}H_{22}O_2$.

Finally, the n.m.r. spectrum of *C* shows signals for four $—CH_3$ groups: two singlets at δ 1.20 and 1.23 ppm, a doublet (J = 3 Hz) at δ 0.98 ppm, and a singlet at δ 1.90 ppm. Note that these data also agree with *C′*.

Devise a structure for *C* that accommodates these data and show how it is formed; explain its ready transformation into *B*.

22

The product *A* of the Mannich reaction of acetophenone with dimethylamine hydrochloride and formaldehyde reacts with phenylmagnesium bromide to give *B*, $C_{17}H_{21}NO$. When *B* is treated with acetic anhydride under reflux, and the reaction mixture is decomposed with water, ether extraction yields only a hydrocarbon *C*, $C_{14}H_{12}$. Examination of the residual solution disclosed the presence of dimethylamine (C_2H_7N).

Compound *C* decolorized bromine in CCl_4. Oxidation of *C* with $KMnO_4$ or CrO_3 yielded *D*, $C_{13}H_{10}O$, a stable compound resistant to further oxidation. Compound *D* formed an oxime *E*, which underwent a Beckmann rearrangement to give *F* (isomeric with *E*). Acid hydrolysis of *F* gave an acid *G*, $C_7H_6O_2$ and an amine *H*, C_6H_7N.

Explain how *C* is formed from *B*, and describe the fate of the carbon atom that is not accounted for in the transformation of *B* into *C*.

23

When chloroacetone is allowed to react with benzaldehyde, in the presence of trimethylamine, the product is *A*, $C_{10}H_{10}O_2$, which shows an IR absorption bond at 1720 cm^{-1}, but no hydroxyl absorption; and has no high intensity UV absorption in the 210–270 nm range.

When *A* is treated with hot, ethanolic sodium acetate it is rearranged into an isomer *B*, which reacts with *o*-phenylenediamine (1,2-diaminobenzene) to give a yellow compound *C*, $C_{16}H_{14}N_2$. When *B* is treated with alkaline hydrogen peroxide it is smoothly cleaved to give two carboxylic acids, *D* and *E*. *D* is $C_2H_4O_2$; *E* is $C_8H_8O_2$. Oxidation of *E* with $KMnO_4$ converts it in good yield to *F*, $C_7H_6O_2$, which is quite stable to further oxidation.

24

Treatment of *o*-hydroxyacetophenone with benzoyl chloride-pyridine gives *A*, $C_{15}H_{12}O_3$. When *A* is allowed to react with powdered KOH in pyridine, followed by acidification, it is converted into *B*, an isomer. *B* is a yellow compound, which gives a deep red color with ferric chloride. *B* is acidic, but is not a carboxylic acid. Treatment of *B* with HCl in acetic acid converts it into *C*, $C_{15}H_{10}O_2$, a colorless, neutral compound with a carbonyl stretching frequency in the IR at 1650 cm^{-1} (in $CHCl_3$). Although *C* is neutral and insoluble in alkali, when it is heated in 10% ethanolic KOH it is first reconverted to *B*, and upon further heating in alkali is transformed into *o*-hydroxyacetophenone and benzoic acid.

Account for the low value of the IR carbonyl stretching frequency (acetophenone shows a C=O bond at 1692 cm^{-1}, *o*-methoxyacetophenone at 1667 cm^{-1}).

25

Reaction of resorcinol with phenylacetonitrile in ether with zinc chloride and HCl gives a crystalline salt *A*, $C_{14}H_{14}NO_2{}^+Cl^-$. When *A* is heated in water it is transformed into *B*, $C_{14}H_{12}O_3$, which is alkali-soluble and gives a deep red color with ferric chloride. Treatment of *B* with one molar equivalent of methyl iodide and excess anhydrous K_2CO_3 in dry acetone gives *C*, $C_{15}H_{14}O_3$, which is also alkali-soluble and gives a deep ferric chloride color. When powdered sodium is added to a solution of *C* in ethyl formate, a vigorous reaction ensues and a solid (*D*) forms. Addition of the solid *D* to a solution of HCl in acetic acid causes it to be transformed into a colorless compound *E*, $C_{15}H_{12}O_3$. *E* is insoluble in alkali and gives no color with ferric chloride. When heated with 10% alcoholic KOH, *E* is converted into *C* and formic acid.

Answers

PART 1

Exercise 1.1

1. (*a*) $C_{20}H_{28}O$ C, 84.50; H, 9.85; O, 5.65
   (*b*) $C_{12}H_{12}$ C, 92.30; H, 7.70
   (*c*) $C_6H_6O_4$ C, 50.70; H, 4.23; O, 45.07
   (*d*) $C_6H_{12}O_6$ C, 40.00; H, 6.67; O, 53.33
   (*e*) C_5H_5N C, 76.00; H, 6.34; N, 17.66
   (*f*) CCl_4 C, 7.80; Cl, 92.20
   (*g*) CH_2Cl_2 C, 14.14; H, 2.36; Cl, 83.50
   (*h*) C_5H_4NBr C, 38.00; H, 2.53; N, 8.88; Br, 50.50
   (*i*) C_2H_5NO C, 40.60; H, 8.48; N, 23.80; O, 27.10
   (*j*) $C_{10}H_7N_2O_2Br$ C, 45.00; H, 2.62; N, 10.48; Br, 29.90; O, 12.00
   (*k*) $C_4H_{10}SO$ C, 45.30; H, 9.40; S, 30.20; O, 15.10

2. (*a*) $C_6H_{12}O_6 \longrightarrow 6\,CO_2$
   180 $6 \times 44 = 264$
   $3.60\text{ mg} \longrightarrow 5.28\text{ mg } CO_2$
   (*b*) $C_8H_8NOBr \longrightarrow 8\,CO_2$
   214 $8 \times 44 = 352$

$$4.60 \longrightarrow \frac{4.60}{2.14} \times 3.52 = 7.57\text{ mg } CO_2$$

   (*c*) $C_6H_{12}O \longrightarrow 6\,CO_2$
   100 264

   $3.45\text{ mg} \longrightarrow 3.45 \times 2.64\text{ mg } CO_2$

(*d*) $C_{12}H_{22}O_{11}$; 4.84 mg $\longrightarrow$ 10.56 mg CO_2

(*e*) $C_{10}H_8$; 128 mg $\longrightarrow$ 440 mg CO_2; $\frac{440}{128} \times 5.11 = 17.5$ mg CO_2

3. (*a*) Density of gasoline = approx. 0.70. One U.S. gallon = 3780 ml. One U.S. gallon of gasoline = 2650 g
   Mol wt C_7H_{16} = 100; 2650 g = 26.5 moles
   1 mole C_7H_{16} gives 7 moles CO_2 + 8 moles H_2O
   Yield of $CO_2 = 26.5 \times 7 \times 44 = 8160$ g CO_2
   $H_2O = 26.5 \times 8 \times 18 = 3820$ g H_2O

   (*b*) Ethanol (C_2H_6O); mol wt 46. One U.S. ounce = 28.35 g. Thus, one U.S. ounce of ethanol = 0.615 mole.
   1 g mole ethanol $\longrightarrow 2 \times 44$ g CO_2 and 3×18 g H_2O
   So, 1 ounce ethanol $\longrightarrow 2 \times 44 \times 0.615 = 54.2$ g CO_2
   And $3 \times 18 \times 0.615 = 33.3$ g H_2O

   (*c*) Cellulose = $(C_6H_{10}O_5)_x$; mol wt = $(162)_x$
   162 g cellulose $\longrightarrow 6 \times 44 = 264$ g $CO_2 + 5 \times 18$ g H_2O
   1.62 mg cellulose $\longrightarrow$ 2.64 mg CO_2 + 0.90 g H_2O
   3.34 mg cellulose = 5.28 mg CO_2 + 1.80 g H_2O

   (*d*) Sucrose ($C_{12}H_{22}O_{11}$); mol wt = 242
   242 g sucrose $\longrightarrow 12 \times 44$ g $CO_2 + 11 \times 18$ g H_2O

   $$100 \text{ g sucrose} \longrightarrow \frac{12 \times 44}{2.42} = 218 \text{ g } CO_2$$

   $$\longrightarrow \frac{11 \times 18}{2.42} = 81.7 \text{ g } H_2O$$

   (*e*) 12 g diamond $\longrightarrow$ 44 g CO_2

   $$2.15 \text{ mg diamond} \longrightarrow \frac{2.15}{1.20} \times 4.40 = 7.88 \text{ mg } CO_2$$

Exercise 1.2

1. (*a*) C, 40.12; H, 6.75; O, 53.13
   40.12/12 = 3.43   3.43/3.32 = 1.03
   6.75/1 = 6.75   6.75/3.32 = 2.13
   53.13/16 = 3.32   3.32/3.32 = 1.00
   Empirical formula: CH_2O
   Calculated for CH_2O: C, 40.00; H, 6.67

   (*b*) $C_7H_{12}O$ (*d*) C_6H_7N (*f*) C_8H_9NO
   (*c*) $C_{13}H_{20}O_4$ (*e*) C_5H_5NO

2. (*a*) $C_{13}H_{20}O_2$ (*c*) C_6H_5OBr
   (*b*) $C_5H_8O_3$ (*d*) $C_{12}H_{10}S_2$

Exercise 1.3

1. (*a*) $CH_3CH(OH)CH_3$: C_3H_8O
   (*b*) $(CH_3CH_2CH_2CH_2)_2O$: $C_8H_{18}O$
   (*c*) $CH_3(CH_2)_7CH_3$: C_9H_{20}
   (*d*) $(CH_3)_2CHCH_2OCH_2CH_2OCH_2CH_3$: $C_8H_{18}O_2$
   (*e*) $(HOCH_2)_4C$: $C_5H_{12}O_4$
   (*f*) cholesterol: $C_{27}H_{46}O$
   (*g*) β-carotene: $C_{40}H_{56}$
   (*h*) adamantane: $C_{10}H_{16}$

Those with no rings or double bonds conform to C_nH_{2n+2} (that is, (*a*), (*b*), (*c*), (*d*), (*e*)). The others contain 2 fewer hydrogen atoms for each ring or double bond.

2. (*a*) C_2H_6O: CH_3CH_2OH, CH_3OCH_3
   (*b*) $C_4H_{10}O$: $CH_3CH_2CH_2CH_2OH$, $CH_3CH_2OCH_2CH_3$
   (*c*) $C_6H_{14}O_6$: $HOCH_2CH(OH)CH(OH)CH(OH)CH(OH)CH_2OH$, $HOCH_2CH(OH){-}C(OH)(CH_2OH){-}CH(OH)CH_2OH$
   (*d*) C_5H_{12}: $CH_3(CH_2)_3CH_3$, $(CH_3)_2CHCH_2CH_3$
   (*e*) $C_8H_{18}O_3$: $CH_3(CH_2)_3CH(OH)CH_2CH(OH)CH_2OH$, $HOCH_2CH_2CH_2CH_2OCH_2CH_2CH_2CH_2OH$
   (*f*) $C_5H_{12}O_4$: $HOCH_2CH(OH)CH_2CH(OH)CH_2OH$, $CH_3CH(OH)CH(OH)CH(OH)CH_2OH$
   (*g*) C_3H_8O: $CH_3OCH_2CH_3$, $CH_3CH_2CH_2OH$
   (*h*) $C_3H_8O_2$: $HOCH_2CH_2CH_2OH$, $CH_3CH(OH)CH_2OH$ 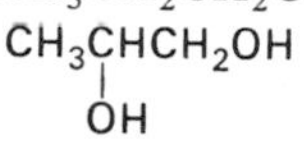
   (*i*) $C_2H_6O_2$: $HOCH_2CH_2OH$, $CH_3CH(OH)_2$
   (*j*) $C_{10}H_{22}O$: $(CH_3CH_2CH_2CH_2CH_2)_2O$, $CH_3(CH_2)_8CH_2OH$
3. General formula for a saturated amine is $C_nH_{2n+3}N$ (e.g., $(CH_3CH_2CH_2)_2NH = C_6H_{15}N$)
   (*a*) C_2H_7NO: $HOCH_2CH_2NH_2$, CH_3CH_2NHOH
   (*b*) $C_3H_{10}N_2O$: $H_2NCH_2CH_2OCH_2NH_2$, $H_2NCH_2CH_2CH_2NHOH$
   (*c*) $C_4H_{11}NO$: $HOCH_2CH_2CH_2CH_2NH_2$, $HOCH(CH_3)CH_2CH_2NH_2$
   (*d*) $C_5H_{11}N$: $CH_3CH{=}CHCH_2CH_2NH_2$, N H
   (*e*) C_5H_9NO: N H O, N O CH_3
4. Saturated, noncyclic compounds are (*b*), (*e*), (*f*), (*h*).
5. Possible structures for 4(*a–b–h–i–j*)
   (*a*) Cyclohexane
   (*b*) $(HOCH_2CH_2CH_2CH_2)_2O$
   (*h*) $HOCH_2CH_2CH_2CH_2NH_2$
   (*i*) Pyridine
   (*j*) N H

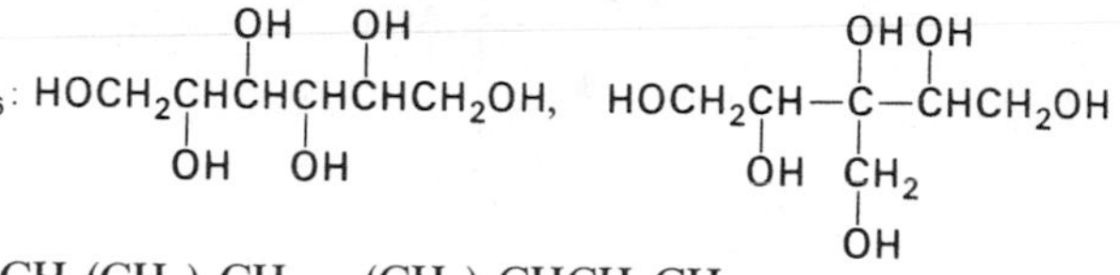

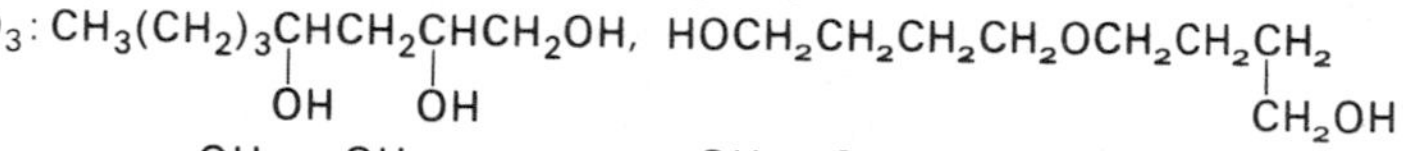

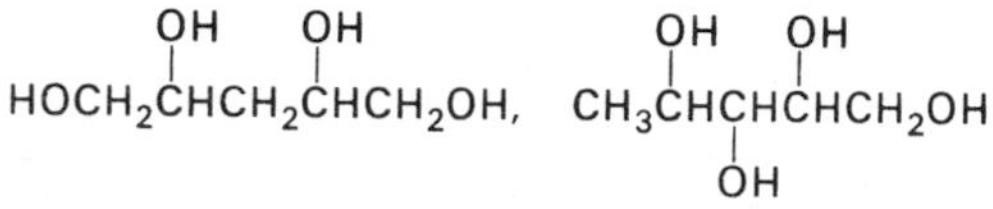

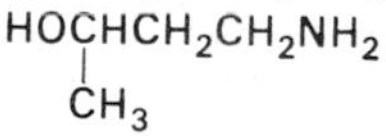

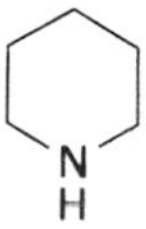

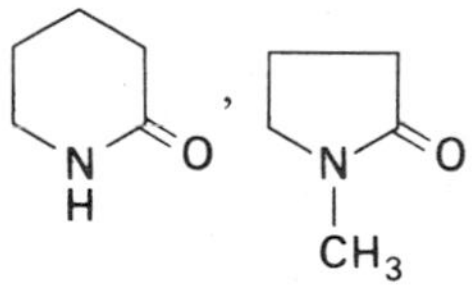

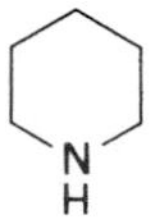

Exercise 1.4

1. $C_7H_{10}O$; contains 3 rings or double bonds $\left(\frac{16-10}{2}=3\right)$. Possible structures include

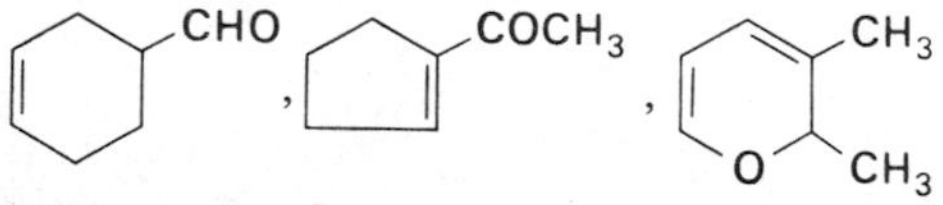

CHO, $COCH_3$, CH_3 O CH_3, $CH_3C{\equiv}CCH{=}CHCH_2OCH_3$,

$(CH_3)_2CH$ O

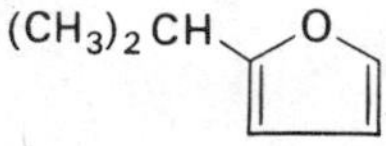

2. (*b*) C_6H_7N; saturated amine = $C_6H_{15}N$; thus, 4 rings or double bonds $\left(\frac{15-7}{2}\right)$.

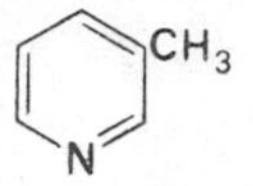

(*g*) $C_2H_2O_2$; saturated C_2-compound is C_2H_6; thus, 2 rings or double bonds; thus $C_2H_2O_2$ can be O=CH—CH=O.

(*h*) $C_2H_2O_4$ can be oxalic acid, HOOC—COOH.

(*i*) $C_4H_6O_2$ contains (10 − 6)/2 = 2 rings or double bonds; thus, it can be $CH_3CH{=}CH{-}CHO$.

PART 2

Exercise 2.1

(*a*) 109.5° (*c*) 180° (*e*) 120°
(*b*) 120° (*d*) 109.5° (*f*) Br—C—Br, 180°; H—C—H, 120°

Exercise 2.2

(*a*) SiF_4, 109.5°; as in methane, CH_4
(*b*) CO_2, 180°; O=C=O
(*c*) H_3O^+, near 105°; as in NH_3
(*d*) NCl_3, near 105°, as in NH_3
(*e*) $(CH_3)_2C{=}O$, about 120°
(*f*) $Ni(CO)_4$, 109.5°
(*g*) $Fe(CO)_5$, bipyramid
(*h*) PCl_6^-, octahedral
(*i*) $ZnCl_4^=$, tetrahedral
(*j*) Monomeric $AlCl_3$, plane trigonal, as in BF_3
(*k*) $Si(CH_3)_4$, tetrahedral
(*l*) SiO_2, tetrahedral, in polymeric matrix
(*m*) $POCl_3$, approximately tetrahedral
(*n*) $Hg(CH_3)_2$, linear, 180°
(*o*) $Ag(NH_3)_2^+$, linear, 180°
(*p*) $N(CH_3)_4^+$, tetrahedral
(*q*) $CH_3MgCl{\cdot}2\,Et_2O$, tetrahedral around Mg
(*r*) $(CH_3)_3\overset{+}{N}{-}\overset{-}{B}F_3$, both N and B tetrahedral
(*s*) $(CH_3)_3NO$, approximately tetrahedral around N

Exercise 2.3

(*a*) 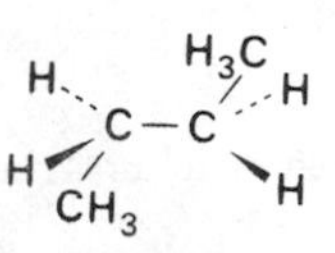 (*b*) 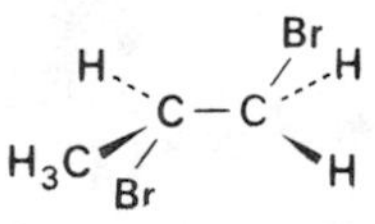 (*c*)

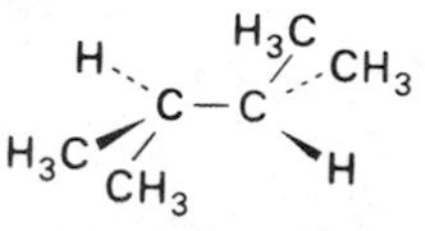

(*d*) H H C H···C—C···H H H

(*e*) H_2C—CH_2 H_2C—CH_2—CH_2

(*f*) H_2C—CH_2—CH_2 H_2C—CH_2—CH_2

or

H HH H—C—C—C—H H—C—C—C—H HH H

Exercise 2.4

All the compounds assume staggered ring conformations with minimal interactions between adjacent H atoms. The following structures are most probable:

cyclooctane cyclodecane

Cyclooctadecane (a ring containing eighteen $-CH_2-$ groups) could assume a variety of essentially equivalent conformations, with the possibility of facile interconversion.

Exercise 2.5

1. Compounds *a* and *d* would be highly strained and have not been prepared; *e* and *f* can assume stable and unstrained conformations; for instance:

(*e*)

Compounds *b* and *c* are strained but possible of existence.

2. Cyclooctene can exist as enantiomeric (optical isomeric) forms (see Part 3):

(±)

Note: By making models of these you will observe that they are not superimposable. They are nonidentical mirror images.

PART 3

Exercise 3.1 (a–d)

(*a*) H H C=C H_3C CH_3 H CH_3 C=C H_3C H

cis-2-butene *trans*-2-butene

(*b*)

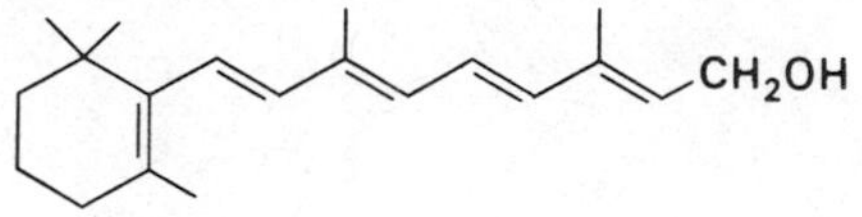

CH_2OH

(*d*)

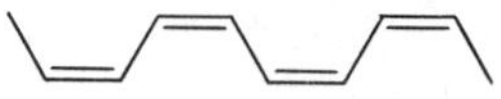

all *cis*-

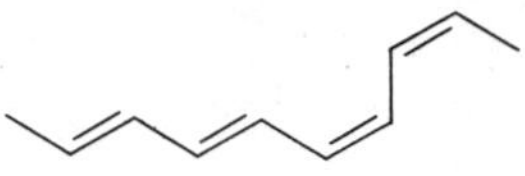

cis-cis-trans-trans

Exercise 3.2

(*a*) H, OH, OH, H

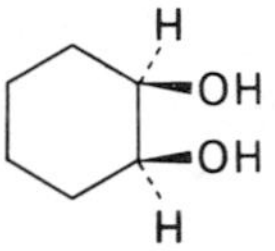

(*d*) H, COOH, COOH, H

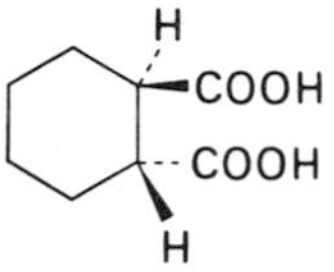

(*b*) H, OH, H, OH, OH, H

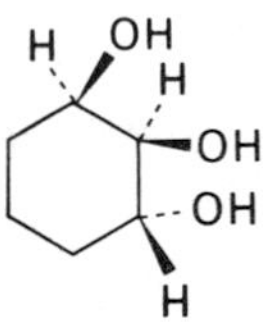

(*e*) H, OH, H, CH_3

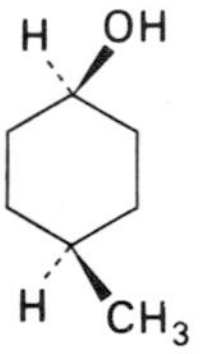

(*c*) H, COOH, COOH, H

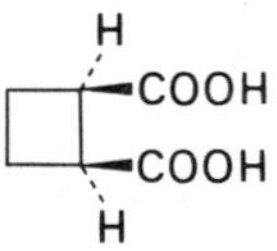

(*f*) H, H_3C, C, CO, O, H_3C, C, CO, H

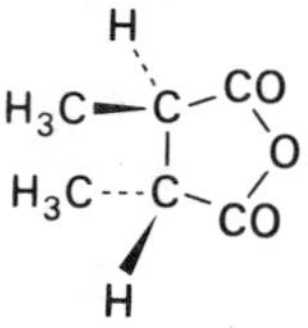

Exercise 3.3

1. (*a*) CH_3, H_3C, H, H   H, H, CH_3, CH_3 (Identical)

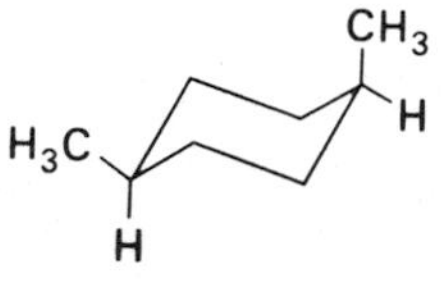

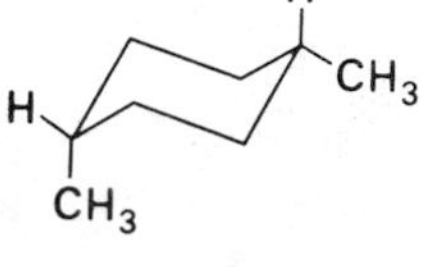

(*b*) * H, H_3C, CH_3, H CH_3, H, H, CH_3 (*More stable)

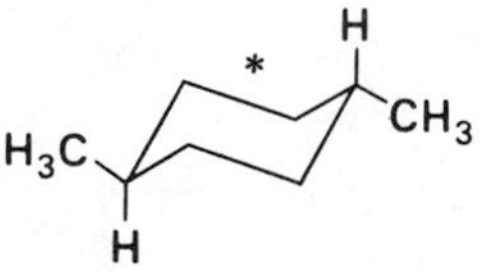

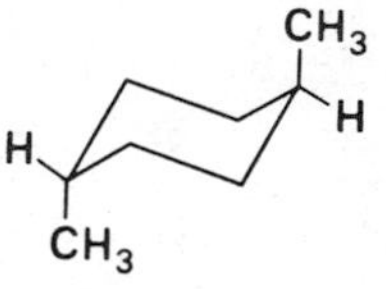

(*c*) Br, H, Br, H   H, *, Br, H, Br   (*More stable)

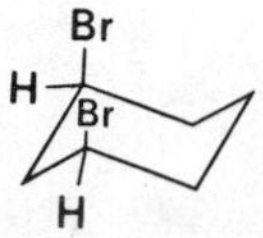

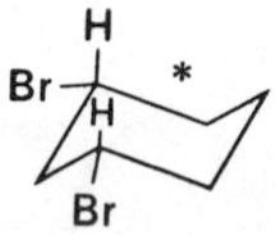

(d) (*More stable)

(e) (Only one form as in (a), with one $COOCH_3$ group axial and the other $COOCH_3$ group equatorial)

2\.

Conformational change permits intramolecular reaction.

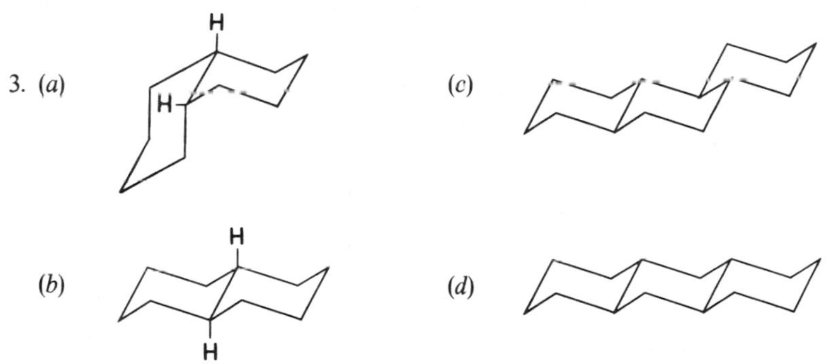

Exercise 3.4

The following compounds can exist in enantiomeric forms: *a, b, g, h, i, k, m, n, o, s.* The others possess elements of symmetry.

Exercise 3.5

(*a*) If *i* is called (+), *ii* is (−), *iii* is (+), *iv* is (−), *v* is (−).
(*b*) Compounds *i* and *iv* are enantiomers, and *iii* and *i* are enantiomers; therefore *iii* and *iv* are identical; *ii* and *iv* are enantiomers; therefore *i* and *ii* are identical.
(*c*) Solution in text.

Exercise 3.6

(*a*) Four ((±)-*cis*, (±)-*trans*)
(*b*) Four (two (±) pairs)
(*c*) Two (one (±) pair)
(*d*) none
(*e*) none
(*f*) Eight (three asymmetric carbon atoms)
(*g*) Two
(*h*) none
(*i*) none
(*j*) none
(*k*) Two
(*l*) Two (note: the bridge is necessarily 1,4-*cis*, so the apparent asymmetry at C-1 and C-4 does not exist).

Exercise 3.7

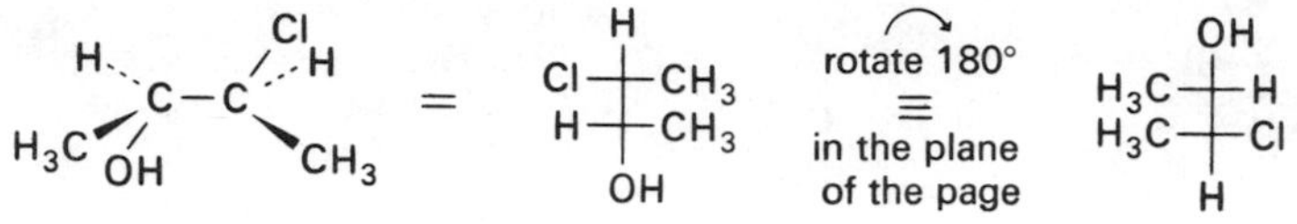

= (by Fischer transformations)

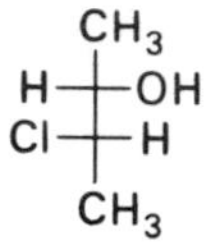

rotate 180° ≡ in the plane of the page

Exercise 3.8

(*a*) ;

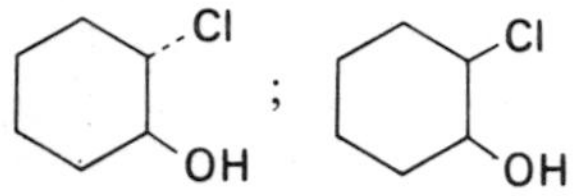

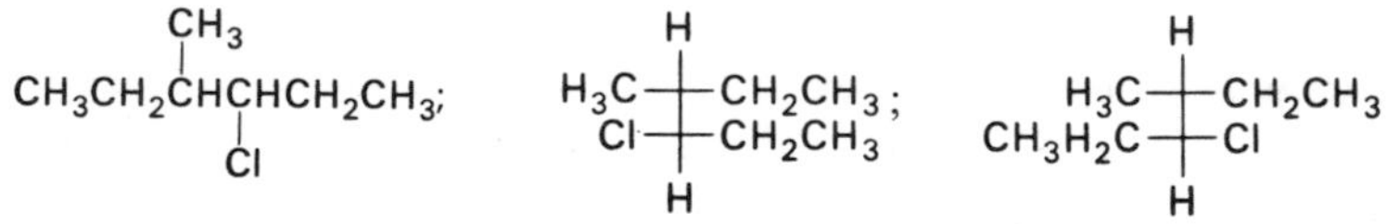

(*b*) $CH_3CH_2\overset{CH_3}{\overset{|}{C}}H\underset{Cl}{\underset{|}{C}}HCH_2CH_3$;

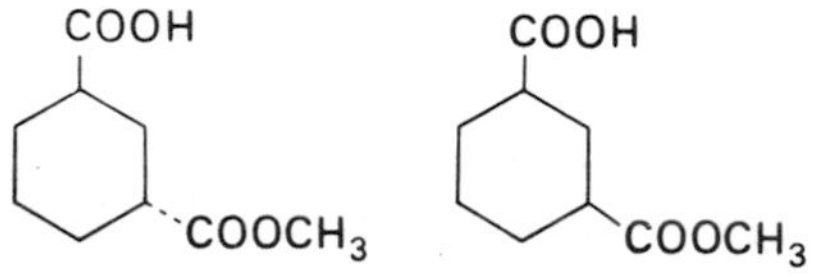

(*c*)

(*d*) $CH_3CH_2CH_2\overset{OH}{\overset{|}{C}}H\underset{OH}{\underset{|}{C}}H\overset{OH}{\overset{|}{C}}HCH_3$;

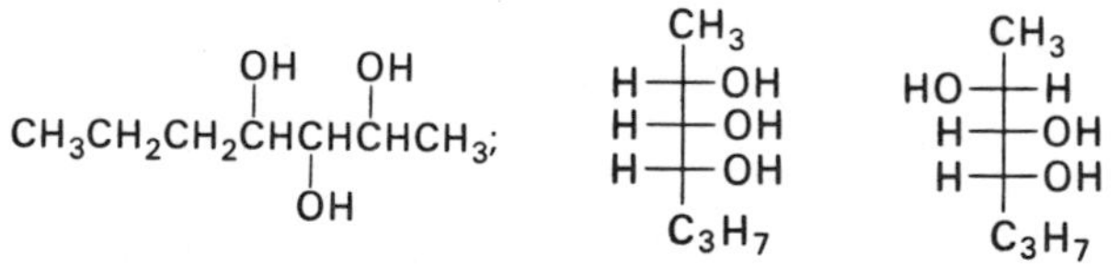

(*e*) *Cis*-$CH_3CH_2CH{=}CHCH_2CH_3$ + HOCl 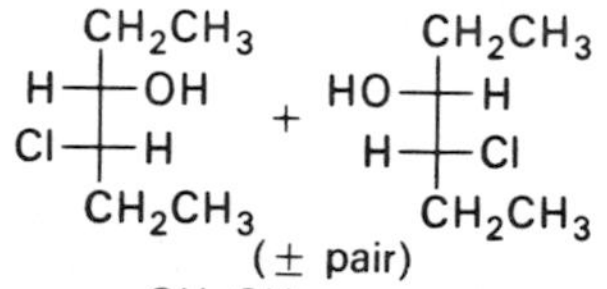 ⟶ (± pair)

Trans-3-hexene + HOCl ⟶ (± pair)

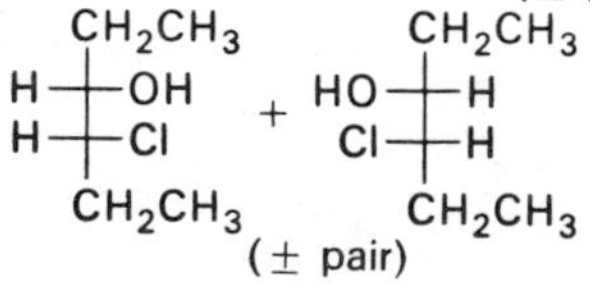

Exercise 3.9

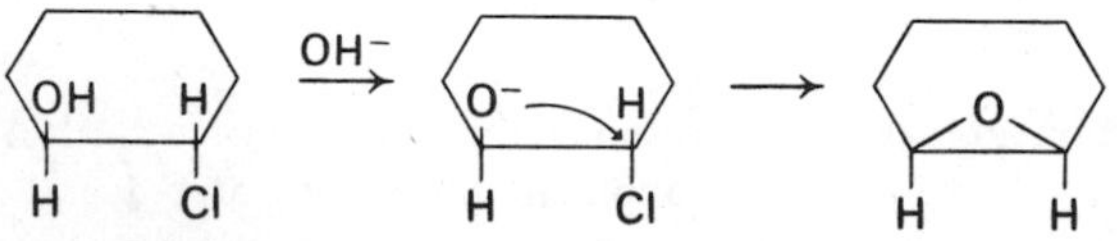

$\xrightarrow{OH^-}$ ⟶

trans-2-chloro-cyclohexanol

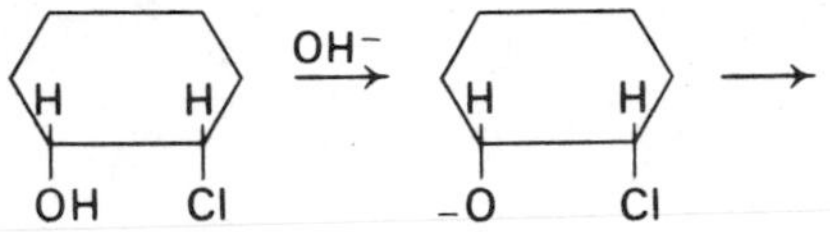

—O⁻ cannot attack $\rangle$C—Cl from rear-side with inversion-displacement.

cis-2-chloro-cyclohexanol $\xrightarrow{OH^-}$ ⟶ no oxide

Exercise 3.10

$R_2C{=}C{=}CH_2$ = — plane normal to page, or plane of page, is plane of symmetry.

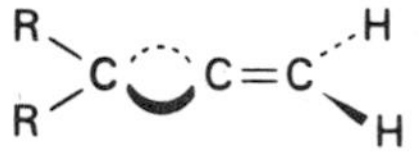

$RCH{=}C{=}CH_2$ = — plane of page is plane of symmetry.

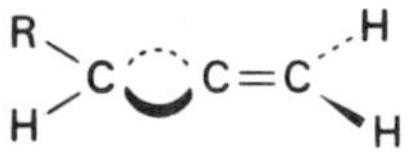

$R_2C{=}C{=}CHR$ = — plane normal to page is plane of symmetry.

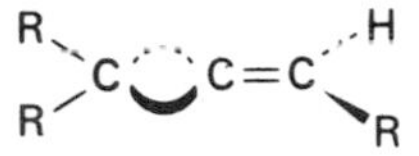

Note: if $R_2C{=}$ is RR′C—, there is no plane of symmetry, and two enantiomers are possible.

$R_2C{=}C{=}CR_2$ = — plane normal to page or plane of page is plane of symmetry.

In R′RC=C=CRR″, enantiomers are possible.

Exercise 3.11

Resolvable biphenyls are (*b*) and (*d*). Compounds *a* and *c* have a plane of symmetry; compound *e* can rotate around central bond because of small size of 2′ and 6′ F atoms.

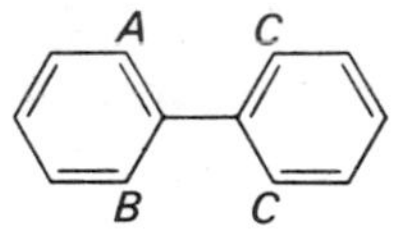

compound *C* 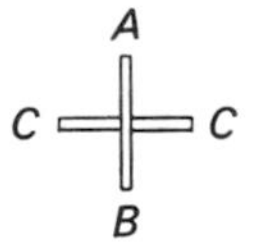 — end view — plane of symmetry through *A*, *B*, and between *C*, *C*.

Exercise 3.12 (use partial answer in text)

Exercise 3.13

(*a*) Answer in text

(*b*) $H(H_3C)C{=}C(H)CH_3 + Br_2 \longrightarrow$ H_3C–CHBr–CHBr–CH_3 + H_3C–CHBr–CHBr–CH_3

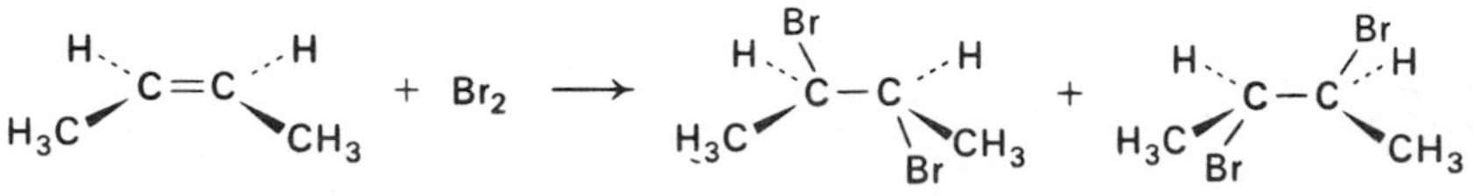

enantiomers

(*c*) H_3C, CH_3 cyclohexene $\xrightarrow{Br_2}$ dibromide + dibromide

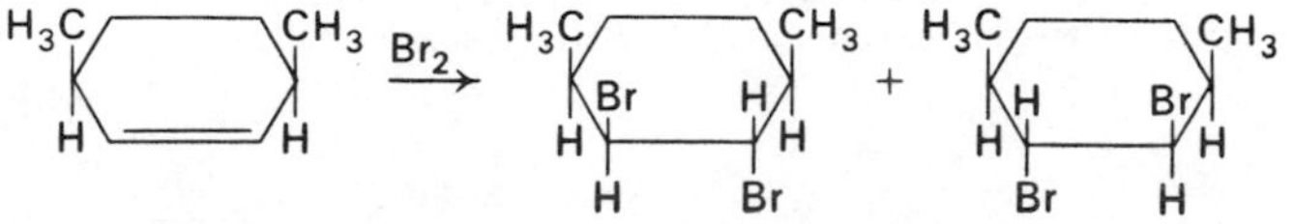

enantiomers

(*d*) See (*b*)

(*e*) Vinylcyclohexane (H, $CH=CH_2$ on the ring carbon) $+ Br_2 \longrightarrow$ 1-(1,2-dibromoethyl)cyclohexane, drawn as two enantiomers: H, C···Br, CH_2Br (wedge), H + H, C···Br, H (wedge), CH_2Br

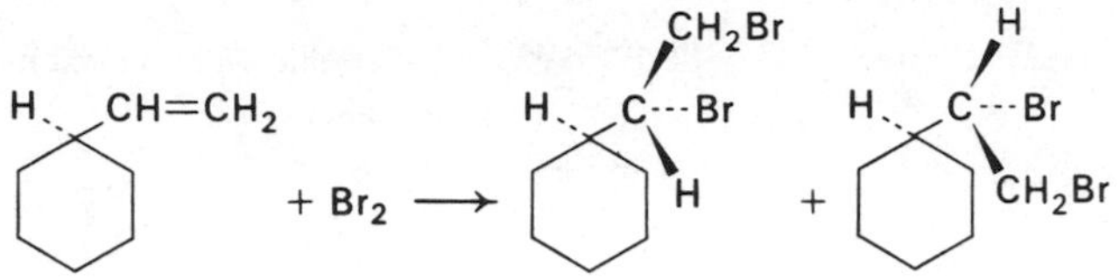

enantiomers
(only one asymmetric carbon atom)

PART 4

Exercise 4.1

$$CH_3CHO + B:^- \rightleftharpoons {}^-:CH_2CHO + BH$$

$$CH_3CHO + {}^-:CH_2CHO \rightleftharpoons CH_3-\underset{\displaystyle O_-}{\underset{|}{C}}HCH_2CHO$$

$$CH_3-\underset{\displaystyle O_-}{\underset{|}{C}}HCH_2CHO + BH \rightleftharpoons CH_3-\underset{\displaystyle OH}{\underset{|}{C}}HCH_2CHO$$

Exercise 4.2

(*a*) $CH_3CHO + CH_3CH_2CHO \longrightarrow CH_3\overset{\displaystyle OH}{\overset{|}{C}}HCH_2CHO + CH_3\overset{\displaystyle OH}{\overset{|}{C}}H\underset{\displaystyle CH_3}{\underset{|}{C}}HCHO$

$$+ CH_3CH_2\overset{\displaystyle OH}{\overset{|}{C}}HCH_2CHO + CH_3CH_2\overset{\displaystyle OH}{\overset{|}{C}}H\underset{\displaystyle CH_3}{\underset{|}{C}}HCHO$$

(*b*) $CH_3COCH_3 + CH_3CHO \longrightarrow CH_3\underset{\displaystyle OH}{\underset{|}{C}}HCH_2CHO + CH_3\underset{\displaystyle OH}{\underset{|}{C}}HCH_2COCH_3$

$$+ (CH_3)_2\overset{\displaystyle OH}{\overset{|}{C}}CH_2COCH_3$$

(*c*) $C_6H_5COCH_3 + C_6H_5CHO \longrightarrow C_6H_5COCH{=}CHC_6H_5$

(*d*) $C_6H_5CH{=}CHCOCH_3 + C_6H_5CHO \longrightarrow C_6H_5CH{=}CHCOCH{=}CHC_6H_5$

(*e*) Cyclohexanone $+ C_6H_5CHO \longrightarrow$ 2,6-bis(benzylidene)cyclohexanone ($\phi CH{=}$ and ${=}CH\phi$ on the carbons flanking C=O)

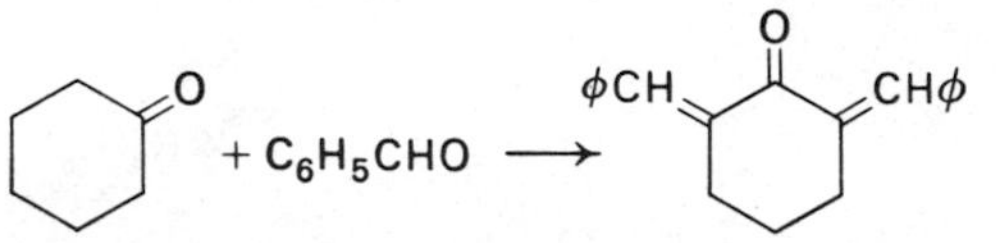

(ϕ is a shorthand designation for the phenyl group C_6H_5;
ϕCHO is the same as C_6H_5CHO (benzaldehyde))

Note that condensation at an α-carbon atom that is substituted gives rise to an aldol that cannot be further stabilized by loss of water:

$$\text{2-methylcyclohexanone} + C_6H_5CHO \rightleftharpoons \text{2-methyl-2-}(CH(OH)C_6H_5)\text{cyclohexanone}$$

The equilibrium for this condensation is so unfavorable that for practical purposes the product is not isolable.

Exercise 4.3

(*a*) $(H_3C)(CH_3CH_2)CHCHO \xrightarrow[OH^-]{HCHO} (H_3C)(CH_2OH)C(CH_3CH_2)(CHO) \xrightarrow{HCHO} (H_3C)(CH_2OH)C(CH_3CH_2)(CH_3CH_2)$

(*b*) $CH_3CHO \xrightarrow[OH^-]{2\ HCHO} CH_3{-}C(CH_2OH)_2{-}CHO \xrightarrow{HCHO} CH_3{-}C(CH_2OH)_3$

(*c*) $CH_3CHO \xrightarrow{OH^-} CH_3CH(OH)CH_2CHO \xrightarrow{reduce} CH_3CH(OH)CH_2CH_2OH$

(*d*) 2,6,6-trimethylcyclohex-1-enyl–CHO $\xrightarrow[OH^-]{CH_3COCH_3}$ 2,6,6-trimethylcyclohex-1-enyl–CH=CH–CO–CH_3

(*e*) 2 furyl–CHO + $CH_3COCH_3 \xrightarrow{OH^-}$ furyl–CH=CHCOCH=CH–furyl

(*f*) $C_6H_5CH_2CN + C_6H_5CHO \xrightarrow{OH^-} (C_6H_5)(CN)C{=}CHC_6H_5$

(*g*) $CH_3COCH_2CH_2COCH_3 \xrightarrow{OH^-}$ 3-methylcyclopent-2-enone { "inner" or cyclic aldol condensation

(*h*) $C_6H_5CH_2COCH_2C_6H_5 + C_6H_5CO{-}COC_6H_5 \xrightarrow{OH^-}$ 2,3,4,5-tetraphenyl-4-hydroxycyclopent-2-enone (O=, ϕ, ϕ, ϕ, OH) $\xrightarrow{-H_2O}$ tetraphenylcyclopentadienone (O=, ϕ, ϕ, ϕ, ϕ)

Exercise 4.4

(*a*) 3-Methyl-1-indanone will condense with benzaldehyde to give the 2-benzal derivative; 2-Methyl-1-indanone will not give a benzal derivative.

(*b*) 2,3-Dimethylcyclohexanone will give a benzal derivative (at 6—); 2,6-dimethylcyclohexanone will not.

(*c*) 3-Methyl-2-heptanone will give a benzal derivative; diisopropyl ketone will not.

Note: Although condensations of the above types can be carried out with other aldehydes, benzaldehyde (or *p*-methoxybenzaldehyde) is generally used for practical reasons of ready accessibility and absence of undesirable side reactions.

Exercise 4.5

(*a*) $CH_3COCH_3 + EtO^- \rightleftarrows CH_3COCH_2:^- + EtOH$

$$CH_3COCH_2:^- + CH_3COOEt \rightleftarrows \underset{\displaystyle COCH_3}{\underset{|}{CH_3COCH_2}} + EtO^-$$

(*b*) $C_6H_5COCH_2CH_3 + MeO^- \rightleftarrows C_6H_5CO\ddot{C}HCH_3{}^- + MeOH$

$$\underset{\displaystyle CH_3}{\underset{|}{C_6H_5COCH}}:^- + HCOOMe \rightleftarrows \underset{\displaystyle CH_3}{\underset{|}{C_6H_5COCH}}-CHO + MeO^-$$

(cyclohexanone) + $EtO^- \rightleftarrows$ (cyclohexanone carbanion, $:^-$ at α-carbon) + EtOH

(cyclohexanone carbanion, $:^-$) + $O=C(OEt)_2 \rightleftarrows$ (2-COOEt-cyclohexanone) + EtO^-

(*d*) $C_6H_5COCH_3 + EtO^- \rightleftarrows C_6H_5COCH_2:^- + EtOH$

$$C_6H_5COCH_2:^- + \underset{\displaystyle COOEt}{\underset{|}{COOEt}} \rightleftarrows \underset{\displaystyle COOEt}{\underset{|}{C_6H_5COCH_2CO}} + EtO^-$$

(*e*) $CH_3CH_2COOMe + MeO^- \rightleftarrows CH_3\ddot{C}H^-COOMe + MeOH$

$$\underset{\displaystyle COOMe}{\underset{|}{CH_3CH}}:^- + \underset{\displaystyle CH_2CH_3}{\underset{|}{COOMe}} \rightleftarrows \underset{\displaystyle COOMe}{\underset{|}{CH_3CH}}-COCH_2CH_3 + MeO^-$$

(*f*) and (*g*) have solutions similar to (*b*) and (*d*).

(*h*) The α-carbon atom of $C_6H_5CH_2CN$ is "active":

$$C_6H_5CH_2CN + EtO^- \rightleftarrows C_6H_5\ddot{C}H^-CN + EtOH$$

$$C_6H_5\ddot{C}H^-CN + CH_3COOEt \rightleftarrows \underset{\displaystyle COCH_3}{\underset{|}{C_6H_5CHCN}} + EtO^-$$

Exercise 4.6

(*a*) $C_6H_5COCH_2CH_3 + CO(OEt)_2(NaOEt)$

(*b*) $2\,C_6H_5CH_2COOEt(NaOEt)$

(*c*)

$$CH_3-\overset{\displaystyle CH_2COOEt}{\overset{|}{CH}}-\underset{\displaystyle CH_3}{\underset{|}{C}}HCH_2COOEt \quad (NaOEt)$$

(*d*) cyclopentanone (C=O) + HCOOEt(NaOEt)

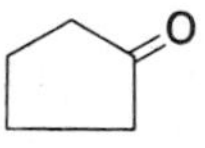

(*e*) cyclopentanone (C=O) + $(COOEt)_2$(NaOEt)

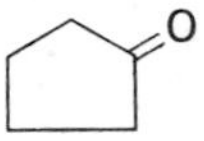

(*f*) $C_6H_5CH_2COCH_2C_6H_5 + (COOEt)_2(NaOEt)$

(*g*) $EtOOCCH_2CH_2N(CH_3)CH_2CH_2COOEt$(NaOEt)

(*h*) 2 $EtOOCCH_2CH_2COOEt$(NaOEt)

(*i*) $EtOOCCH_2CH_2CH_2COOEt + (COOEt)_2$(NaOEt)

(*j*) Refer to answers to (*a*) and (*k*).

(*k*) $C_6H_5CH_2CH_2COOEt + (COOEt)_2$(NaOEt)

(*l*) $o\text{-}C_6H_4(COOEt)(OCOCH_3)$ (NaH); or $o\text{-}C_6H_4(OH)(COCH_3)$ + $CO(OEt)_2$(Na)

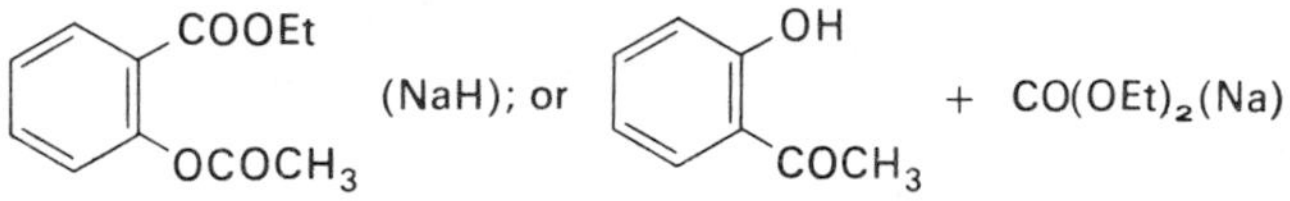

Exercise 4.7

(E-1) $C_6H_5CHO + (CH_3CO)_2O \xrightarrow{OAc^-} C_6H_5CH{=}CHCOOH$

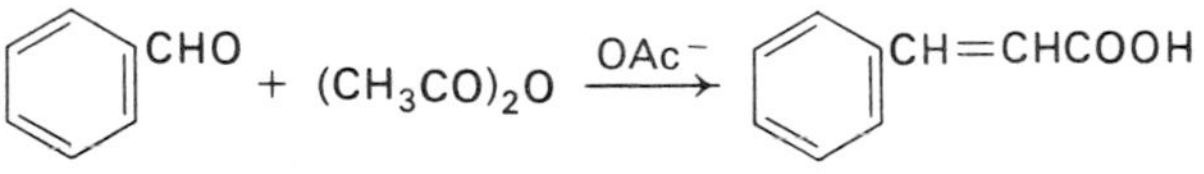

overall reaction

Steps

(*i*) $CH_3COOCOCH_3 + OAc^- \rightleftarrows CH_3COOCOCH_2{:}^- + HOAc$

(ii) 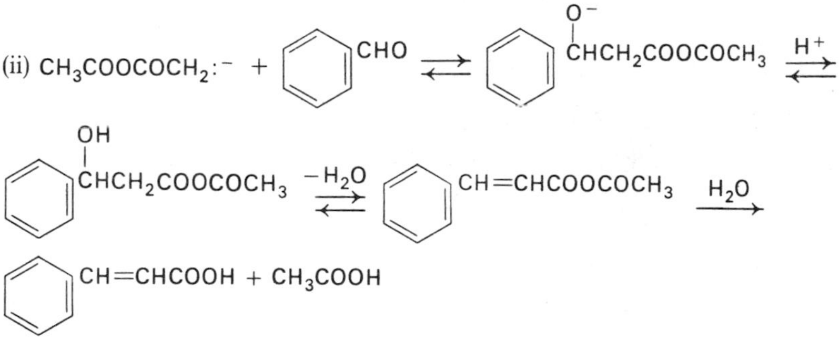 $CH_3COOCOCH_2{:}^- + C_6H_5CHO \rightleftarrows C_6H_5CH(O^-)CH_2COOCOCH_3 \underset{}{\overset{H^+}{\rightleftarrows}}$

$C_6H_5CH(OH)CH_2COOCOCH_3 \overset{-H_2O}{\rightleftarrows} C_6H_5CH{=}CHCOOCOCH_3 \xrightarrow{H_2O}$

$C_6H_5CH{=}CHCOOH + CH_3COOH$

(E-2)

Steps

(*i*) $C_6H_5CH_2COOH + Ac_2O \rightleftarrows C_6H_5CH_2COOAc + HOAc$

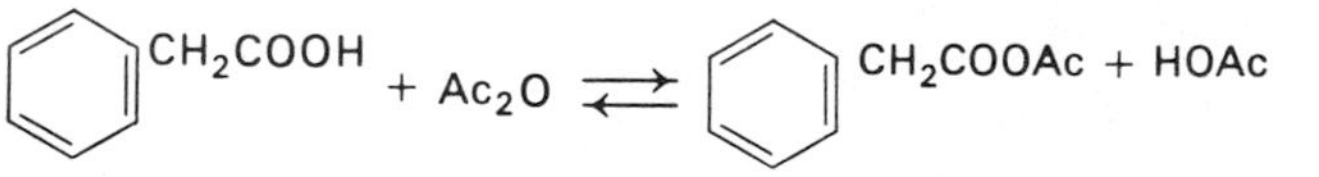

(*ii*) $C_6H_5CH_2COOAc + AcO^- \rightleftarrows C_6H_5\ddot{C}^-HCOOAc$

(*iii*) $C_6H_5\ddot{C}^-HCOOAc$ + $C_6H_4(CO)_2O$ $\rightleftarrows$ $C_6H_4[C(O^-)(CH(\phi)COOAc)\text{–}O\text{–}CO]$

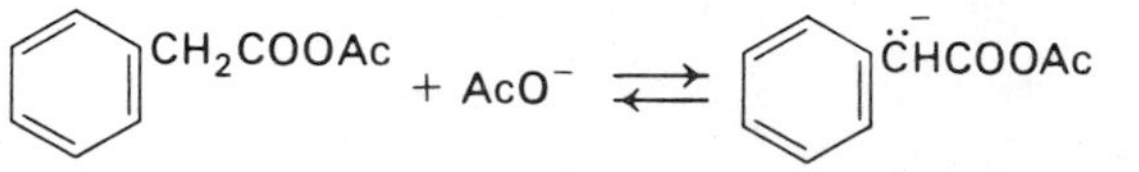

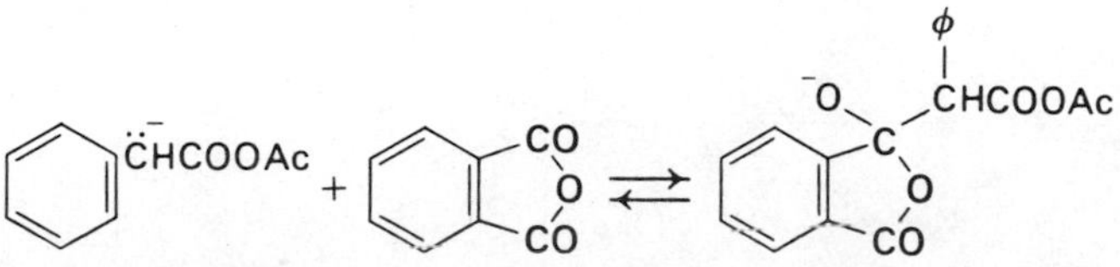

(*iv*) [structure: phthalide with ^-O and $CHCOOAc$ (φ) substituents] + HOAc ⇌ [structure: HO, CHCOOH (φ) phthalide]

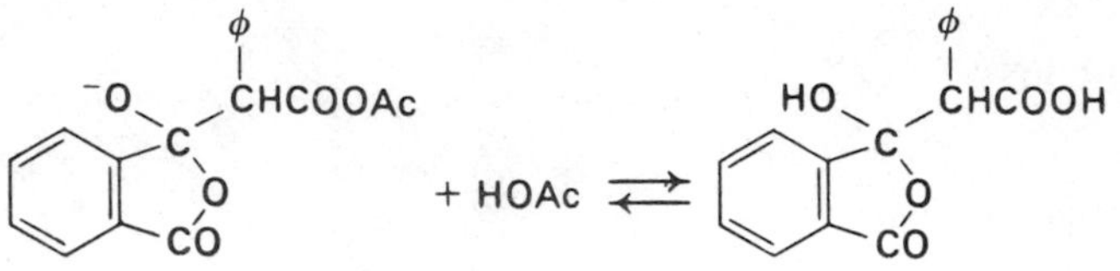

(*v*) [structure: HO, CHCOOH (φ) phthalide] $\xrightarrow[-CO_2]{-H_2O}$ [structure: benzylidene phthalide, $CH\phi$]

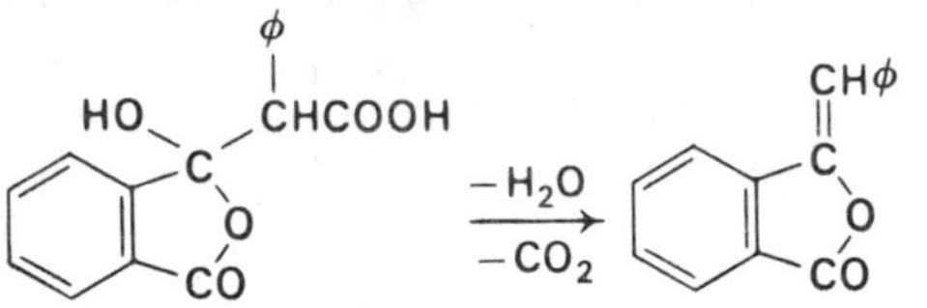

(E-3), See (E-1), above

Exercise 4.8

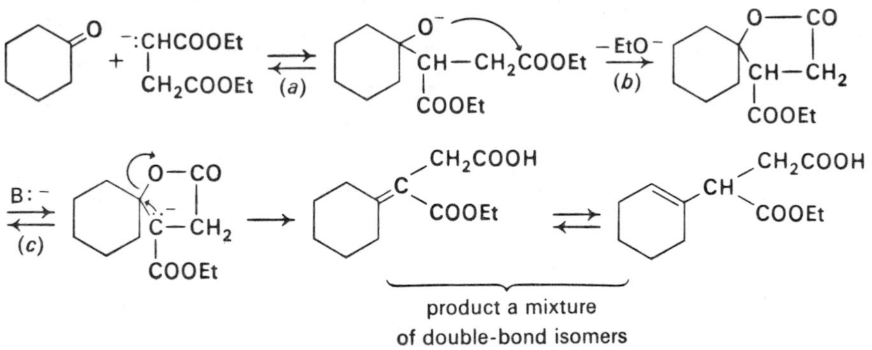

cyclohexanone + $^-$:CHCOOEt / CH_2COOEt $\underset{(a)}{\rightleftharpoons}$ [O$^-$, CH—CH_2COOEt, COOEt adduct] $\xrightarrow[(b)]{-EtO^-}$ [β-lactone: O—CO, CH—CH_2, COOEt]

$\underset{(c)}{\overset{B:^-}{\rightleftharpoons}}$ [O—CO, C$^-$—CH_2, COOEt] ⟶ [cyclohexylidene C, CH_2COOH, COOEt] ⇌ [cyclohexenyl CH, CH_2COOH, COOEt]

product a mixture of double-bond isomers

Exercise 4.9

(*a*) β-cyclocitral (CHO) + CH_3COCH_3 $\xrightarrow{Ba(OH)_2}$ β-ionone ($CH{=}CHCOCH_3$)

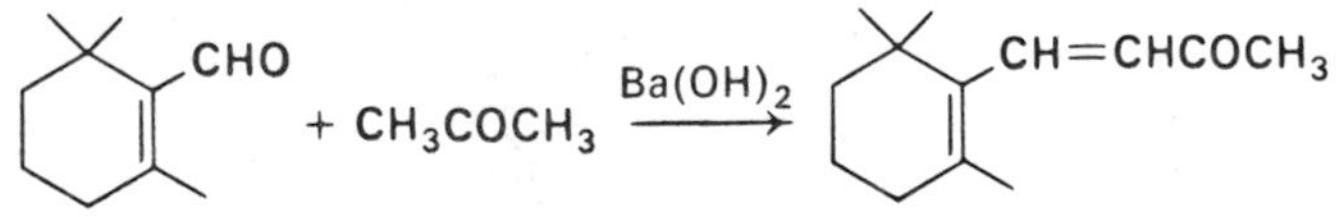

(*b*) $CH_3CHO \xrightarrow{OH^-} CH_3CH(OH)CH_2CHO \xrightarrow{\text{reduce}} CH_3CH(OH)CH_2CH_2OH$

(*c*) $C_6H_5CHO \xrightarrow[\text{pyridine}]{CH_2(COOH)_2} C_6H_5CH{=}CHCOOH \xrightarrow{Br_2} C_6H_5CHBrCHBrCOOH$

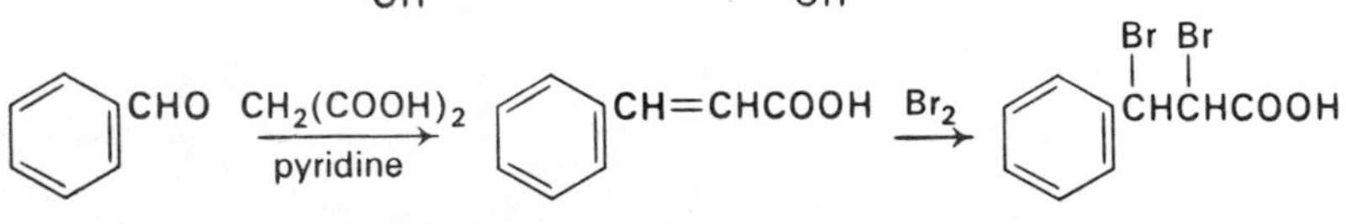

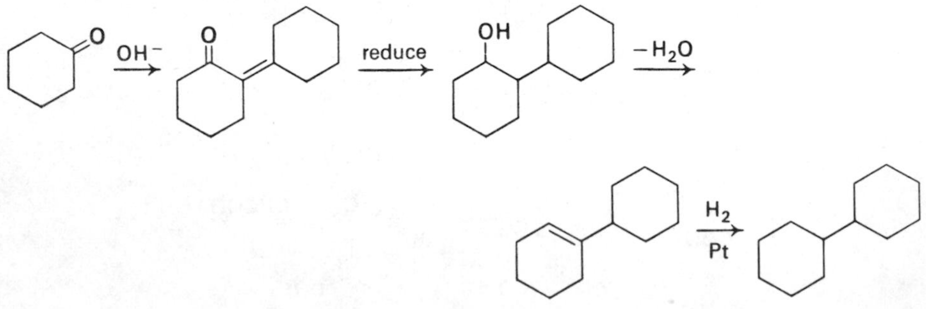

(*d*) cyclohexanone $\xrightarrow{OH^-}$ 2-cyclohexylidenecyclohexanone (O) $\xrightarrow{\text{reduce}}$ 2-cyclohexylcyclohexanol (OH) $\xrightarrow{-H_2O}$ 1-cyclohexylcyclohexene $\xrightarrow[Pt]{H_2}$ bicyclohexyl (*e*) $CH_3COCH_3 \xrightarrow{Ba(OH)_2} (CH_3)_2C(OH)CH_2COCH_3 \xrightarrow[I_2]{-H_2O} (CH_3)_2C{=}CHCOCH_3 \xrightarrow[Pt]{H_2} (CH_3)_2CHCH_2COCH_3$

(*f*) $(CH_3)_2C{=}CHCOCH_3$ (as in (*e*)) $\xrightarrow[NaOH]{Br_2} (CH_3)_2C{=}CHCOOH + HCBr_3$

(*g*) $C_6H_5CHO + CH_3COCH_3 \xrightarrow{OH^-} C_6H_5CH{=}CHCOCH{=}CHC_6H_5$

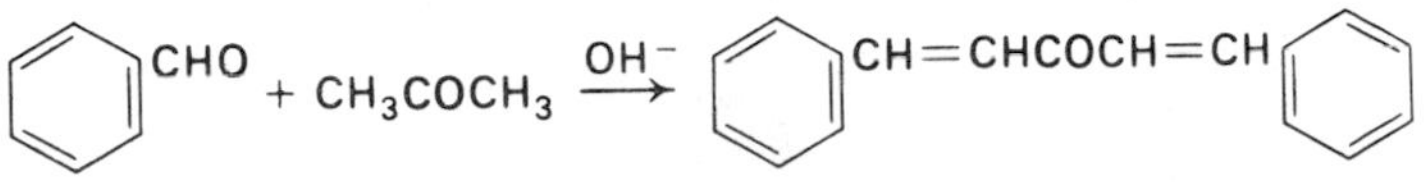

(*h*) $C_6H_5CHO \xrightarrow{\text{(as in (c))}} C_6H_5CH{=}CHCOOH \xrightarrow[Pt]{H_2} C_6H_5CH_2CH_2COOH \xrightarrow{SOCl_2} C_6H_5CH_2CH_2COCl \xrightarrow{AlCl_3}$ 1-indanone

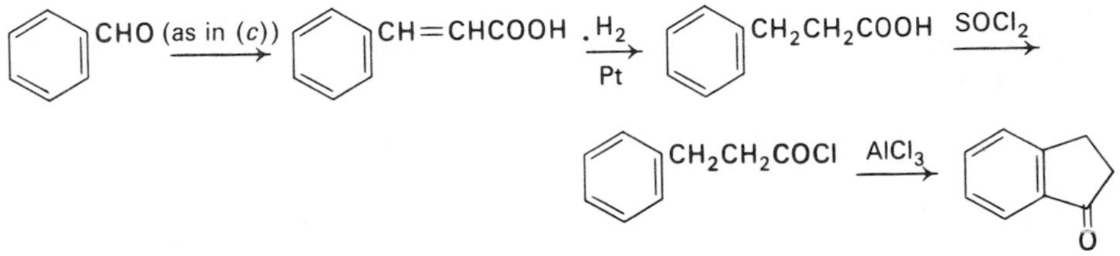

(*i*) $C_6H_5CHO \quad CH_3CH_2COCH_2CH_3 \xrightarrow{OH^-} C_6H_5CH{=}C(CH_3)COCH_2CH_3 \xrightarrow{\text{reduce}} C_6H_5CH_2CH(CH_3)COCH_2CH_3$

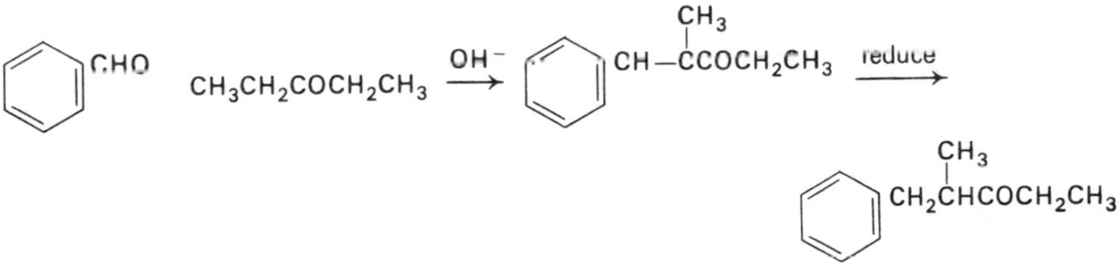

(*j*) $C_6H_5CH_3 \xrightarrow{Cl_2} C_6H_5CH_2Cl \xrightarrow{KCN} C_6H_5CH_2CN \xrightarrow[NaOEt]{CH_3COOEt} C_6H_5CH(CN)COCH_3 \xrightarrow[H^+]{H_2O} \{C_6H_5CH(COOH)COCH_3\} \xrightarrow{-CO_2} C_6H_5CH_2COCH_3$

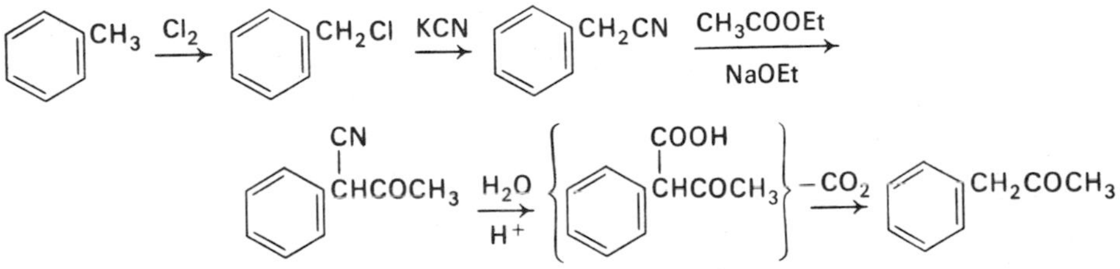

Exercise 4.10

(*a*) $C_6H_5CH{=}CHCOCH_3 + CH_3CH(COOEt)_2 \longrightarrow C_6H_5CH(CH_3{-}C(COOEt)_2)CH_2COCH_3$

(*b*) $CH_3CH{=}CHCOOEt + CH_3COCH_2COOEt \longrightarrow CH_3CH(CH_3COCHCOOEt){-}CH_2COOEt$

(*c*) $(C_6H_5CH_2)_2NH + CH_2{=}CHCN \longrightarrow (C_6H_5CH_2)_2NCH_2CH_2CN$

(*d*) $CH_3OH + CH_3CH_2CH{=}CHCOOCH_3 \longrightarrow CH_3CH_2CH(OCH_3)CH_2COOCH_3$

(*e*) $CH_3OCH_2CH_2COOEt + EtOH \longrightarrow EtOCH_2CH_2COOEt + CH_3OH$

(*f*) cyclohexanone $+ C_6H_5COCH{=}CH_2 \longrightarrow$ 2-($CH_2CH_2COC_6H_5$)cyclohexanone

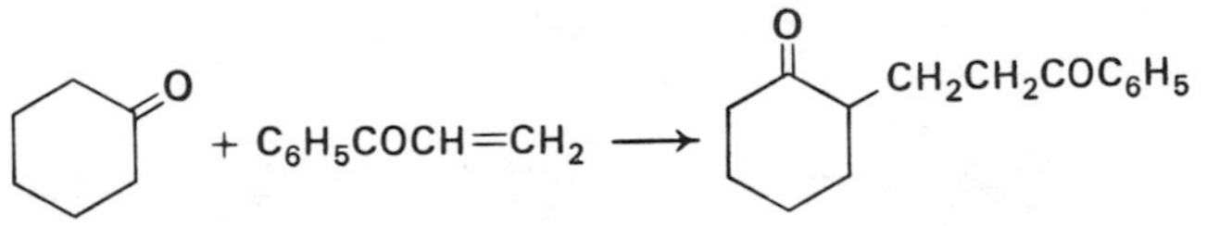

(g) $C_6H_5COCH_2CH_2\overset{+}{N}(CH_3)_3I^- + CH_2(COOEt)_2 \longrightarrow C_6H_5COCH_2CH_2CH(COOEt)_2$

(h) $CH_3CH{=}CHCOOEt + CH_3CH(COOEt)_2 \longrightarrow CH_3CH{-}CH_2COOEt$ with $CH_3{-}C(COOEt)_2$ attached to CH

(i) 2-methylenecyclohexanone + $CH_2(CN)(COOEt) \longrightarrow$ 2-($CH_2CH(CN)COOEt$)cyclohexanone

(j) $C_6H_5CH{=}CHCOC_6H_5 + KCN \longrightarrow C_6H_5CH(CN)CH_2COC_6H_5$

(k) The product is not formed by the addition of the fragments $-CH_3$ and $-CH(COOEt)_2$, but by the following route:

(i) $CH_3CH{=}CHCOOEt + CH_3CH(COOEt)_2 \xrightarrow[\text{Michael}]{\text{"normal"}} CH_3{-}CH(CH_2COOEt){-}C(CH_3)(COOEt)_2$

(ii) $CH_3{-}CH(CH_2COOEt){-}C(CH_3)(COOEt)_2 \xrightarrow{OEt^-} CH_3{-}CH(\overset{-}{C}HCOOEt){-}C(CH_3)(COOEt)_2 \longrightarrow$ [cyclic intermediate: $CH_3{-}CH{-}CHCOOEt$, $CH_3{-}C{-}CO\}OEt^-$, COOEt] $\longrightarrow$

$CH_3{-}CH(CHCOOEt)(CO\text{-}OEt)\ /\ CH_3{-}\overset{-}{C}:(COOEt) \xrightarrow{H^+} CH_3{-}CH{-}CH(COOEt)_2$ with $CH_3{-}CH(COOEt)$ attached

Note that the $CH_3CH(COOEt)COOEt$ [boxed: $CH_3CHCOOEt$ bearing COOEt] unit in the original addend is intact in the final product.

Exercise 4.11

Example G-5:

3,4-dihydroxyphenethyl-NHCH$_3$ + HCHO $\rightleftarrows$ ArCH$_2$CH$_2$N(CH$_3$)CH$_2$OH $\rightleftarrows$

ArCH$_2$CH$_2\overset{+}{N}$(CH$_3$)=CH$_2$ $\longrightarrow$ cyclized cation (HO, HO, H, N$-$CH$_3$) $\xrightarrow{-H^+}$

6,7-dihydroxy-2-methyl-1,2,3,4-tetrahydroisoquinoline (HO, HO, N$-$CH$_3$)

Example G-9:

HCHO + pyrrolidine (NH) $\rightleftarrows$ NCH$_2$OH $\rightleftarrows$ $\overset{+}{N}{=}CH_2$

$$\text{(pyrrolidinyl)}NCH_2C{\equiv}CH + B{:}^- \rightleftharpoons \text{(pyrrolidinyl)}NCH_2C{\equiv}C{:}^-$$

$$\text{(pyrrolidinyl)}NCH_2C{\equiv}C{:}^- + CH_2{=}\overset{+}{N}\text{(pyrrolidine)} \longrightarrow \text{(pyrrolidinyl)}NCH_2C{\equiv}CCH_2N\text{(pyrrolidine)}$$

Exercise 4.12

A methylolamine (1-hydroxy-2-methyl-1,2,3,4-tetrahydroisoquinoline; N—CH_3, OH) readily ionizes in the presence of an acid catalyst:

$$\underset{\text{(i)}}{\text{N—}CH_3\text{, 1-OH}} \underset{}{\overset{H^+}{\rightleftharpoons}} \underset{\text{(ii)}}{\text{N—}CH_3\text{, 1-}{}^+OH_2} \rightleftharpoons \underset{\text{(iii)}}{\text{(3,4-dihydroisoquinolinium) }\overset{+}{N}\text{—}CH_3} + H_2O$$

Reaction of the nucleophilic cyanide ion with the electrophile (iii) gives the 1-cyano compound. As is to be expected, (iii) also reacts with active methylene compounds:

$$\text{(3,4-dihydroisoquinolinium) }\overset{+}{N}\text{—}CH_3 + {}^-{:}CH_2COR \longrightarrow \text{(tetrahydroisoquinoline) N—}CH_3\text{, 1-}CH_2COR$$

Exercise 4.13

(*a*) $$\underset{\text{indole}}{\text{indole (N–H)}} \xrightarrow[Me_2NH]{HCHO} \text{3-}CH_2NMe_2\text{-indole} \xrightarrow{CH_3I} \text{3-}CH_2\overset{+}{N}Me_3I^-\text{-indole} \xrightarrow{KCN} \text{3-}CH_2CN\text{-indole} + NMe_3$$

(*b*) $$\text{phenol (OH)} + HCHO + Me_2NH \longrightarrow \text{2-}CH_2NMe_2\text{-phenol (OH)} \xrightarrow{CH_3I} \text{2-}CH_2\overset{+}{N}Me_3I^-\text{-phenol (OH)}$$

(*c*) $$CH_3COCH_3 \xrightarrow[Me_2NH]{HCHO} CH_3COCH_2CH_2NMe_2 \xrightarrow{CH_3MgI} (H_3C)_2C(OH)CH_2CH_2NMe_2$$

(*d*) (indole-3-yl)$CH_2\overset{+}{N}Me_3I^-$ $\xrightarrow{NaCH(COOEt)_2}$ (indole-3-yl)$CH_2CH(COOEt)_2$ $\xrightarrow[(2)\ \Delta,\ -CO_2]{(1)\ \text{saponify}}$ (indole-3-yl)CH_2CH_2COOH

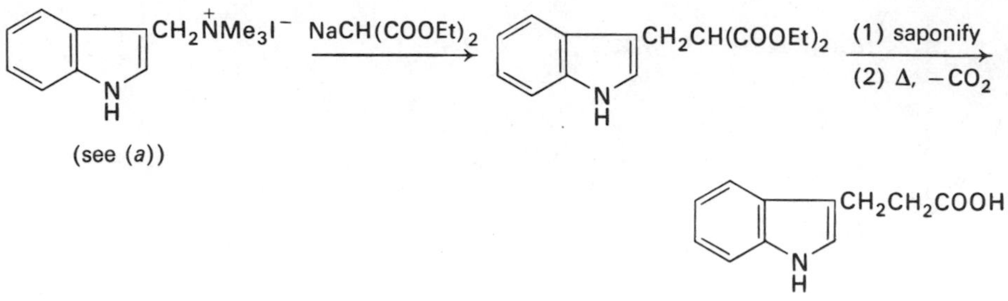

(see (*a*))

(*e*) 3,4-(HO)$_2C_6H_3CH_2CH_2NHCH_3$ $\xrightarrow{CH_3CHO}$ 6,7-dihydroxy-1-CH_3-2-CH_3-1,2,3,4-tetrahydroisoquinoline ($N-CH_3$)

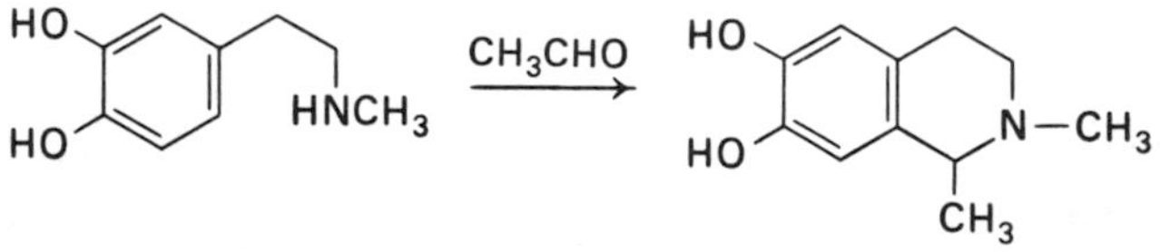

(compare Equation G-5)

(*f*) $C_6H_5COCH_3$ + pyrrolidine (NH) + HCHO $\longrightarrow$ $C_6H_5COCH_2CH_2N$(pyrrolidine) $\xrightarrow[(2)\ -H_2O]{(1)\ C_6H_5MgBr}$ $(C_6H_5)_2C{=}CHCH_2N$(pyrrolidine)

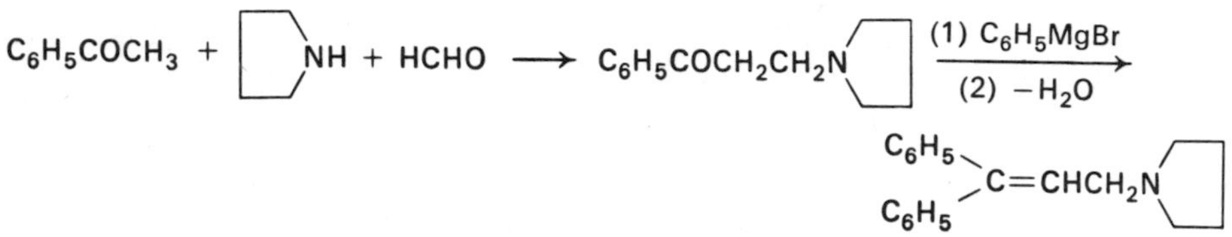

(*g*) (pyrrolidine)$NCH_2C{\equiv}CCH_2N$(pyrrolidine) $\xrightarrow[Pd]{1\text{ mole } H_2}$ (pyrrolidine)$NCH_2CH\overset{cis}{=}CHCH_2N$(pyrrolidine)

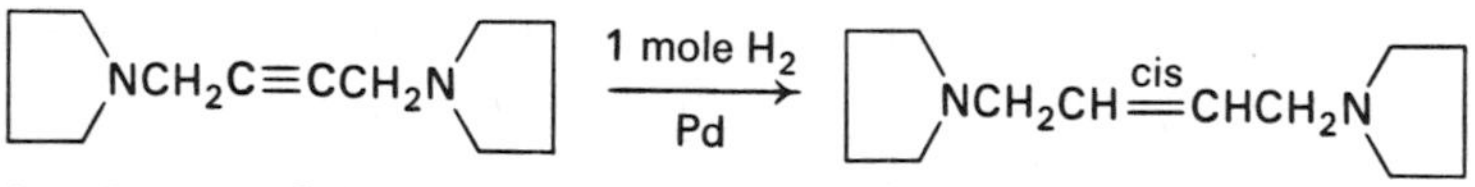

(see Equation G-9)

(*h*) $(H_3C)_2CHNO_2$ $\xrightarrow[Me_2NH]{HCHO}$ $CH_3-C(CH_3)(NO_2)-CH_2NMe_2$ $\xrightarrow[Pt]{H_2}$ $CH_3-C(CH_3)(NH_2)-CH_2NMe_2$

PART 5

Exercise 5.1

(*a*) $CH_3CH_2MgBr + CH_3CHO$; or $CH_3CH_2CHO + CH_3MgI$

(*b*) $CH_3CH_2CH_2CH(CH_3)MgBr + HCHO$; or *via* carbonation of the Grignard reagent and $LiAlH_4$ reduction of the acid

(*c*) $CH_3CH_2COOR + CH_3MgI$ (R can be Me, Et)

(*d*) $CH_3COCH_3 + CH_3CH_2MgBr$; or $CH_3COCH_2CH_3 + CH_3MgI$

(*e*) $(CH_3)_2CHCH_2CH_2MgBr + CO_2$

(*f*) $(CH_3)_2CHCH_2CH_2CHO + (CH_3)_2CHCH_2CH_2MgBr$

(*g*) $CH_3-C(CH_3)_2-CHO + CH_3CH_2MgBr$; or $CH_3-C(CH_3)_2-MgCl + CH_3CH_2CHO$

(*h*) $C_6H_5COCH_2CH_2NMe_2 + CH_3MgBr$; or $CH_3COCH_2CH_2NMe_2 + C_6H_5MgBr$

(*i*) $CH_3OCH_2CH_2CHO + CH_3CH_2MgBr$; why not $CH_3OCH_2CH_2MgBr + CH_3CH_2CHO$?

(*j*) $CH_3OCH_2COOEt + C_6H_5MgBr$

(*k*) $Br-C_6H_4-CHO + C_6H_5MgBr$

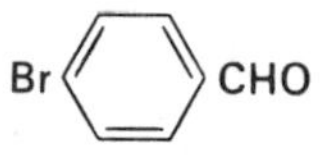

(*l*) $BrCH_2CH_2COOEt + CH_3MgI$

(*m*) $EtOOC-C_6H_4-COOEt$ (para) $+ CH_3CH_2MgBr$

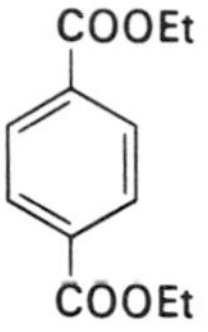

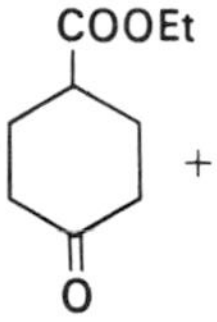

(*n*) 4-oxocyclohexane-COOEt $+ CH_3MgI$

(*o*) $C_6H_5CH_2CH_2COOEt + CH_3CH_2MgBr$

(*p*) $BrCH_2COOEt + NaCH(COOEt)_2 \longrightarrow EtOOCCH_2CH(COOEt)_2 \xrightarrow[(2)\ \Delta,\ -CO_2]{(1)\ \text{saponify}}$

$HOOCCH_2CH_2COOH \xrightarrow[HCl]{EtOH} EtOOCCH_2CH_2COOEt \xrightarrow{CH_3CH_2MgBr}$

$CH_3CH_2-C(OH)(CH_2CH_3)-CH_2CH_2-C(OH)(CH_2CH_3)-CH_2CH_3$

Exercise 5.2

(*a*) $CH_3CH_2CH(CH_3)CHO + HCHO \xrightarrow{OH^-}$

(*b*) $C_6H_5CHO + C_6H_5COCH_3 \xrightarrow{OH^-} C_6H_5CH{=}CHCOC_6H_5 \xrightarrow[Pt]{H_2}$

$C_6H_5CH_2CH_2COC_6H_5 \xrightarrow{CH_3MgI}$

(*c*) $C_6H_5CHO + C_6H_5CH_2COC_6H_5 \xrightarrow{OH^-}$

(*d*) $C_6H_5COOH \xrightarrow[H^+]{CH_3OH} C_6H_5COOCH_3 \xrightarrow{CH_3MgI}$

$$C_6H_5\underset{CH_3}{\overset{OH}{|}}C\,CH_3 \xrightarrow{H^+} C_6H_5\underset{CH_3}{|}C{=}CH_2$$

(*e*) Addition of CH_3MgI to $C_6H_5\underset{CH_2COOEt}{|}CHCH_2COOEt$,

which is prepared by Michael reaction starting with $C_6H_5CH{=}CHCOOEt$.

(*f*) $C_6H_5CHO + CH_3NO_2 \xrightarrow{OH^-} C_6H_5CH{=}CHNO_2$,

then $LiAlH_4$ reduction.

(*g*) Addition of CH_3MgI to $(CH_3)_2CHCH_2COCH_3$, then dehydration.

(*h*) Pimelic acid (ester) $\xrightarrow{NaOEt}$ 2-(COOEt)cyclohexanone (=O, COOEt) $\xrightarrow[(2)\ -CO_2]{(1)\ \text{saponify}}$

cyclohexanone (=O) $\xrightarrow[(2)\ -H_2O]{(1)\ CH_3MgI}$ 1-methylcyclohexene (CH_3)

(*i*) $C_6H_5COCH_3 \xrightarrow{\text{Mannich}} C_6H_5COCH_2CH_2N$(pyrrolidine ring) then addition of CH_3MgI.

(*j*) cyclohexanol (OH) $\xrightarrow{CrO_3}$ cyclohexanone (O) $\xrightarrow[NaH]{CO(OEt)_2}$

2-carbethoxycyclohexanone (O, COOEt) $\xrightarrow[(2)\ \text{saponify};\ -CO_2]{(1)\ CH_2{=}CHCOOEt}$ cyclohexanone with CH_2CH_2COOH (O, CH_2CH_2COOH)

(*k*) $HOOCCH_2CH_2COOH \longrightarrow \text{ester} \xrightarrow{CH_3MgI} (CH_3)_2\overset{OH}{\overset{|}{C}}CH_2CH_2\overset{OH}{\overset{|}{C}}(CH_3)_2 \xrightarrow{H^+}$

2,2,5,5-tetramethyltetrahydrofuran (H_3C, H_3C, O, CH_3, CH_3)

(*l*) Pimelic acid $\longrightarrow$ ester $\xrightarrow{NaOEt}$ 2-carbethoxycyclohexanone (O, COOEt) $\xrightarrow{CH_2{=}CHCOCH_3}$

2-carbethoxy-2-(3-oxobutyl)cyclohexanone (O, $CH_2CH_2COCH_3$, COOEt)

(*m*) $2\,CH_2{=}CHCOOEt - CH_3NH_2 \longrightarrow EtOOCCH_2CH_2N(CH_3)CH_2CH_2COOEt \xrightarrow{NaOEt}$

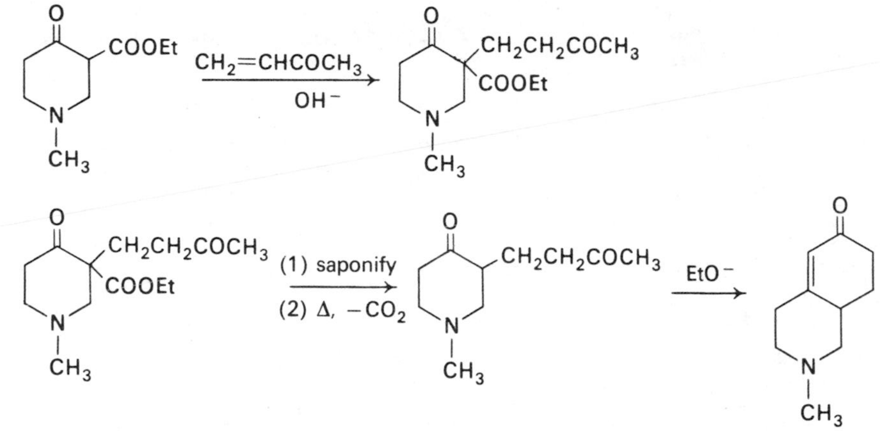

O, COOEt, N, CH_3 $\xrightarrow[OH^-]{CH_2{=}CHCOCH_3}$ O, $CH_2CH_2COCH_3$, COOEt, N, CH_3

O, $CH_2CH_2COCH_3$, COOEt, N, CH_3 $\xrightarrow[(2)\ \Delta,\ -CO_2]{(1)\ \text{saponify}}$ O, $CH_2CH_2COCH_3$, N, CH_3 $\xrightarrow{EtO^-}$ O, N, CH_3

(*n*) (cyclopentanone)=O + $BrCH_2COOEt \xrightarrow{Zn}$ CH_2COOEt, OH $\xrightarrow{LiAlH_4}$ CH_2CH_2OH, OH

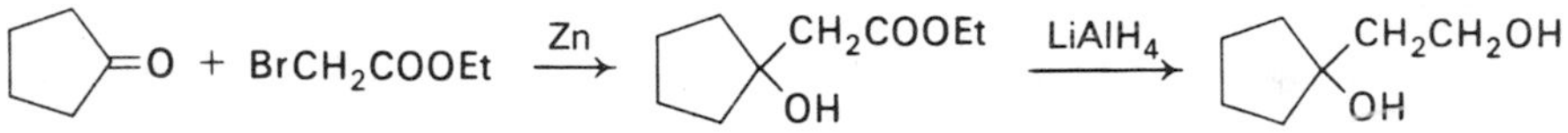

(*o*) $C_6H_5COCH_3 \xrightarrow[NaOEt]{ClCH_2COOEt}$ $C_6H_5C(CH_3)(-O-)CHCOOEt \xrightarrow[(2)\ H^+,\ \Delta]{(1)\ \text{saponify}}$

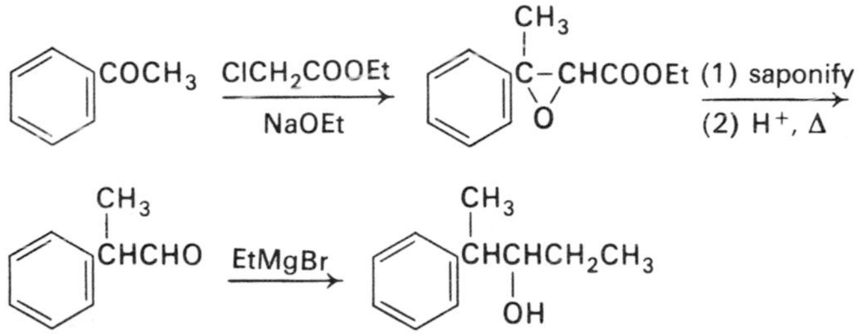

$C_6H_5CH(CH_3)CHO \xrightarrow{EtMgBr} C_6H_5CH(CH_3)CH(OH)CH_2CH_3$

(*p*) $C_6H_5COCH_3 \xrightarrow[Me_2NH]{HCHO} C_6H_5COCH_2CH_2NMe_2 \xrightarrow[(2)\ -H_2O]{(1)\ C_6H_5MgBr}$

$(C_6H_5)_2C{=}CHCH_2NMe_2 \xrightarrow[Pt]{H_2} (C_6H_5)_2CHCH_2CH_2NMe_2$

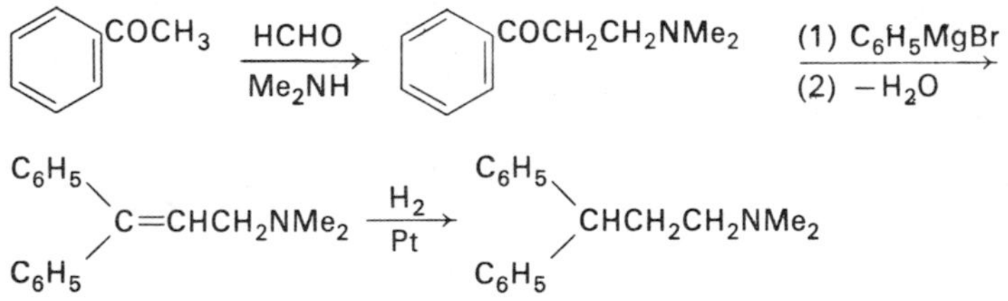

PART 6

Exercise 6.1

(*a*) Alkylate $CH_2(COOEt)_2$ with $(CH_3)_2CHCH_2Br$; saponify, decarboxylate

(*b*) Alkylate $CH_2(COOEt)_2$ with CH_3COCH_2Br; saponify, decarboxylate to give

$CH_3COCH_2CH_2COOH$; then $\xrightarrow{LiAlH_4} CH_3CH(OH)CH_2CH_2CH_2OH$

(*c*) Alkylate $CH_2(COOEt)_2$ with $BrCH_2CH_2CH_2CH_2Br \longrightarrow$

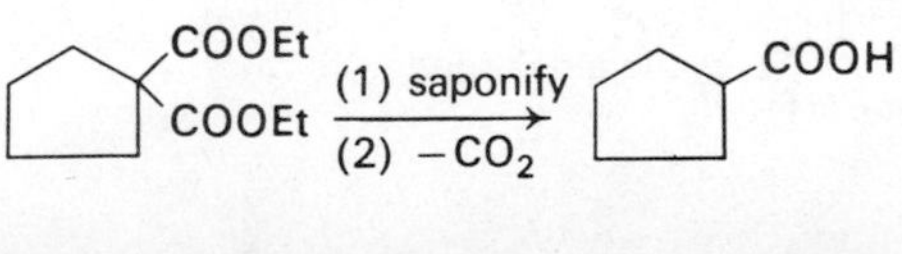

(cyclopentane)(COOEt)(COOEt) $\xrightarrow[(2)\ -CO_2]{(1)\ \text{saponify}}$ (cyclopentane)COOH (*d*) Alkylate $CH_2(COOEt)_2$ with $C_6H_5CH_2Cl$, then with $CH_3I \longrightarrow C_6H_5CH_2C(CH_3)(COOEt)_2$; then saponify, decarboxylate, and esterify $\longrightarrow C_6H_5CH_2CH(CH_3)COOEt \xrightarrow{NH_3} C_6H_5CH_2CH(CH_3)CONH_2 \xrightarrow{LiAlH_4} C_6H_5CH_2CH(CH_3)CH_2NH_2$

(*e*) Alkylate $CH_2(COOEt)_2$ with $C_6H_5CH_2Cl$, to give finally $C_6H_5CH_2CH_2COOH$; reduce ($LiAlH_4$) to $C_6H_5CH_2CH_2CH_2OH$; convert to *p*-toluenesulfonate, alkylate malonic ester to give $C_6H_5CH_2CH_2CH_2CH(COOEt)_2$

Alkylate $CH_2(COOEt)_2$ with $C_6H_5CH_2Cl$, then with CH_3I to give finally $C_6H_5CH_2CH(CH_3)COOH$. Convert to acid chloride, cyclize ($AlCl_3$) to 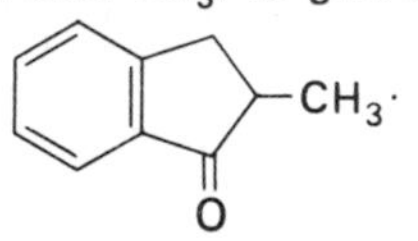 Then add CH_3MgI, and dehydrate.

Exercise 6.2

(*b*) $CH_2{=}CHCH_2Br + CH_3COCH_2COOEt \xrightarrow{NaOEt} CH_2{=}CHCH_2CH(COOEt)COCH_3 \xrightarrow[(2)\ \Delta,\ -CO_2]{(1)\ \text{saponify}} CH_2{=}CHCH_2CH_2COCH_3$

(*c*) $C_6H_5CH_2Cl + CH_3COCH_2COOEt \xrightarrow[(a)]{\text{as in}} C_6H_5CH_2CH_2COCH_3 \xrightarrow{CH_3MgI}$

$C_6H_5CH_2CH_2C(CH_3)(OH){-}CH_3 \xrightarrow[(-H_2O)]{H^+,\ \Delta} C_6H_5CH_2CH{=}C(CH_3)CH_3$

(*d*)

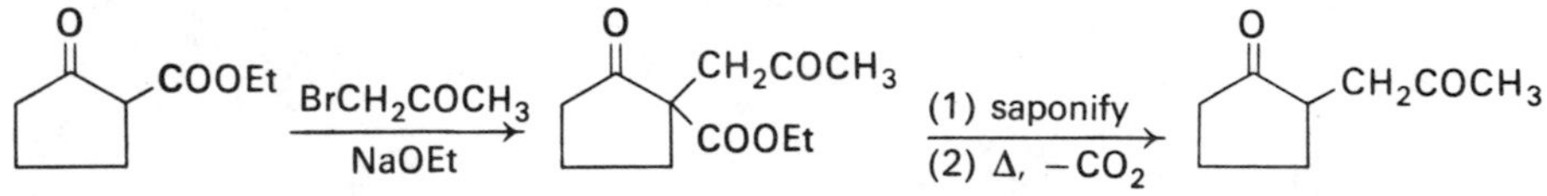

from ethyl adipate

(*e*)

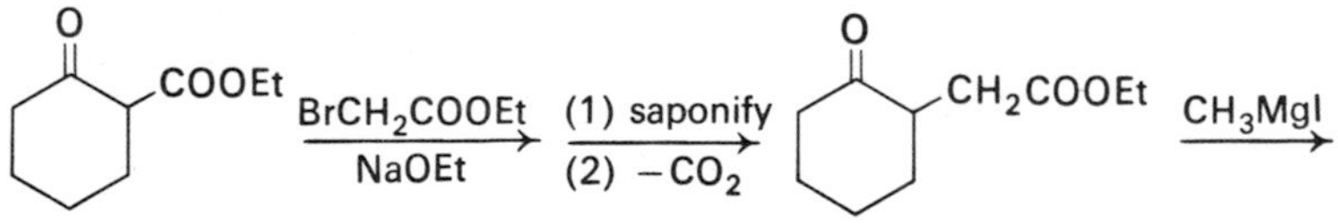

from ethyl pimelate

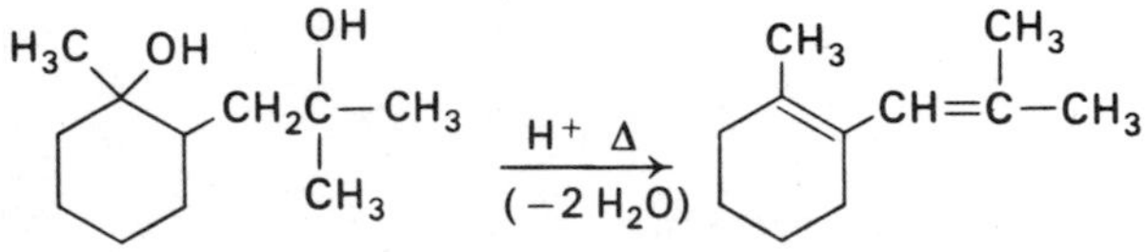

(*f*)

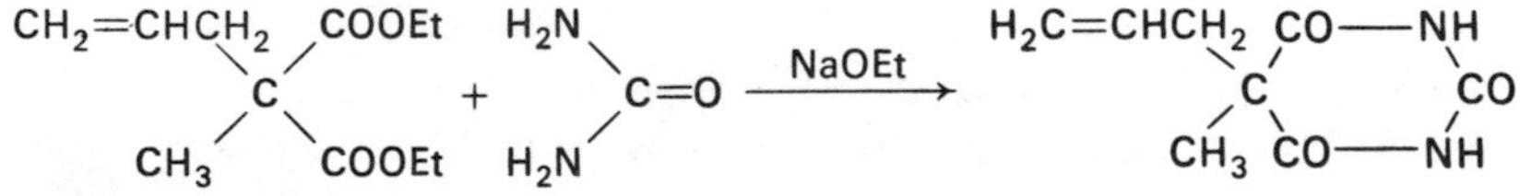

by alkylation of malonic ester

(*g*) alkylate acetoacetic ester with $BrCH_2COCH_3$ to give finally

$CH_3COCH_2CH_2COCH_3 \xrightarrow{OH^-}$ 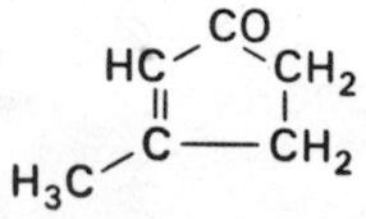 cyclic aldol condensation (*h*) $CH_2{=}CHCOOMe + CH_2(COOEt)_2 \xrightarrow{OH^-} MeOOCCH_2CH_2C(COOEt)_2CH_2CH_2COOMe \xrightarrow[(2)\ \Delta,\ -CO_2]{(1)\ \text{saponify}}$

$HOOCCH_2CH_2CH(COOH)CH_2CH_2COOH$

(*i*) alkylate ethyl cyclopentanone-2-carboxylate with $BrCH_2CH_2CH_2Br$

(*j*) $(CH_3)_2C{=}CHCOOEt + CH_2(COOEt)_2 \xrightarrow{NaOEt} \xrightarrow[(2)\ \text{decarboxylate}]{(1)\ \text{saponify}}$

$(CH_3)_2C(CH_2COOH)CH_2COOH \xrightarrow[HCl]{EtOH} (CH_3)_2C(CH_2COOEt)CH_2COOEt \xrightarrow[NaOEt]{(COOEt)_2}$

(cyclic diketo diester: $(CH_3)_2C$ ring with two HC(COOEt) and two CO) $\xrightarrow[NaOEt]{CH_3I}$ (ring with two C(H_3C)(COOEt) and two CO, $(CH_3)_2C$) $\xrightarrow[(2)\ \Delta,\ -CO_2]{(1)\ \text{saponify}}$ (ring with two HC(CH_3) and two CO, $(CH_3)_2C$)

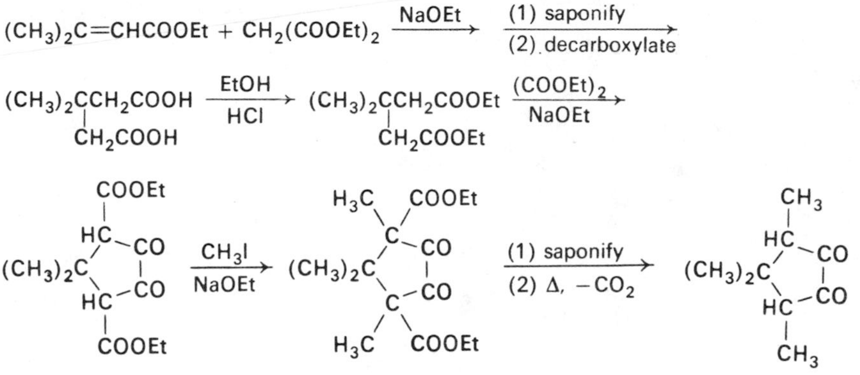

(*k*) see text

(*l*) alkylate acetoacetic ester with $(CH_3)_2C{=}CHCH_2Br$. Prepare $(CH_3)_2C{=}CHCH_2OH$ by $LiAlH_4$ reduction of $(CH_3)_2C{=}CHCOOEt$. Prepare $(CH_3)_2C{=}CHCOOH$ by haloform reaction from $(CH_3)_2C{=}CHCOCH_3$.

Exercise 6.3

(*a*) $CH_3COCH_3 + HNR_2 \longrightarrow CH_3{-}C({=}CH_2){-}NR_2 \xrightarrow{CH_2=CHCH_2Br}$

$CH_3{-}C(CH_2CH_2CH{=}CH_2){=}NR_2{}^+Br^- \xrightarrow{H^+/H_2O} CH_3COCH_2CH_2CH{=}CH_2$

(*Note:* HNR_2 can be pyrrolidine).

(*b*) $CH_3CH_2COCH_2CH_3 + HNR_2 \longrightarrow CH_3CH_2C(NR_2){=}CHCH_3 \xrightarrow{CH_3COCH_2Br}$

$CH_3CH_2C({=}\overset{+}{N}R_2Br^-){-}CH(CH_3)CH_2COCH_3 \xrightarrow[H_2O]{H^+} CH_3CH_2COCH(CH_3)CH_2COCH_3$

(*c*) 2-methylcyclohexanone (H_3C, O) $\xrightarrow{HNR_2}$ enamine (H_3C, NR_2) $\xrightarrow{CH_3OCH_2Cl}$ iminium salt (H_3C, $NR_2^+\ Cl^-$, CH_2OCH_3) $\xrightarrow{H^+/H_2O}$

2-methyl-6-(methoxymethyl)cyclohexanone (H_3C, O, CH_2OCH_3)

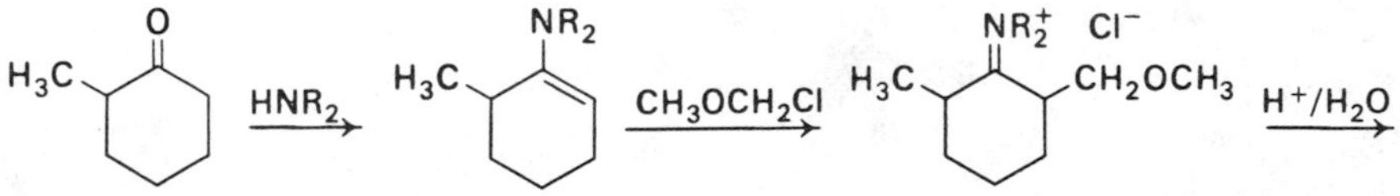

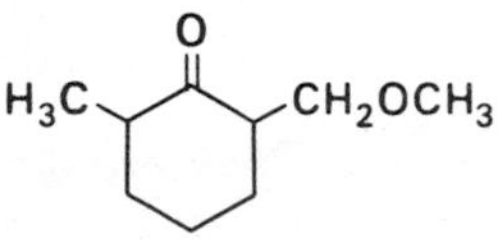

(*d*) Cyclopentanone $\xrightarrow{HNR_2}$ enamine (NR_2) $\xrightarrow[(2)\ H^+/H_2O]{(1)\ C_6H_5CH_2Cl}$ 2-($CH_2C_6H_5$)cyclopentanone $\xrightarrow[(2)\ -H_2O]{(1)\ CH_3MgI}$ 1-CH_3-2-$CH_2C_6H_5$-cyclopentene

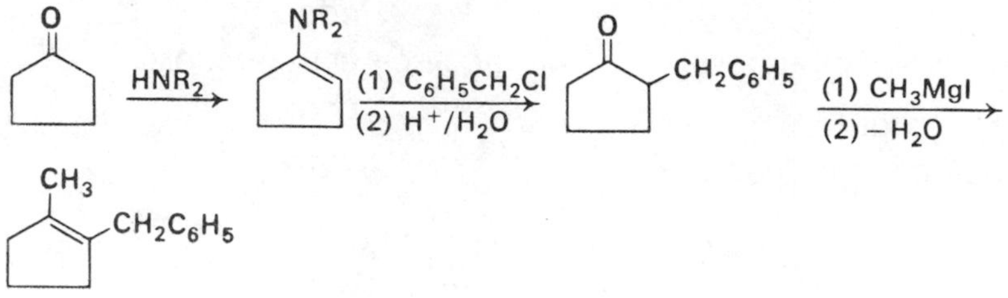

(*e*) Product of (*d*) $\xrightarrow{O_3}$ $CH_3CO(CH_2)_3COCH_2C_6H_5$ $\xrightarrow{OH^-}$ 3-CH_3-2-C_6H_5-cyclohex-2-enone

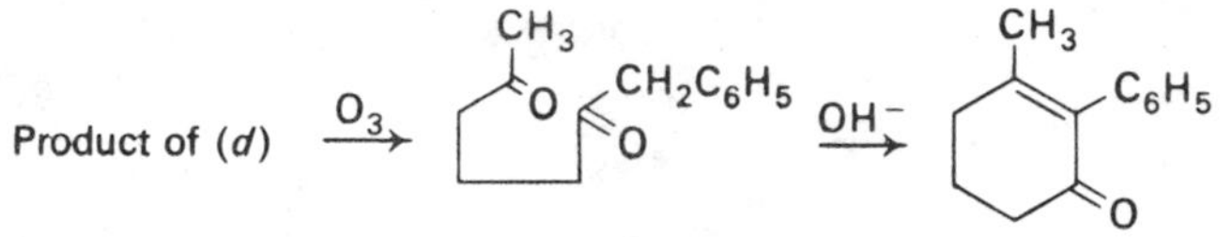

(*f*) $(CH_3)_2CHCOCH_3 \xrightarrow{HNR_2} (CH_3)_2CHC(NR_2){=}CH_2 \xrightarrow[(2)\ H^+/H_2O]{(1)\ CH_2{=}CHCH_2Br}$
$(CH_3)_2CHCOCH_2CH_2CH{=}CH_2$

(*g*) Cyclohexanone $\xrightarrow[+C_6H_5CH_2Cl]{\textit{via}\ \text{enamine}}$ 2-($CH_2C_6H_5$)cyclohexanone $\xrightarrow[(2)\ -H_2O]{(1)\ LiAlH_4}$ 1-($CH_2C_6H_5$)cyclohexene

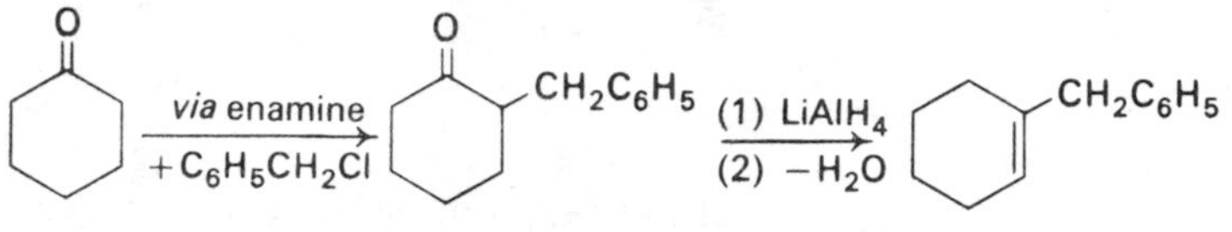

(*Note:* It would be more direct to prepare this olefin by addition of $C_6H_5CH_2MgCl$ to cyclohexane, followed by dehydration.)

Exercise 6.5

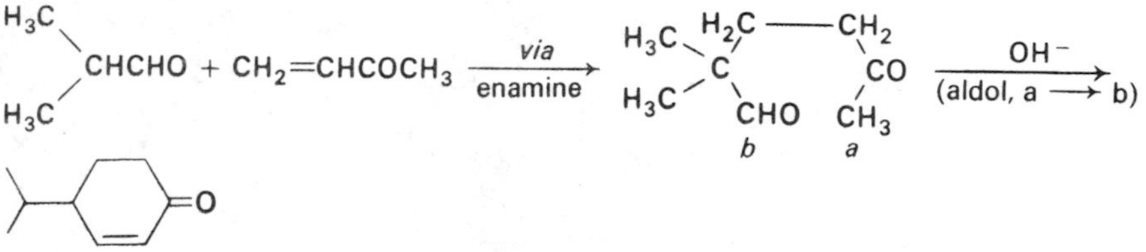

$(H_3C)_2CHCHO + CH_2{=}CHCOCH_3 \xrightarrow[\text{enamine}]{\textit{via}} (H_3C)_2C(CHO)CH_2CH_2COCH_3$ (CHO = *b*, CH_3 = *a*) $\xrightarrow[(\text{aldol, a} \rightarrow \text{b})]{OH^-}$ 4-isopropyl... 4,4-dimethylcyclohex-2-enone (structure shown: $(CH_3)_2C$ ring with C=C and C=O)

(*Note:* The ketoaldehyde could as well be prepared by a direct Michael addition as by the enamine route.)

Exercise 6.6

In all cases, start with preparation of $CH_2(COOEt)(COOR)$ or $CH_2(COOR)_2$ where R = *t*-butyl, benzyl, or tetrahydropyranyl.

(*a*) $C_6H_5COCl + M(HC(COOEt)_2) \rightarrow C_6H_5COCH(COOR)(COOEt) \xrightarrow{\text{remove R as described in text}}$
$C_6H_5COCH(COOEt)COOH \xrightarrow[-CO_2]{\Delta} C_6H_5COCH_2COOEt$

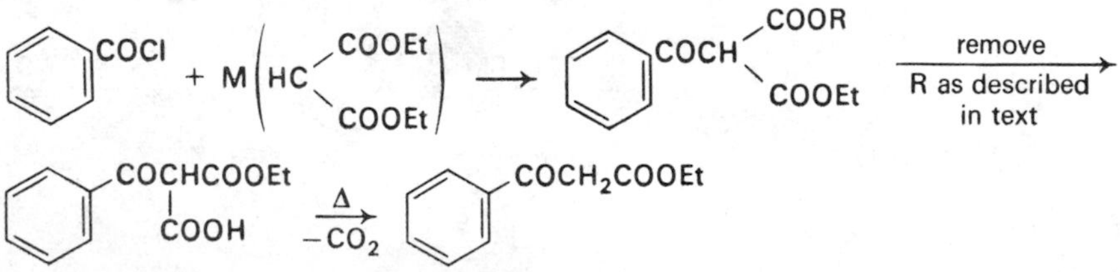

Note: $M(CH(COOR)_2) = CH_3\overset{+}{O}Mg\overset{-}{C}H(COOR)_2$

(*b*) $CH_2(COOH)(COOEt) \xrightarrow{\text{saponify}} CH_2(COOH)_2 \longrightarrow CH_2(COOR)_2$ (R as above)

$$CH_2(COOR)_2 \xrightarrow[\text{MeOH}]{\text{Mg}} MeOMgCH(COOR)_2 \xrightarrow{CH_3CH_2CH_2COCl}$$

$$CH_3CH_2CH_2COCH(COOR)_2 \longrightarrow CH_3CH_2CH_2COCH(COOH)_2 \xrightarrow[-2\,CO_2]{\Delta}$$

$$CH_3CH_2CH_2COCH_3$$

(*c*) *Via* $CH_2{=}CHCH_2COCl + MeOMgCH(COOR)(COOEt)$ (R as above)

(*d*) Starting with $CH_2(COOR)_2$ (R as above),
(1) Acylate with $(CH_3)_2CHCOCl$;
(2) Alkylate with $CH_2{=}CHCH_2Br$,

to give $(H_3C)_2CHCOC(COOR)_2{-}CH_2CH{=}CH_2$, then remove R, decarboxylate.

Exercise 6.7

(*a*) *Via* enamine of cyclopentanone + CH_3CH_2COCl
(*b*) *Via* enamine of cyclohexanone + $(CH_3)_2CHCOCl$
(*c*) *Via* 2-($COCH_2CH_2CH_3$)cyclohexanone
(*d*) *Via* enamine of cyclohexanone + $ClCOCH_2CH_2CH_2COCl$ ⟶

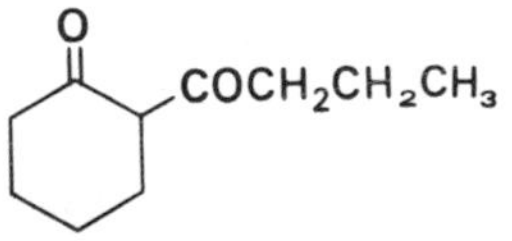

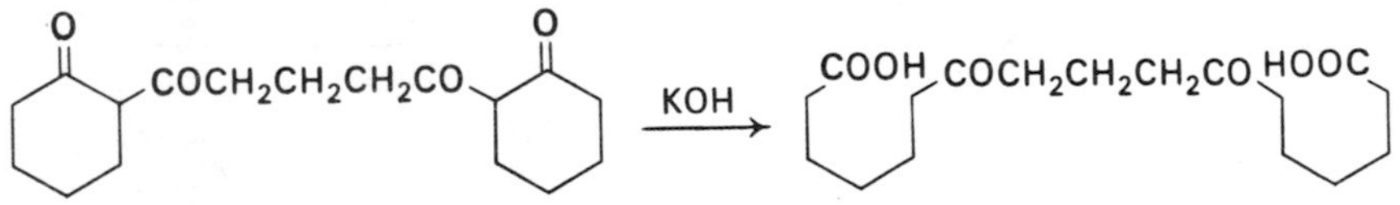

2-oxocyclohexyl–$COCH_2CH_2CH_2CO$–2-oxocyclohexyl $\xrightarrow{KOH}$ $HOOC(CH_2)_5COCH_2CH_2CH_2CO(CH_2)_5COOH$ (COOH … $COCH_2CH_2CH_2CO$ … HOOC, ring-opened)

(*e*) *Via* enamine of $CH_3CH_2CH_2COCH_2CH_2CH_3$ + ClCOOEt.
(*f*) *Via* enamine of diethyl ketone + benzoyl chloride.
(*g*) *Via* enamine of diethyl ketone + phenylacetyl chloride.
(*h*) *Via* enamine of 1-tetralone and propionyl chloride, followed by alkaline cleavage.
(*i*) *Via* enamine of 2-tetralone + acetyl chloride.

PART 7

Exercise 7.1

A nitrogen ylide, $R\bar{C}H{-}\overset{+}{N}R_3$, cannot be represented by the formula $RCH{=}NR_3$, for nitrogen lacks the fifth (stable) orbital required to accommodate ten valence electrons.

Exercise 7.2

(*a*) 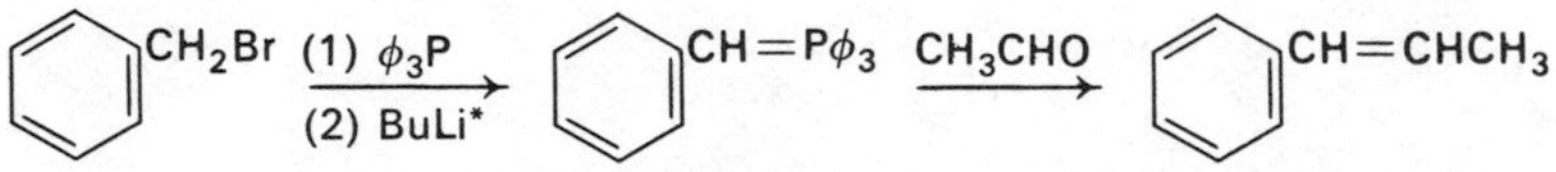
CH₂Br
(1) φ₃P
(2) BuLi*
CH=Pφ₃
CH₃CHO
CH=CHCH₃

* *Note:* The base employed to remove the proton from $RCH_2—\overset{+}{P}\phi_3$ varies with the character of the $—CH_2—$ group. Alkyllithium reagents are often used, but in some cases sodium ethoxide is sufficient.

(*a, alternative*) Wittig reagent from CH_3CH_2Br, with benzaldehyde.

(*b*) Reaction of $\phi_3P=CHCH_2CH_2CH=P\phi_3$ (from 1,4-dibromobutane) with cyclohexanone.

(*c*) 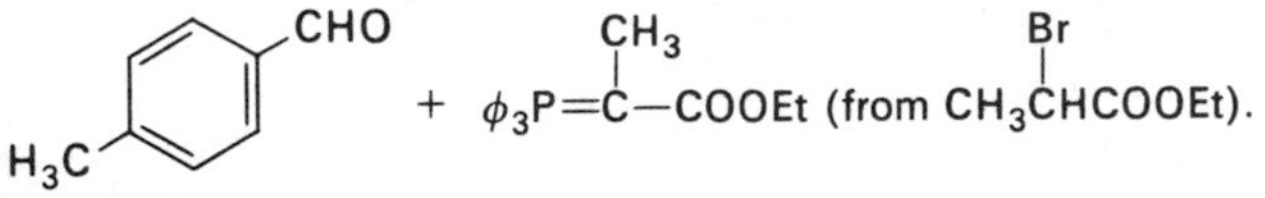
CHO
H₃C
+ φ₃P=C—COOEt (from CH₃CHCOOEt).
CH₃
Br

(*d*) 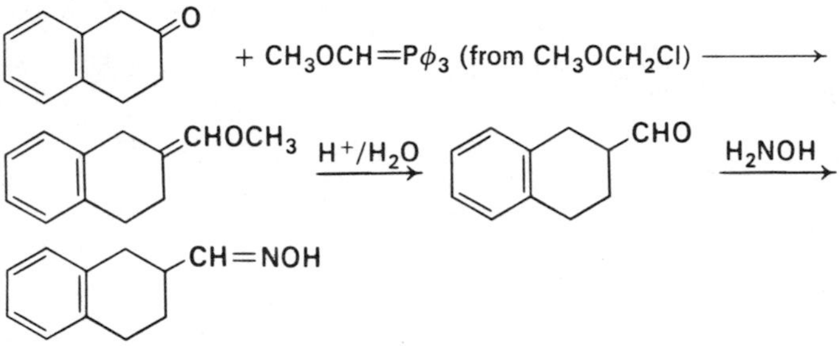
O
+ CH₃OCH=Pφ₃ (from CH₃OCH₂Cl)
CHOCH₃
H⁺/H₂O
CHO
H₂NOH
CH=NOH

(*e*) 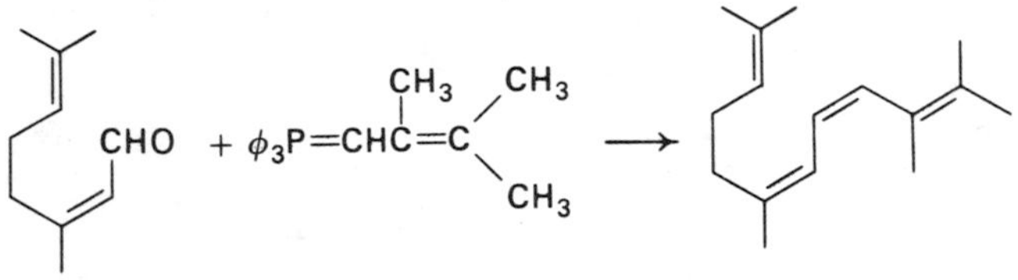
CHO
+ φ₃P=CHC=C
CH₃
CH₃
CH₃

geranial

(*f*) 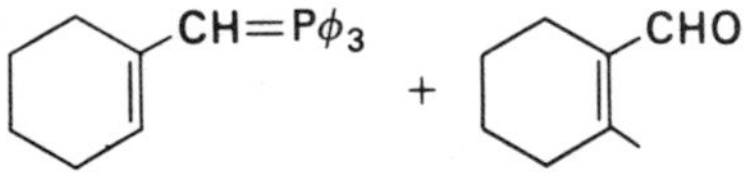
CH=Pφ₃
+
CHO

Exercise 7.3

(*a*) 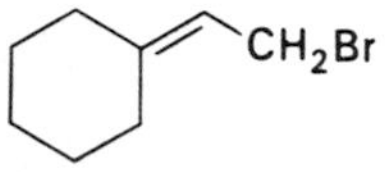
CH₂Br
is required. This could be prepared *via*

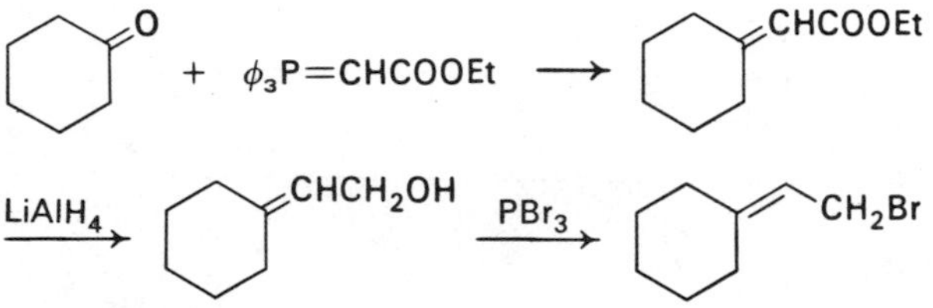
O
+ φ₃P=CHCOOEt
CHCOOEt
LiAlH₄
CHCH₂OH
PBr₃
CH₂Br

followed by the Wittig synthesis:

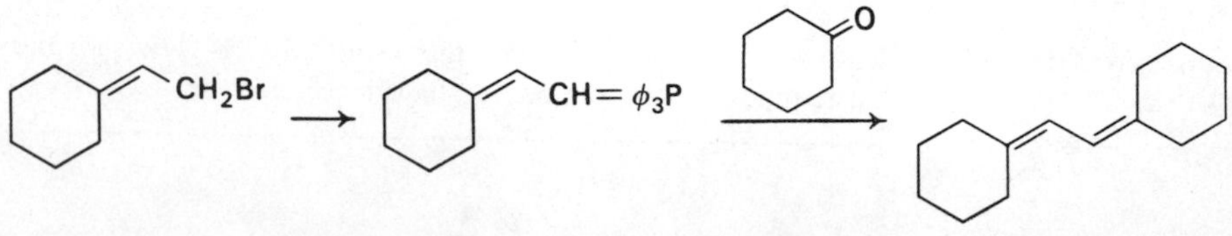
CH₂Br
CH=φ₃P
O

PART 8

Exercise 8.1

(*a*) $CH_3CH_2CH_2OCOCH_3 \xrightarrow[\text{HBr}]{\text{conc.}} CH_3CH_2CH_2Br$

(*b*) $(CH_3)_3COH \xrightarrow[\text{HCl}]{\text{conc.}} (CH_3)_3CCl$

(*c*) $C_6H_5CH_2OCH_3 \xrightarrow[\text{HBr}]{\text{conc.}} C_6H_5CH_2Br$

(*d*) $CH_2{=}CHCH_2OH \xrightarrow[\text{PBr}_3]{\text{HBr or}} CH_2{=}CHCH_2Br$

(*e*) $C_6H_5CHO + C_6H_5CH_2MgCl \longrightarrow C_6H_5CH(OH)CH_2C_6H_5 \xrightarrow[\text{HBr}]{\text{conc.}} C_6H_5CH(Br)CH_2C_6H_5$

(*f*) $(CH_3)_3COH \longrightarrow (CH_3)_3CCl \longrightarrow (CH_3)_3CMgCl \xrightarrow{CO_2} (CH_3)_3CCOOH$

(*g*) $CH_3CH_2CH_2OH \xrightarrow{\text{HBr}} CH_3CH_2CH_2Br \longrightarrow CH_3CH_2CH_2MgBr \xrightarrow{CO_2}$

$CH_3CH_2CH_2COOH \xrightarrow{\text{LiAlH}_4} CH_3CH_2CH_2CH_2OH$

(*h*) As in (*g*), but $CH_2{=}CHCH_2MgBr$ replaces $CH_3CH_2CH_2MgBr$

Exercise 8.2

(*a*) $(CH_3)_2CHO^- + CH_3CH_2Br$

(*b*) 1-methylcyclohexyl-O^- ($C_6H_{10}(CH_3)O^-$) $+ CH_3I$

(*c*) $CH_3O^- + CH_2{=}CHCH_2Br$

(*d*) $C_6H_5CH_2O^- + C_6H_5CH_2Cl$

(*e*) $CH_3CH(OH)CH(Cl)CH_3 + NaOH$
(from *cis*-2-butene + HOCl)

(*f*) $C_6H_5O^- + CH_2{=}CHCH_2Br$

(*g*) 2-naphthyl-O^- ($C_{10}H_7O^-$) $+ (CH_3)_2SO_4$

(*h*) $C_6H_5O^- + ClCH_2COOH$

(*i*) $(CH_3)_3C{-}O^- + C_6H_5CH_2Cl$

(*j*) $2\,C_6H_5O^- + BrCH_2CH_2Br$

(*k*) furanoside ring: HOH_2C, O, OCH_3; H, H, H, H; OH, OH $+ CH_3I + Ag_2O$

(*l*) $CH_3CH(Br)COOCH_3 + CH_3O^-$

Exercise 8.3

(*a*) $\left[CH_3{-}N(C_2H_5)_2{-}CH_2C_6H_5\right]^+ Br^-$

(*b*) $C_5H_5\overset{+}{N}{-}CH_2COOEt\ Br^-$

(*c*) $CH_3OC_6H_4\overset{+}{N}(CH_3)_3\quad BrC_6H_4SO_3^-$

(*d*) $(CH_3)_2NCH_2CH{=}CH_2$

(*e*) quinolinium, $N^+{-}CH_3$, $CH_3OSO_3^-$

(*f*) $CH_3C_6H_4N(C_2H_5){-}CH_3$

(*g*) [structure: N,N-dimethyl quaternary ammonium salt of aporphine-type ring system, $\overset{+}{N}(CH_3)_2$] I^-

(*h*) $CH_3O-C_6H_4-\overset{+}{N}(CH_3)_3$ I^-

(*i*) [structure: isoindoline ring with $\overset{+}{N}(CH_3)(CH_2CH_3)$] I^-

(*j*) [structure: indol-3-yl]$-CH_2-\overset{+}{N}(CH_3)_2-CH_2-$[indol-3-yl] I^-

(*k*) The product is formed by the attack of $-CH_2NMe_2$ upon

$$-CH_2\overset{+}{N}Me_3 \longrightarrow -CH_2-\overset{+}{N}Me_2(-CH_2) + NMe_3.$$

Exercise 8.4

(*a*) $(CH_3)_2C{=}CHCH_2OH \xrightarrow{HBr} (CH_3)_2C{=}CHCH_2Br \xrightarrow{KCN} (CH_3)_2C{=}CHCH_2CN \xrightarrow{LiAlH_4}$
$(CH_3)_2C{=}CHCH_2CH_2NH_2$

(*b*) $C_6H_5CH_2OH \xrightarrow{HCl} C_6H_5CH_2Cl \xrightarrow{KCN} C_6H_5CH_2CN \xrightarrow[NaOMe]{CH_2{=}CHCOOCH_3}$

$$C_6H_5-C(CN)(CH_2CH_2COOCH_3)CH_2CH_2COOCH_3$$

(*c*) $CH_2{=}CHCH_2Cl \xrightarrow{KCN} CH_2{=}CHCH_2CN \xrightarrow{H^+/H_2O}$
$CH_2{=}CHCH_2COOH \xrightarrow[(2)\ CH_2{=}CHCH_2OH]{(1)\ SOCl_2} CH_2{=}CHCH_2COOCH_2CH{=}CH_2$

(*d*) $HOCH_2CH_2OH \xrightarrow{HBr} BrCH_2CH_2Br \xrightarrow{KCN} \underset{CH_2CN}{\overset{CH_2CN}{|}} \xrightarrow{LiAlH_4} \underset{CH_2CH_2NH_2}{\overset{CH_2CH_2NH_2}{|}}$

(*e*) [phthalan: $C_6H_4(CH_2)_2O$] $\xrightarrow{HBr}$ $C_6H_4(CH_2Br)_2$ $\xrightarrow{KCN}$ $C_6H_4(CH_2CN)_2$ $\xrightarrow[H^+]{EtOH}$

$C_6H_4(CH_2COOEt)_2 \xrightarrow{NaOEt}$ [indanone ring: C_6H_4 fused with $CH_2-CO-CH(COOEt)$]

Exercise 8.5

(*a*) $CH_3COO^- + C_6H_5CH_2Br$

(*b*) $C_6H_5CH_2Cl + (NH_2)_2C{=}S \longrightarrow C_6H_5CH_2{-}\overset{+}{S}{=}C(NH_2)_2\ Br^- \xrightarrow{H_2O/OH^-}$

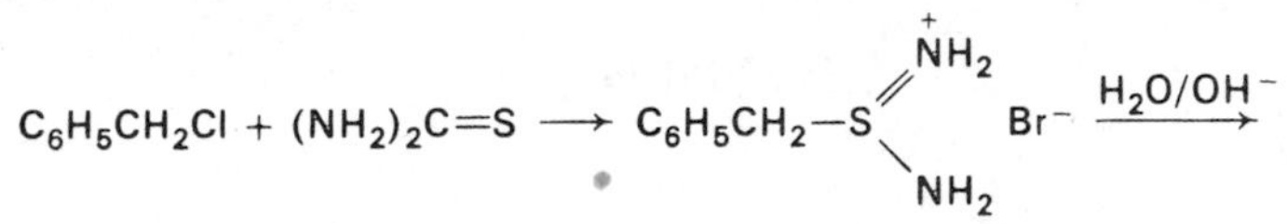

$C_6H_5CH_2SH \xrightarrow[NaOH]{CH_2{=}CHCH_2Cl} C_6H_5CH_2SCH_2CH{=}CH_2$

(*c*) $C_6H_5NHCH_3 + C_6H_5CH_2Br$

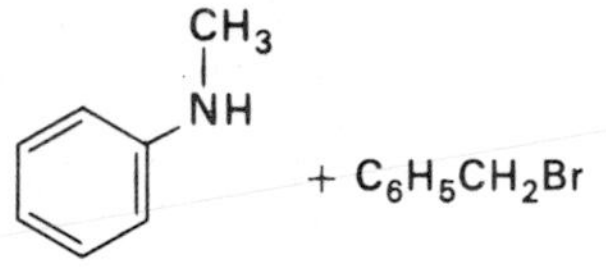

(*d*) $2\ CH_3COO^- + BrCH_2CH_2Br$

(*e*) $BrCH_2CH_2CH_2CH_2CH_2SH + NaOH$

(*f*) $C_6H_5{-}C_6H_4{-}COCH_2Br + (CH_3)_2CHCOO^-$

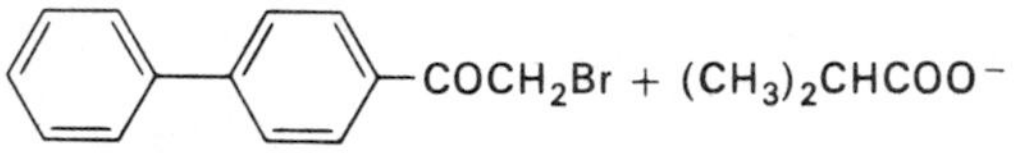

by bromination of the corresponding $-COCH_3$ compound

(*g*) $C_6H_5CH_2Cl + Na_2SO_3$

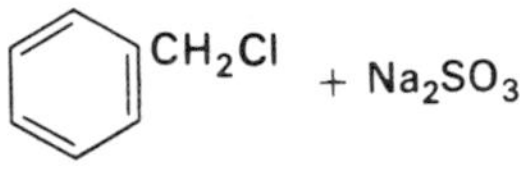

(*h*) $(H_3C)_2CHCH_2Br + (NH_2)_2C{=}S$

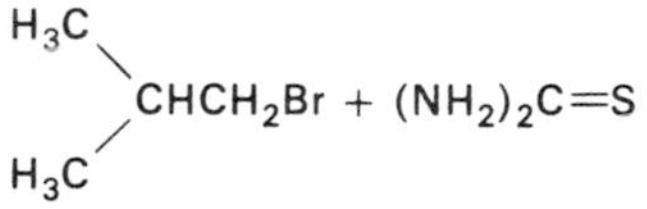

(*i*) $CH_3SCH_3 + CH_3I$

(*j*) pyridine $+ H_2O_2 \longrightarrow C_5H_5\overset{+}{N}{-}O^- \xrightarrow{CH_3I} C_5H_5\overset{+}{N}{-}OCH_3\ I^-$

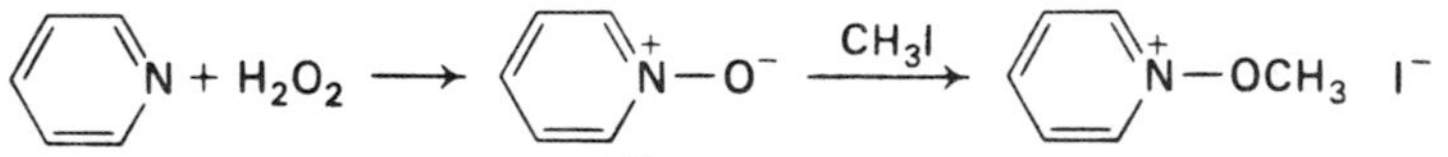

pyridine-N-oxide

Exercise 8.6

(*a*) $C_6H_4(CH_2OH)_2 + RSO_2Cl \longrightarrow C_6H_4(CH_2OSO_2R)(CH_2OH) \xrightarrow{OH^-} C_6H_4(CH_2OSO_2R)(CH_2O^-)$

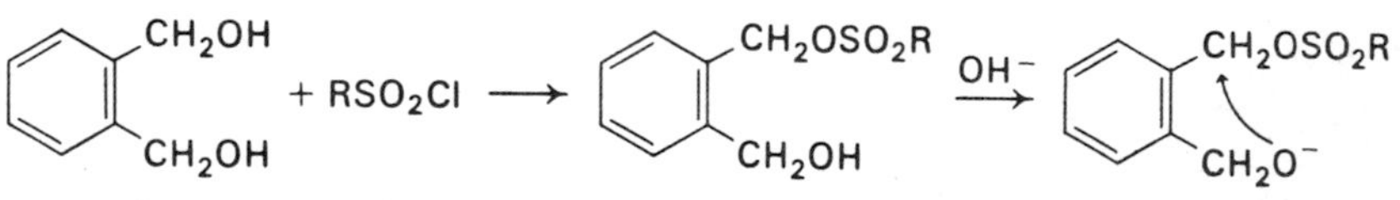

(*b*) cyclohexane with $O{-}C(=O){-}CH_3$ and OSO_2CH_3 $\longrightarrow$ cyclic $O{-}\overset{+}{C}{-}CH_3{-}O$ $\xrightarrow{EtOH}$ cyclic $O{-}C(OEt)(CH_3){-}O$

(*c*) 1-methyl-2-(chloromethyl)pyrrolidinium (H, CH_3, CH_2Cl) $\xrightarrow{OH^-}$ N-CH_3 pyrrolidine with CH_2Cl $\longrightarrow$ $+N{-}CH_2$ bicyclic ion, CH_3, Cl^- $\longrightarrow$ 3-chloro-1-methylpiperidine (Cl, N, CH_3)

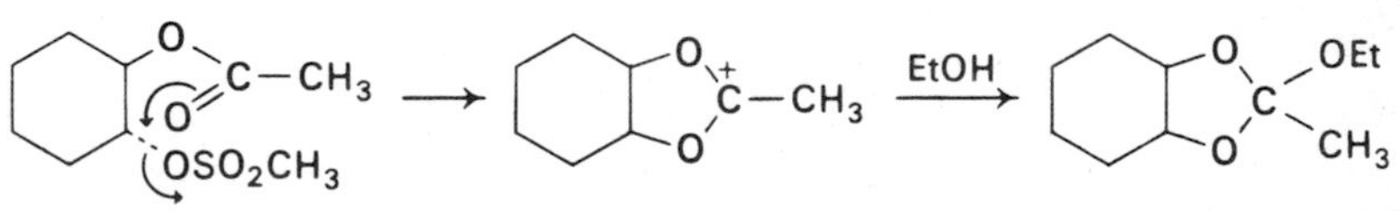

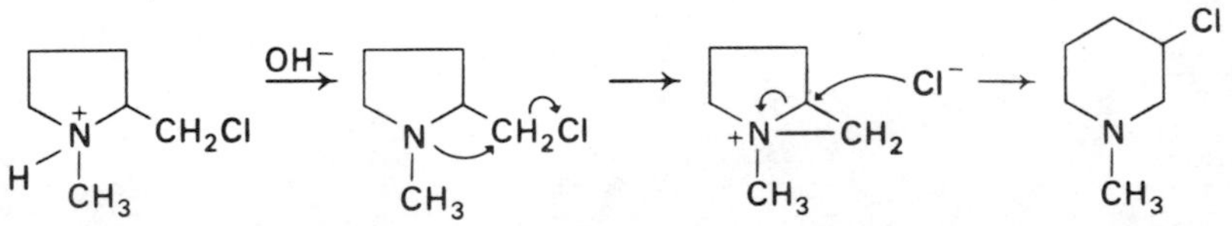

(*d*) $(CH_3)_2CHNO_2 \xrightarrow{OH^-} (CH_3)_2C{=}\overset{\overset{O}{\uparrow}}{N}{-}O^- \xrightarrow{C_6H_5CH_2Br}$

$C_6H_5CH_2{-}O{-}\overset{\overset{O}{\uparrow}}{N}{=}C(CH_3)_2 \xrightarrow[(-H^+)]{OH^-} C_6H_5CH{=}O + {}^-O{-}N{=}C(CH_3)_2$

(*e*) $C_6H_5CH_2OSO_2C_6H_5 + (CH_3)_2S{=}O \longrightarrow$

$C_6H_5CH_2{-}O{-}\overset{+}{S}(CH_3)_2 \; (+\ C_6H_5SO_2O^-) \xrightarrow[(-H^+)]{OH^-} C_6H_5CHO + S(CH_3)_2$

PART 9

Exercise 9.1

(a) Br
(i) CH_3CO N $COCH_3$
(q) O +

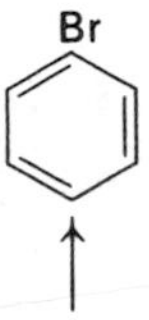

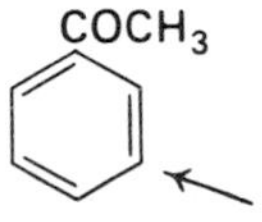

(b) $COCH_3$
(j) $N(COCH_3)_2$ OCH_3
(r) NO_2

(c) CN
(k) OH Br Br
(s) $NHCOCH_3$ $OCOCH_3$

(d) OCH_3 OCH_3
(l) NO_2 NO_2
(t) $\overset{+}{S}(CH_3)_2$

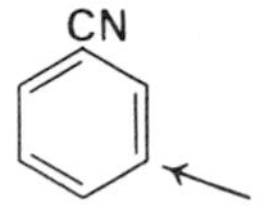

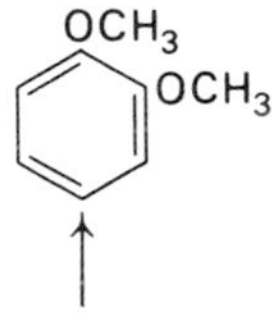

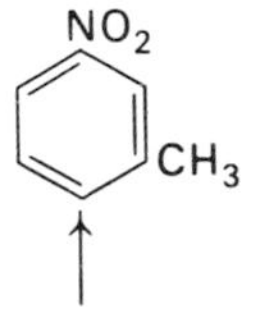

(e) NO_2 CH_3
(m) NO_2 CN
(u) CF_3

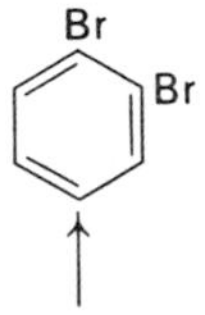

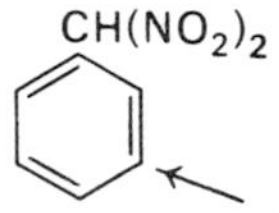

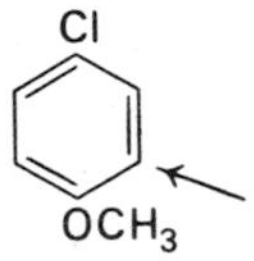

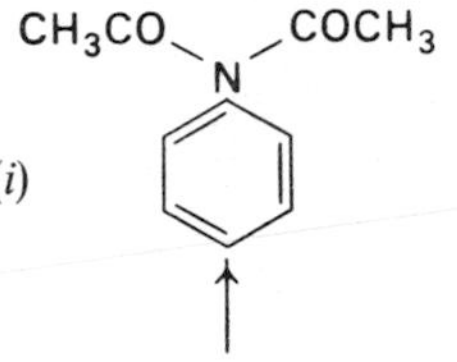

(f) Br Br
(n) $N(CH_3)_2$ OCH_3
(v) CF_3 CH_3

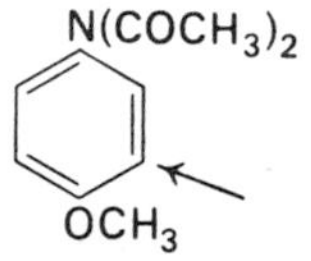

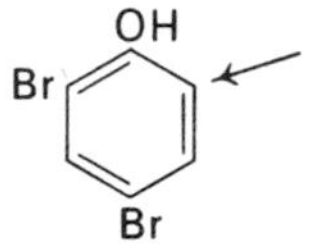

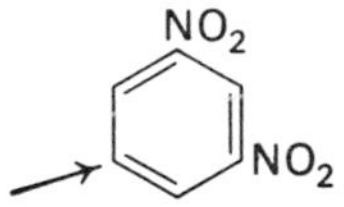

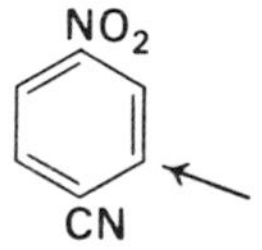

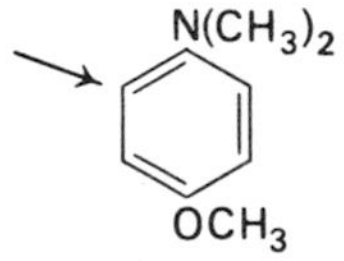

(g) $CH(NO_2)_2$
(o) $\overset{+}{N}(CH_3)_3$
(w) CH_2CH_2COOH

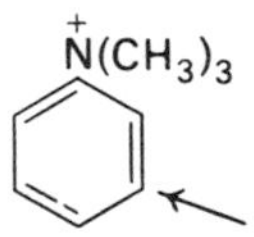

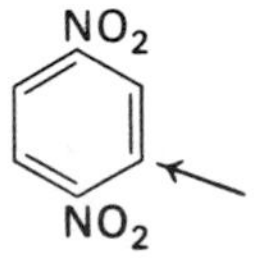

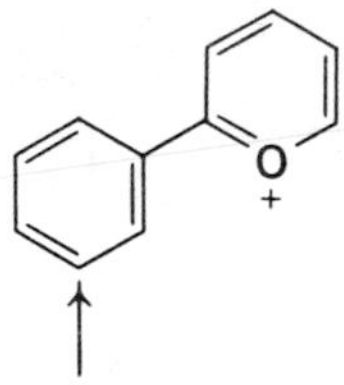

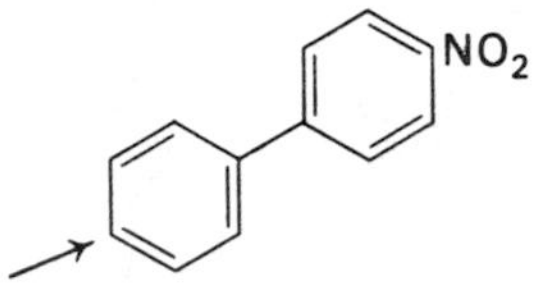

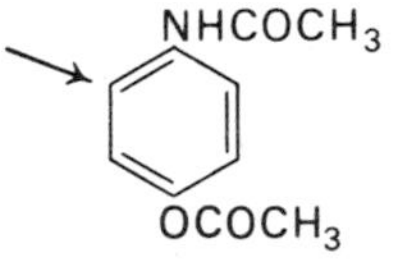

(h) Cl OCH_3
(p) NO_2 NO_2
(x) $COOCH_3$ Br

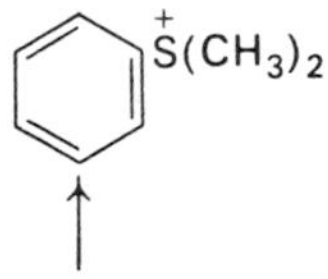

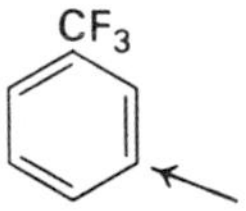

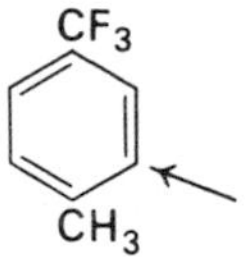

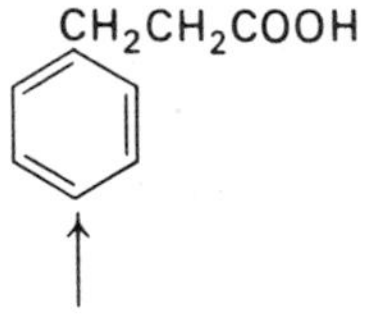

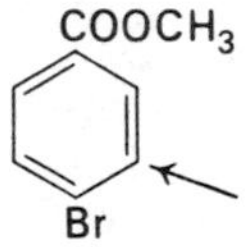

Exercise 9.2

(*a*) CH_3 $\xrightarrow[H_2SO_4]{HNO_3}$ CH_3 NO_2 + CH_3 NO_2

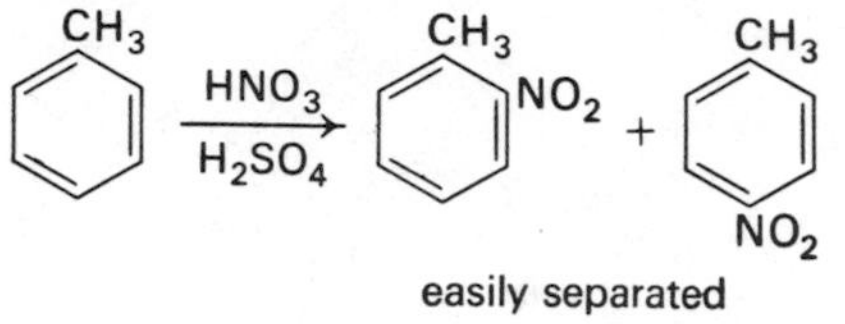

easily separated

(*b*) CH_3 $\xrightarrow[FeBr_3]{Br_2}$ CH_3 Br

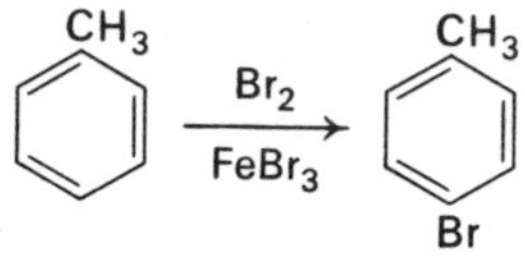

(*c*) CH_3 $\xrightarrow[AlCl_3]{CH_3COCl}$ CH_3 $COCH_3$ $\xrightarrow[H_2SO_4]{HNO_3}$ CH_3 NO_2 $COCH_3$

(*d*) Br $\xrightarrow[FeBr_3]{Br_2}$ Br Br

(*e*) $NHCOCH_3$ $\xrightarrow{Br_2}$ $NHCOCH_3$ Br

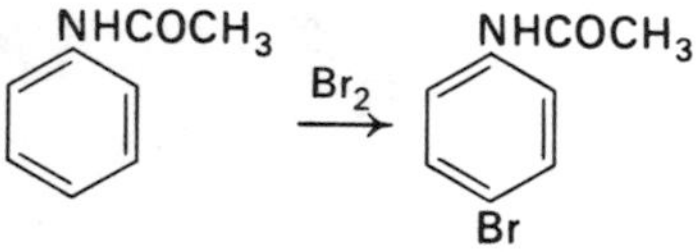

(*f*) CHO $\xrightarrow[H_2SO_4]{HNO_3}$ CHO NO_2 $\xrightarrow[\text{pyridine}]{CH_2(COOH)_2}$ CH=CHCOOH NO_2

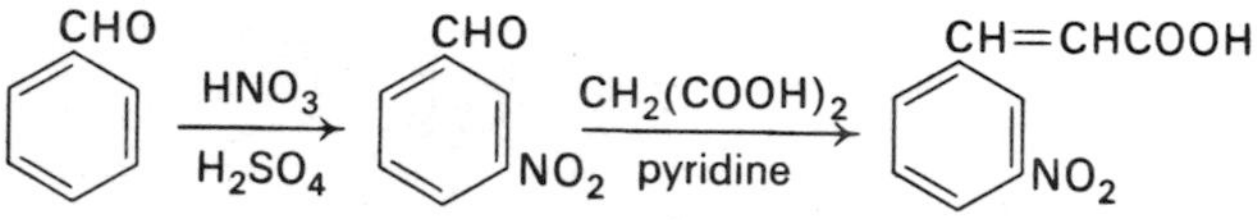

(*g*) CH_3 $\xrightarrow[H_2SO_4]{HNO_3}$ CH_3 NO_2 NO_2

(*h*) OCH_3 $\xrightarrow[H_2SO_4]{HNO_3}$ OCH_3 NO_2 $\xrightarrow[FeBr_3]{Br_2}$ OCH_3 Br NO_2

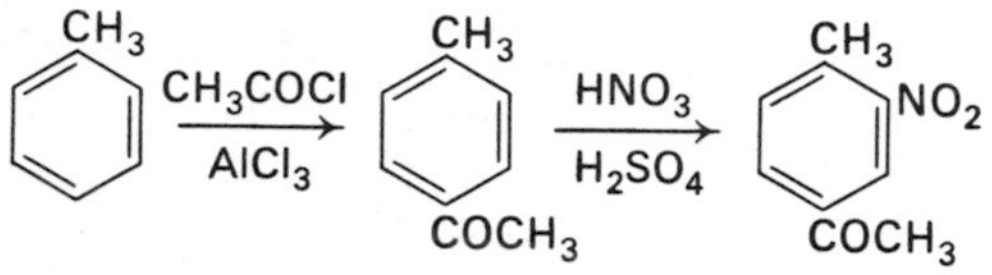

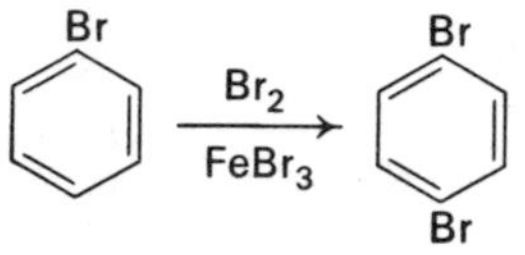

(*i*) CHO $\xrightarrow[\text{reaction}]{\text{Perkin}}$ CH=CHCOOH $\xrightarrow[(2)\ SOCl_2\ (3)\ AlCl_3]{(1)\ H_2/Pt}$ O

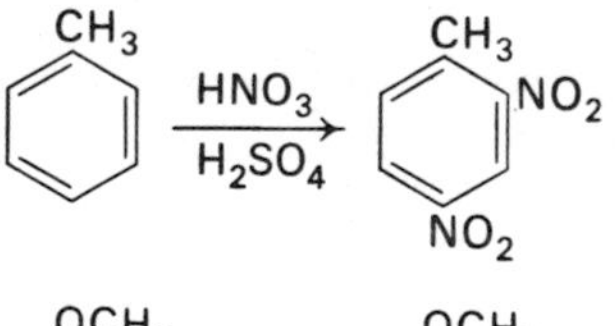

(*j*) CHO $\xrightarrow[(i)]{\text{as in}}$ O $\xrightarrow[NaOH]{\phi CHO}$ =CHϕ O

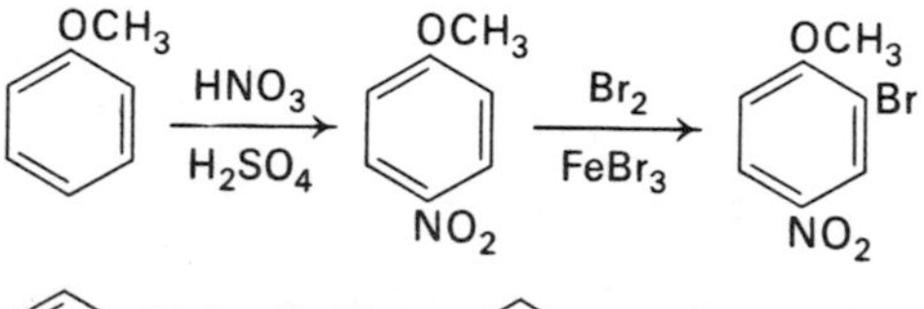

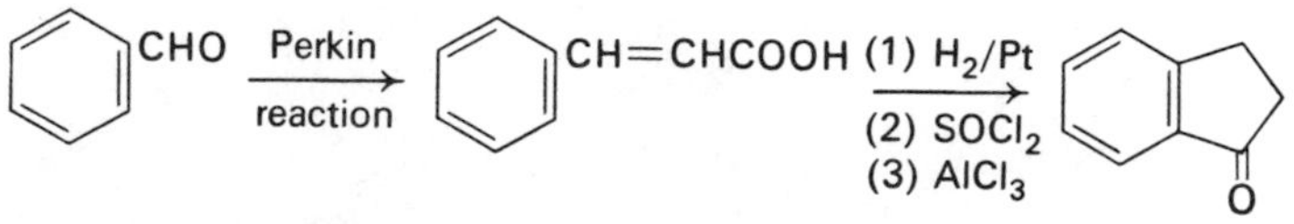

(*k*) OCH_3 $\xrightarrow[AlCl_3]{(CH_3)_2CHCOCl}$ OCH_3 $COCH(CH_3)_2$

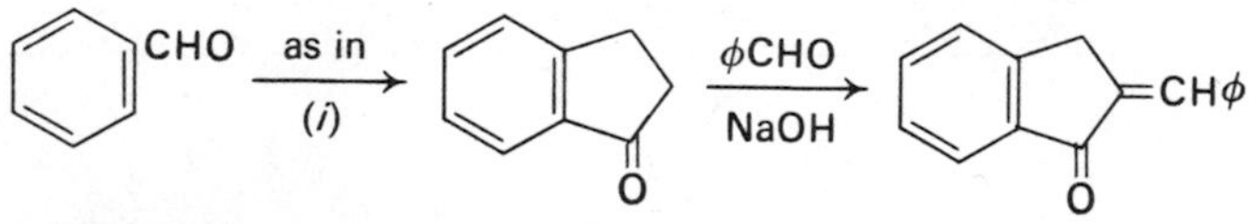

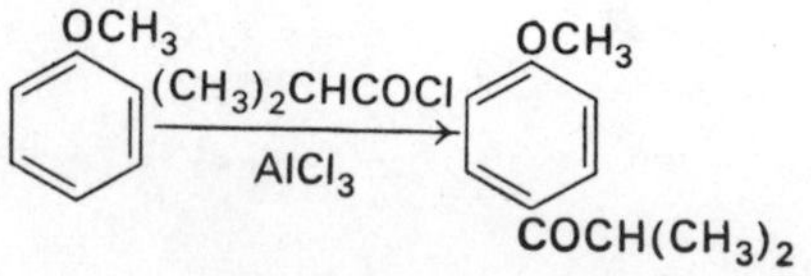

(*l*) Cl (benzene) $\xrightarrow[H_2SO_4]{HNO_3}$ Cl, NO_2, NO_2

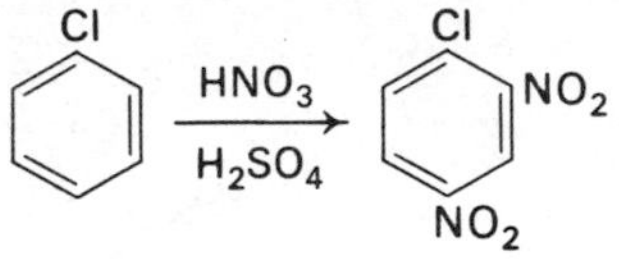

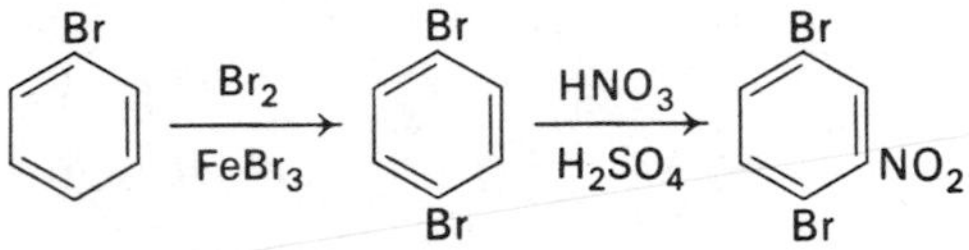

(*m*) Br (benzene) $\xrightarrow[FeBr_3]{Br_2}$ Br, Br $\xrightarrow[H_2SO_4]{HNO_3}$ Br, NO_2, Br

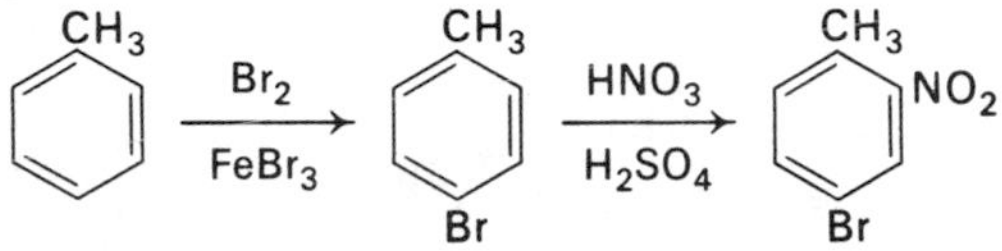

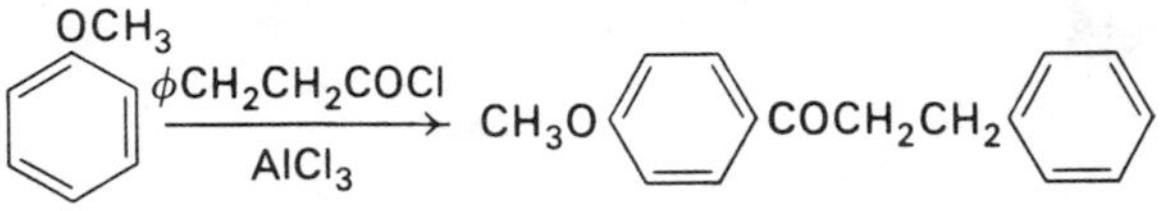

(*n*) CH_3 (benzene) $\xrightarrow[FeBr_3]{Br_2}$ CH_3, Br $\xrightarrow[H_2SO_4]{HNO_3}$ CH_3, NO_2, Br

(*o*) OCH_3 (benzene) $\xrightarrow[AlCl_3]{\phi CH_2CH_2COCl}$ CH_3O–(benzene)–$COCH_2CH_2$–(benzene)

Exercise 9.3

(*a*) Br, NO_2 (benzene) + NaOMe

(*b*) Br, NO_2, NO_2 (benzene) + NaOH

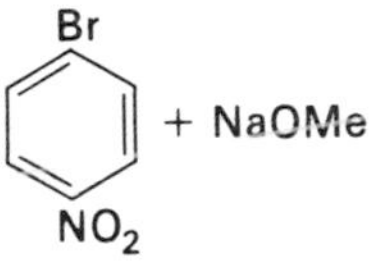

(*c*) Br, NO_2, NO_2 (benzene) + NH_2OH

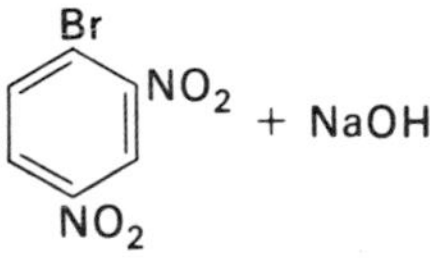

(*d*) Br–(benzene)–NO_2 + (benzene)–NH_2

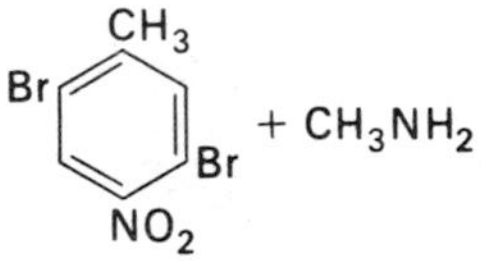

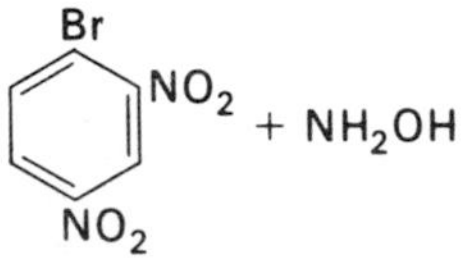

(*e*) Br, Br, NO_2 (benzene) + Me_2NH

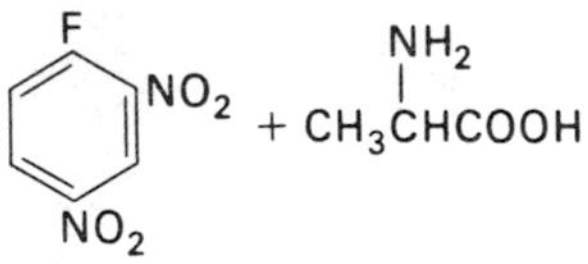

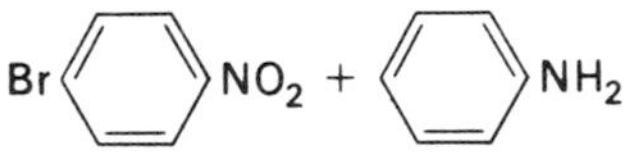

(*f*) CH_3, Br, Br, NO_2 (benzene) + CH_3NH_2

(*g*) F, NO_2, NO_2 (benzene) + $CH_3CH(NH_2)COOH$

(*h*) Br, NO_2 (benzene) + $H_2NCH_2CH_2NH_2$

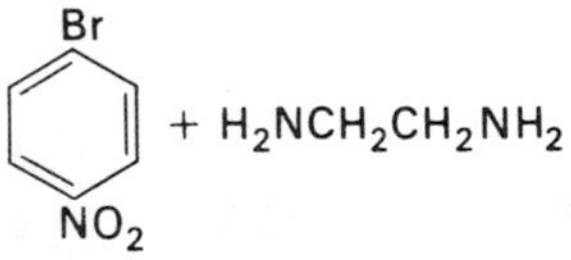

Exercise 9.4

(*a*) $NHCOCH_3$ (benzene) $\longrightarrow$ $NHCOCH_3$, Br $\longrightarrow$ $NHCOCH_3$, NO_2, Br $\longrightarrow$

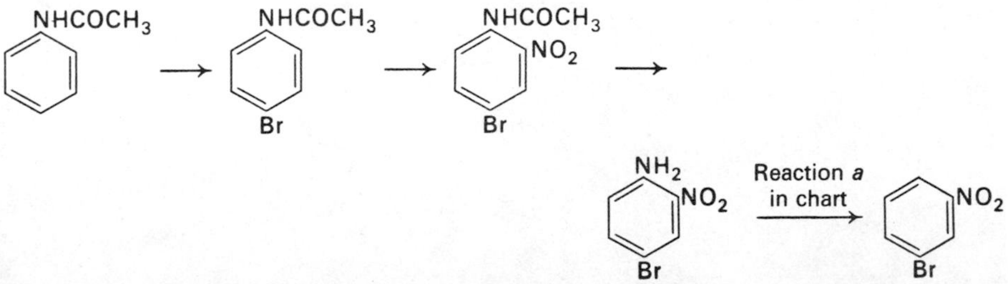

NH_2, NO_2, Br $\xrightarrow{\text{Reaction } a \text{ in chart}}$ NO_2, Br

(b) OCH_3 → OCH_3, NO_2 → OCH_3, NH_2 → OCH_3, $NHCOCH_3$ →

OCH_3, NO_2, $NHCOCH_3$ $\xrightarrow[\text{Exercise } (a)]{\text{as in}}$ OCH_3, NO_2

(c) Br → Br, NO_2 $\xrightarrow[\text{above}]{\text{as}}$ Br, $NHCOCH_3$ → Br, Cl, $NHCOCH_3$ $\xrightarrow[\text{Exercise } (b)]{\text{as in}}$ Br, Cl

(d) $NHCOCH_3$ → $NHCOCH_3$, NO_2 → $NHCOCH_3$, Br, Br, NO_2 →

$NHCOCH_3$, Br, Br, NH_2 → $NHCOCH_3$, Br, Br → NH_2, Br, Br

(e) NH_2 $\xrightarrow{Br_2}$ NH_2, Br, Br, Br $\xrightarrow[-NH_2 \text{ as before}]{\text{remove}}$ Br, Br, Br

(f) Cl → Cl, NO_2 → Cl, $NHCOCH_3$ → Cl, NO_2, $NHCOCH_3$ $\xrightarrow[-NHAc \text{ to } -NH_2 \text{ and remove}]{\text{hydrolyze}}$

Cl, NO_2 → Cl, NH_2

(g) CH_3 → CH_3, NO_2 → CH_3, NH_2 → CH_3, Br, Br, NH_2 → CH_3, Br, Br

(h) $NHCOCH_3$ → $NHCOCH_3$, NO_2 → $NHCOCH_3$, NH_2 → $NHCOCH_3$, Br, Br, NH_2 →

$NHCOCH_3$, Br, Br → NH_2, Br, Br

(*i*) $NHCOCH_3$ → $NHCOCH_3$, NO_2 → NH_2, NO_2 → NH_2, Br, Br, NO_2 → Br, Br, NO_2

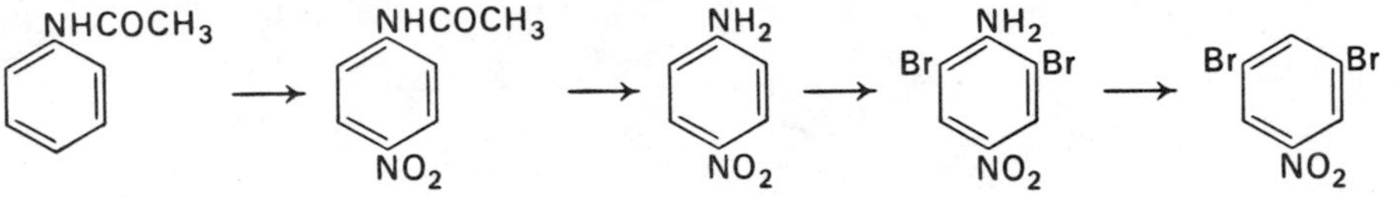

(*Question:* Why cannot 3,5-dibromonitrobenzene be made by direct dibromination of nitrobenzene?)

(*j*) $NHCOCH_3$ → $NHCOCH_3$, NO_2 → $NHCOCH_3$, Br, NO_2 → $NHCOCH_3$, Br, NH_2 →

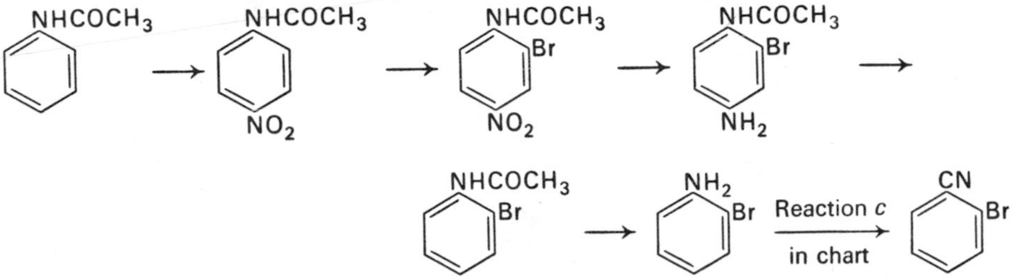

$NHCOCH_3$, Br → NH_2, Br —Reaction *c* in chart→ CN, Br

(*k*) $NHCOCH_3$, Br, NO_2 (from (*j*)) → $NHCOCH_3$, Br, NH_2 → $NHCOCH_3$, Br, CN —as before→ Br, CN

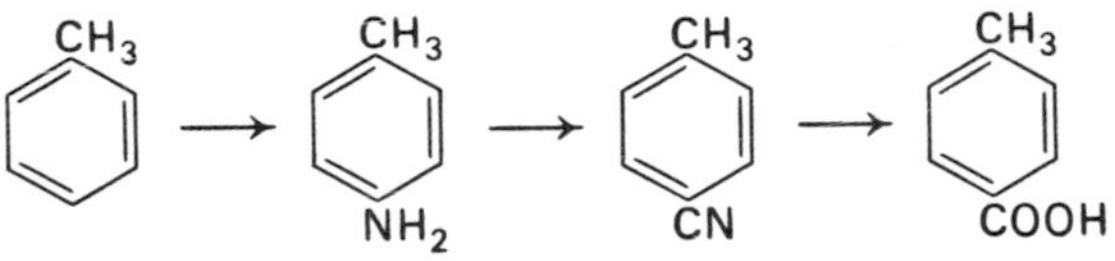

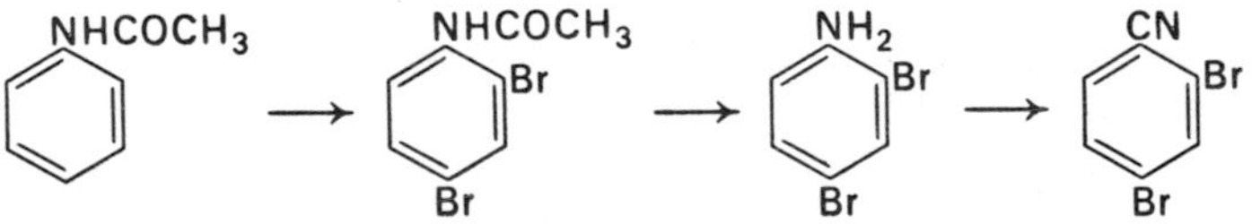

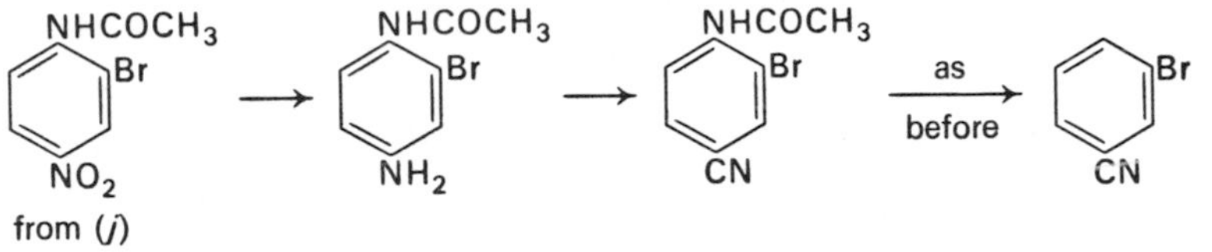

Exercise 9.5*

(*a*) CH_3 → CH_3, NH_2 → CH_3, CN → CH_3, COOH

(*b*) $NHCOCH_3$ → $NHCOCH_3$, Br, Br → NH_2, Br, Br → CN, Br, Br

(*c*) CH_3 → CH_3, $NHCOCH_3$ → CH_3, Br, $NHCOCH_3$ → CH_3, Br, NH_2 →

CH_3, Br, CN —CH_3MgI→ CH_3, Br, $COCH_3$

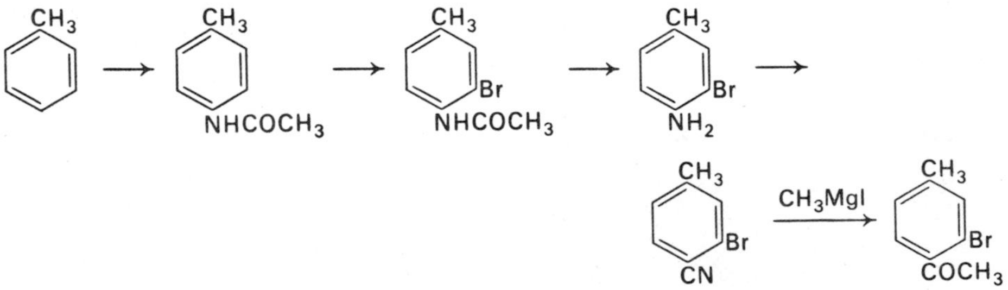

* Some intermediate steps (for example, $-NO_2 \longrightarrow -NH_2 \longrightarrow -NHAc$; $NHAc \longrightarrow NH_2$; $CN \longrightarrow COOH$, and so forth) are omitted from these equations.

(*d*) CH_3 → CH_3, NO_2 → CH_3, Br, NO_2 → CH_3, Br, NH_2 → CH_3, Br → COOH, Br

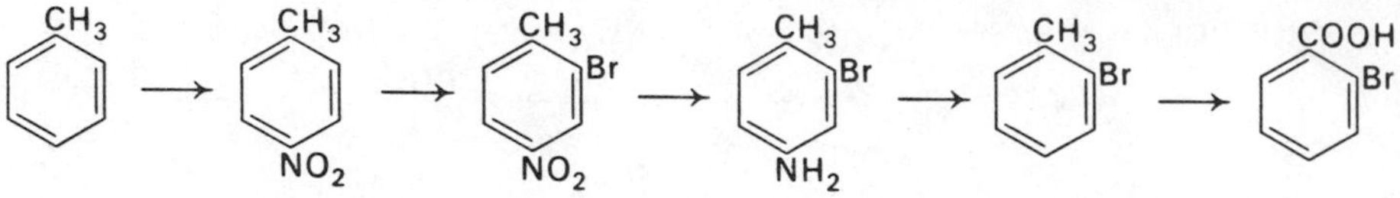

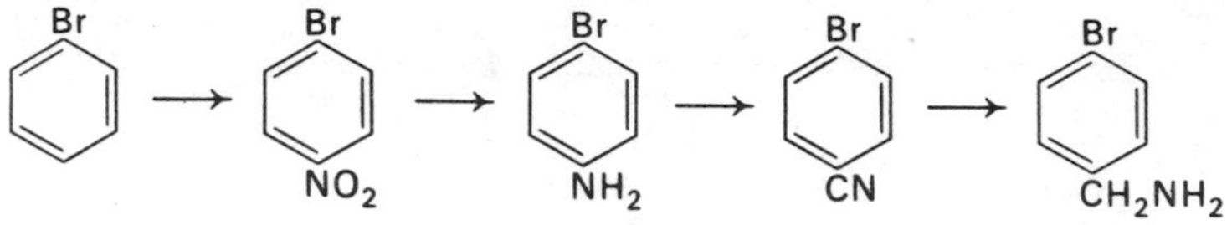

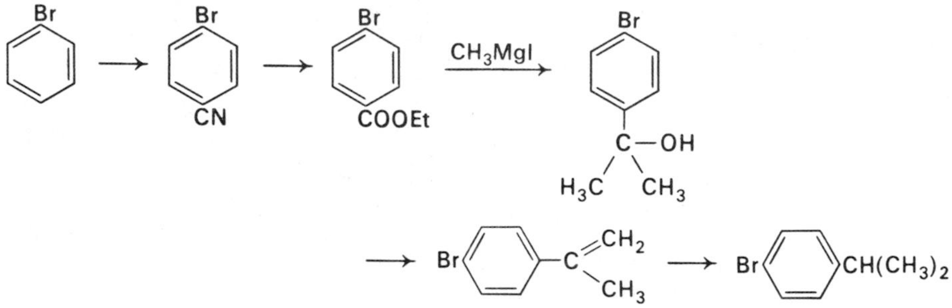

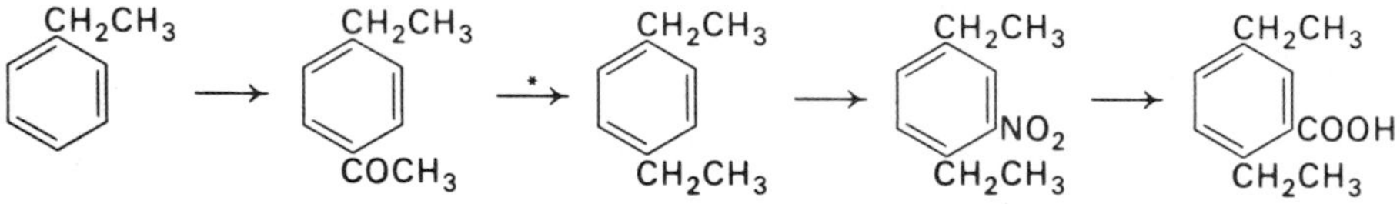

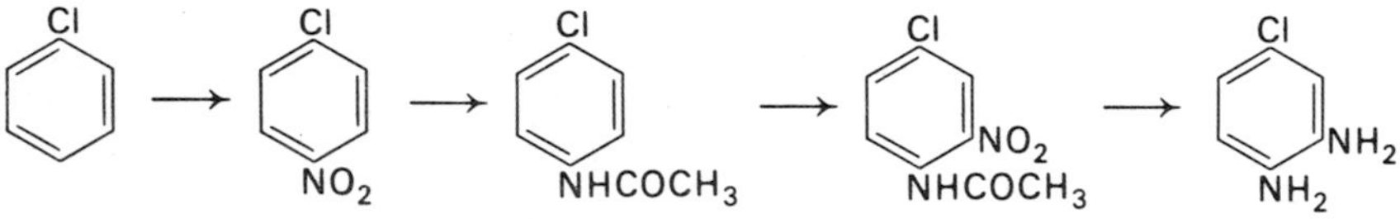

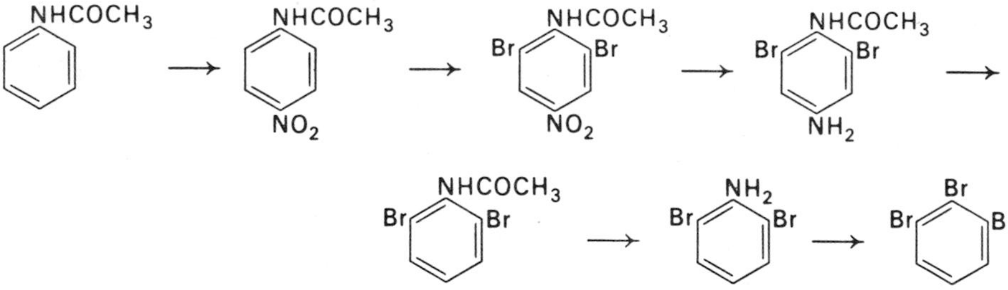

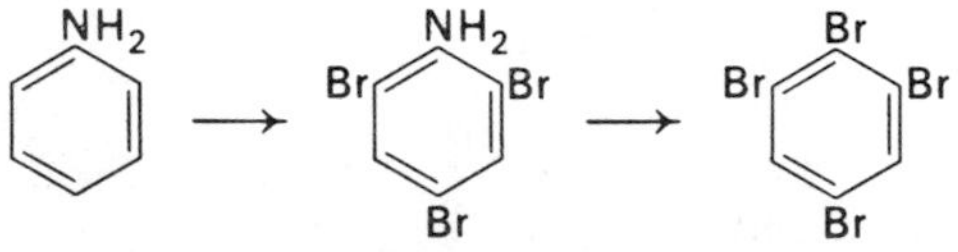

(*e*) Br → Br, NO_2 → Br, NH_2 → Br, CN → Br, CH_2NH_2

(*f*) Br → Br, CN → Br, COOEt $\xrightarrow{CH_3MgI}$ Br, C—OH, H_3C, CH_3

→ Br, C, CH_2, CH_3 → Br, $CH(CH_3)_2$

(*g*) CH_2CH_3 → CH_2CH_3, $COCH_3$ →* CH_2CH_3, CH_2CH_3 → CH_2CH_3, NO_2, CH_2CH_3 → CH_2CH_3, COOH, CH_2CH_3

(*h*) Cl → Cl, NO_2 → Cl, $NHCOCH_3$ → Cl, NO_2, $NHCOCH_3$ → Cl, NH_2, NH_2

(*i*) $NHCOCH_3$ → $NHCOCH_3$, NO_2 → $NHCOCH_3$, Br, Br, NO_2 → $NHCOCH_3$, Br, Br, NH_2 →

$NHCOCH_3$, Br, Br → NH_2, Br, Br → Br, Br, Br

(*j*) NH_2 → NH_2, Br, Br, Br → Br, Br, Br, Br

(*k*) CH_3 → CH_3, NO_2 → CH_3, Br, NO_2 → COOH, Br, NO_2

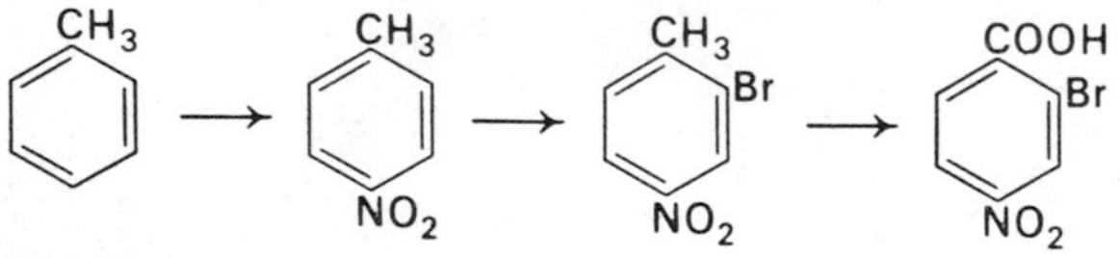

* the reduction $-COCH_3 \rightarrow -CH_2CH_3$ may be accomplished directly (Clemmensen), or by indirect means.

Exercise 9.6

(a) $C_6H_5CHO \xrightarrow{CH_3MgI} C_6H_5CH(OH)CH_3 \xrightarrow{CrO_3} C_6H_5COCH_3$

(b) $C_6H_5COOH \longrightarrow C_6H_5COCl \xrightarrow{Me_2NH} C_6H_5CONMe_2 \xrightarrow{LiAlH_4} C_6H_5CH_2NMe_2$

(c) p-dibromobenzene $\longrightarrow$ 2,5-dibromoacetophenone (Br, Br, $COCH_3$) $\xrightarrow{RCO_3H}$ 2,5-dibromophenyl acetate ($OCOCH_3$) $\longrightarrow$ 2,5-dibromophenol (OH)

(d) $C_6H_5COOH \longrightarrow C_6H_5CH_2OH \longrightarrow C_6H_5CH_2Cl \longrightarrow$

$C_6H_5CH_2MgCl \xrightarrow{CH_3COOEt} CH_3\text{—}C(OH)(CH_2C_6H_5)_2 \xrightarrow{-H_2O} CH_3\text{—}C(CH_2C_6H_5)\text{=}CHC_6H_5$

(e) $C_6H_5NH_2 \longrightarrow C_6H_5CN \qquad C_6H_5CN + C_6H_5MgBr \longrightarrow$

$C_6H_5\text{—}CO\text{—}C_6H_5 \xrightarrow{LiAlH_4} C_6H_5\text{—}CH(OH)\text{—}C_6H_5$

(f) toluene (CH_3) $\longrightarrow$ p-nitrotoluene (CH_3, NO_2) $\longrightarrow$ p-toluidine (CH_3, NH_2) $\xrightarrow[\text{in chart}]{\text{Reaction } b}$ p-cresol (CH_3, OH) $\longrightarrow$ 2,6-dibromo-4-methylphenol (CH_3, Br, Br, OH)

(g) o-xylene (CH_3, CH_3) $\longrightarrow$ nitro-o-xylene (CH_3, CH_3, NO_2) $\xrightarrow[\text{and } -CN]{via\ -NH_2}$ (CH_3, CH_3, COOH)

$\xrightarrow[AlCl_3]{CH_3COCl}$ (CH_3, CH_3, $COCH_3$) $\xrightarrow[NaOH]{Br_2}$ (CH_3, CH_3, COOH)

(h) $C_6H_5COCH_3$ $\longrightarrow$ m-nitroacetophenone ($COCH_3$, NO_2) $\xrightarrow{RCO_3H}$ m-nitrophenyl acetate ($OCOCH_3$, NO_2) $\longrightarrow$ m-nitrophenol (OH, NO_2)

(*Note:* The Friedel-Crafts reaction of nitrobenzene with acetyl chloride cannot be used to prepare *m*-nitroacetophenone.)

(*i*) $CH_3C_6H_4COCH_3$ $\xrightarrow[(32)]{Eq}$ $CH_3C_6H_4NHCOCH_3$ → $CH_3C_6H_3(Br)NHCOCH_3$ → $CH_3C_6H_3(Br)NH_2$ → $CH_3C_6H_3(Br)Cl$

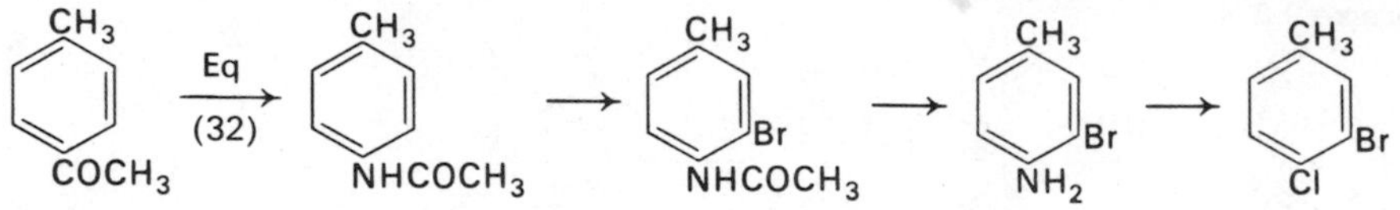

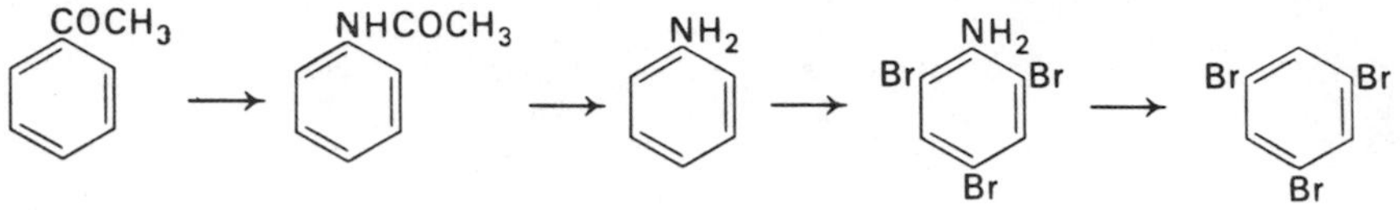

(*j*) $C_6H_5COCH_3$ → $C_6H_5NHCOCH_3$ → $C_6H_5NH_2$ → $NH_2C_6H_2Br_3$ → $C_6H_3Br_3$

Exercise 9.7

(*a*) $C_6H_4(COOH)_2$ (COOH, COOH)

(*b*) $(CH_3O)_2C_6H_3COOH$ (OCH$_3$, CH_3O, COOH)

(*c*) $C_6H_3(COOH)_3$ (COOH, COOH, COOH)

(*d*) C_6H_5COOH

(*e*) $C_6H_3Br(COOH)_2$ (COOH, Br, COOH)

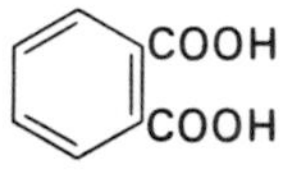

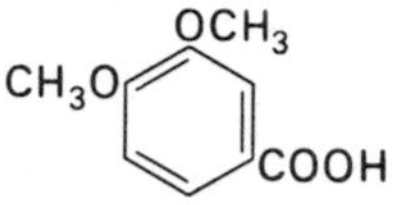

(*f*) ClC_6H_4COOH (Cl, COOH)

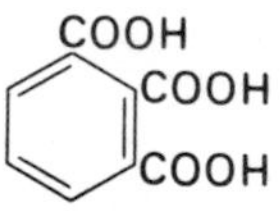

(*g*) $C_6H_4(COOH)COC_6H_5$ (COOH, COC_6H_5)

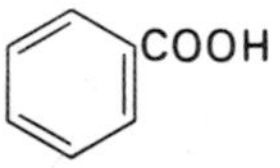

(*h*) $C_6H_3(COOH)_3$ (COOH, COOH, COOH)

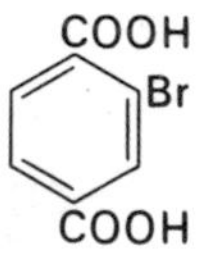

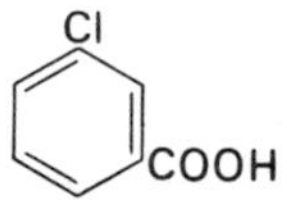

(*i*) $NO_2C_6H_3(COOH)_2$ (NO_2, COOH, COOH)

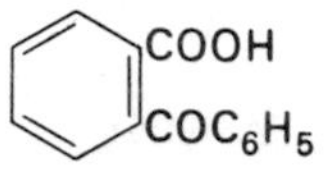

(*j*) $C_6H_4(COOH)_2$ (COOH, COOH)

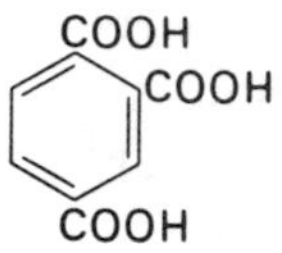

Exercise 9.8

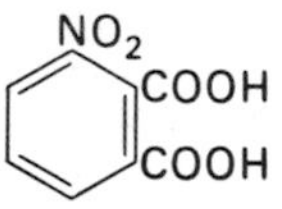

(*a*) C_6H_5OH $\xrightarrow[\text{(Equation 41)}]{\text{Kolbe reaction}}$ $C_6H_4(OH)COOH$ $\xrightarrow[\text{NaOH (excess)}]{Me_2SO_4}$ $C_6H_4(OCH_3)COOH$

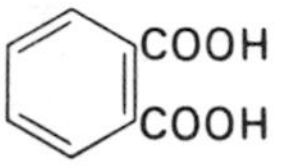

(*b*) HOC_6H_4OH $\xrightarrow[ZnCl_2]{HCN}$ $HOC_6H_3(OH)CHO$ $\xrightarrow[NaOAc]{(CH_3CO)_2O}$ HO-coumarin (O, =O)

(*c*) HOC_6H_4OH + $C_6H_5N_2^+Cl^-$ → $HOC_6H_3(OH)N{=}N{-}C_6H_5$ $\xrightarrow{Na_2S_2O_4}$ $HOC_6H_3(OH)NH_2$

(*d*) HOC_6H_4OH $\xrightarrow[NaOH]{CO_2}$ $HOC_6H_3(OH)COOH$

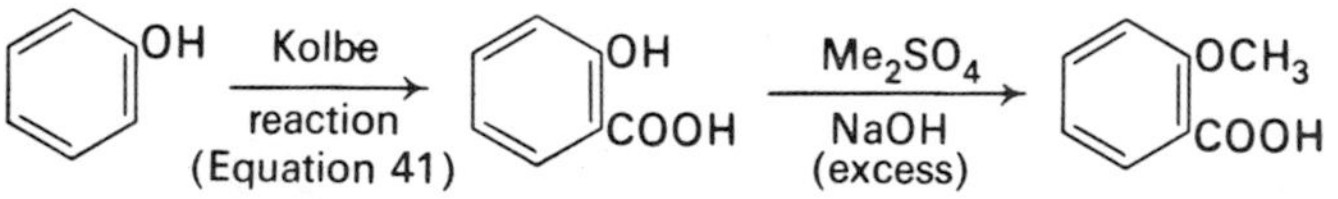

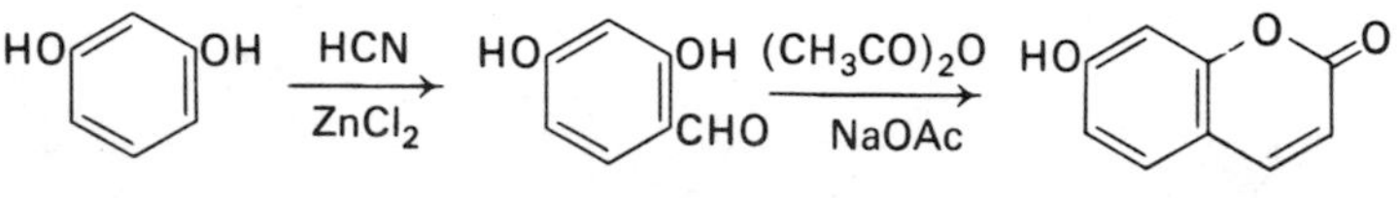

(*e*) HOC_6H_4OH + $(CH_3)_2CHCH_2CN$ $\xrightarrow[HCl]{ZnCl_2}$ $HOC_6H_3(OH)COCH_2CH(CH_3)_2$

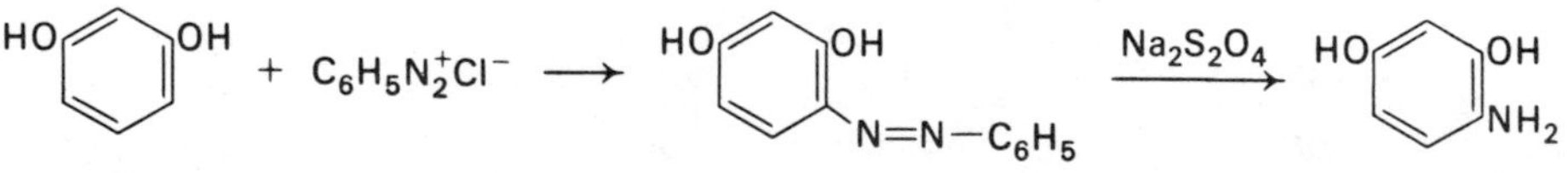

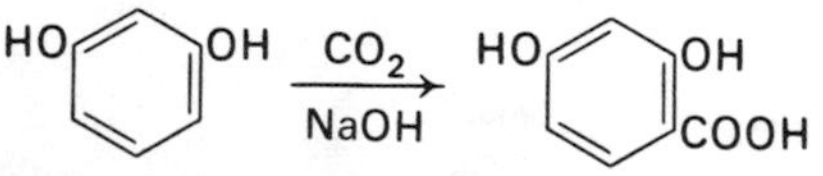

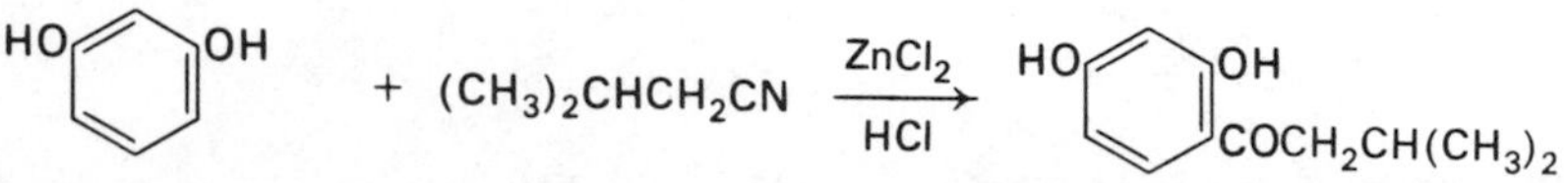

(*f*) HO–C6H4 (phenol) + $C_6H_5N_2^+Cl^-$ ⟶ HO–C6H4–N=N–C_6H_5 $\xrightarrow[(c)]{\text{as in}}$

HO–C6H4–NH_2 $\xrightarrow{(CH_3CO)_2O}$ HO–C6H4–$NHCOCH_3$

(*g*) C6H5OH $\xrightarrow[NaOH]{CHCl_3}$ o-HOC6H4CHO $\xrightarrow{CH_3MgI}$ o-HOC6H4CH(OH)CH_3

(*h*) HO, OH, OH-benzene $\xrightarrow[ZnCl_2]{ClCH_2CN}$ HO, OH, OH-C6H2$COCH_2Cl$ $\xrightarrow{*}$ HO, OH-benzofuran-3-one

(*i*) HO–C6H4–OH $\xrightarrow[ZnCl_2/HCl]{CH_3CN}$ HO, OH-C6H3$COCH_3$ $\xrightarrow[NaOH]{H_2O_2}$ HO-C6H3(OH)(OH)

(*j*) o-HOC6H4CHO $\xrightarrow[NaOH]{H_2O_2}$ o-C6H4(OH)(OH) $\xrightarrow[NaOH]{(CH_3)_2SO_4}$ o-C6H4(OCH_3)(OCH_3)

* A mild base, such as sodium acetate, is sufficient.

PART 10

Exercise 10.1

(*a*) $CH_3CH_2OH \longrightarrow CH_3CHO + 2\,H^+ + 2\,e$

(*b*) $(CH_3)_2CHOH \longrightarrow (CH_3)_2C{=}O + 2\,H^+ + 2\,e$

(*c*) $CH_3CH{=}CHCH_3 + 2\,H_2O \longrightarrow CH_3CH(OH){-}CH(OH)CH_3 + 2\,H^+ + 2\,e$

(*d*) $CH_3COOH + 2\,H_2O \longrightarrow 2\,CO_2 + 8\,H^+ + 8\,e$

(*e*) $CH_3CHOHCOCH_3 + H_2O \longrightarrow CH_3CHO + CH_3COOH + 2\,H^+ + 2\,e$

(*f*) cyclohexene $+ 2H_2O \longrightarrow$ cyclohexane-1,2-dione $+ 6\,H^+ + 6\,e$

(*g*) cyclohexene $+ H_2O \longrightarrow$ cyclohex-2-enone $+ 4\,H^+ + 4\,e$

(*h*) C6H5$CH_2CH_2CH_2$C6H5 $+ 6\,H_2O \longrightarrow CO_2 + 2$ C6H5COOH $+ 16\,H^+ + 16\,e$

(*i*) phenol (OH on benzene ring) $+ H_2O \longrightarrow$ p-benzoquinone (O=C ring C=O) $+ 4 H^+ + 4 e$

(*j*) tetralin $+ H_2O \longrightarrow$ 1-tetralone (C=O on ring) $+ 4 H^+ + 4 e$

(*k*) p-$CH_3C_6H_4COCH_3$ $+ 5 H_2O \longrightarrow$ p-$C_6H_4(COOH)_2$ $+ CO_2 + 14 H^+ + 14 e$

(*l*) 2-methylindane ($-CH_3$) $+ 8 H_2O \longrightarrow$ o-$C_6H_4(COOH)_2$ $+ 2 CO_2 + 22 H^+ + 22 e$

(*m*) 3-phenylindene (C_6H_5) $+ 5 H_2O \longrightarrow$ o-$C_6H_4(COOH)(COC_6H_5)$ $+ CO_2 + 12 H^+ + 12 e$

(*n*) $CH_3CH{=}CH_2 + 4 H_2O \longrightarrow CH_3COOH + CO_2 + 10 H^+ + 10 e$

(*o*) $CH_3CH{=}CH{-}CH{=}CH{-}CH{=}CH{-}CH_3 + 12 H_2O \longrightarrow$

$$2 CH_3COOH + 4 CO_2 + 28 H^+ + 28 e$$

Exercise 10.2

(*a*) The product of reduction of CrO_3 in H_2SO_4 is $Cr_2(SO_4)_3$

(*i*) $CH_3CH(OH)CH_2CH_3 \longrightarrow CH_3COCH_2CH_3 + 2 H^+ + 2 e$

(*ii*) $CrO_3 + 6 H^+ + 3 e \longrightarrow Cr^{3+} + 3 H_2O$

Multiply (*i*) by 3, (*ii*) by 2; add (*i*) and (*ii*):

$$3 CH_3CH(OH)CH_2CH_3 + 2 CrO_3 + 12 H^+ + 6 e \longrightarrow CH_3COCH_2CH_3 + 6 H^+ + 6 e + 2 Cr^{3+} + 6 H_2O$$

Cancelling:

$$3 CH_3CH(OH)CH_2CH_3 + 2 CrO_3 + 6 H^+ \longrightarrow 3 CH_3COCH_2CH_3 + 2 Cr^{3+} + 6 H_2O$$

Add 3 $SO_4^{=}$ to each side:

$$3 CH_3CH(OH)CH_2CH_3 + 2 CrO_3 + 3 H_2SO_4 \longrightarrow 3 CH_3COCH_2CH_3 + Cr_2(SO_4)_3 + 6 H_2O$$

(*b–e*) can be answered in a similar way.

(*f*) (*i*) $CH_3CH{=}CHCOOH + 4\,H_2O \longrightarrow CH_3COOH + 2\,CO_2 + 10\,H^+ + 10\,e$

(*ii*) $MnO_4^- + 4\,H^+ + 3\,e \longrightarrow MnO_2 + 2\,H_2O$

Multiply (*i*) by 3, (*ii*) by 10, add (*i*) and (*ii*):

$$3\,CH_3CH{=}CHCOOH + 12\,H_2O + 10\,MnO_4^- + 40\,H^+ + 30\,e \longrightarrow$$
$$3\,CH_3COOH + 6\,CO_2 + 30\,H^+ + 30\,e + 10\,MnO_2 + 20\,H_2O$$

Cancelling:

$$3\,CH_3CH{=}CHCOOH + 10\,MnO_4^- + 10\,H^+ \longrightarrow 3\,CH_3COOH$$
$$+ 6\,CO_2 + 10\,MnO_2 + 8\,H_2O.$$

Add 10 K^+ and 10 OH^- to each side:

$$3\,CH_3CH{=}CHCOOH + 10\,KMnO_4 + 10\,H_2O \longrightarrow 3\,CH_3COOH + 6\,CO_2$$
$$+ 10\,MnO_2 + 10\,KOH + 8\,H_2O.$$

Finally, 10 H_2O (left side) and 8 H_2O (right side); cancel, leaving 2 H_2O on left side, no H_2O on right side

Check: $\left.\begin{array}{l} H: 18 + 4 = 22 \\ O: 6 + 40 + 2 = 48 \end{array}\right\}$ left $\quad \left.\begin{array}{l} H: 12 + 10 = 22 \\ O: 6 + 12 + 20 + 10 = 48 \end{array}\right\}$ right

---

Exercise 10.3

(*a*) $CH_3COCH_2CH_3$

(*b*) $CH_3COCOCH_3$

(*c*) CH_2CH_2CHO

(*d*) O CH_3

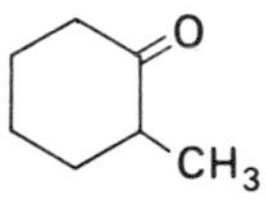

(*e*) O

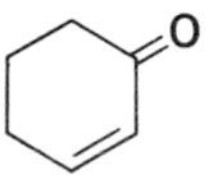

(*f*) $CH_3{-}C(CH_3)(OH)CH_2COCH_3$

(*g*) The product is the 3-keto compound.

(*h*) $CH_3OCH_2CH_2CH_2COCH_3$

(*i*) $CH_3C{\equiv}CCOC{\equiv}CCH_3$

(*j*) $CH_3C{\equiv}CCOCH_3$

Exercise 10.4

(*a*) $CH_3CHO + H_2O \longrightarrow CH_3COOH + 2\,H^+ + 2\,e$

$MnO_4^- + 4\,H^+ + 3\,e \longrightarrow MnO_2 + 2\,H_2O.$

Thus, 3 CH_3CHO require 2 $KMnO_4$.

(*b*) $(CH_2CHO)_2 + 2\,H_2O \longrightarrow (CH_2COOH)_2 + 4\,H^+ + 4\,e$

$MnO_4^- + 4\,H^+ + 3\,e \longrightarrow MnO_2 + 2\,H_2O$

$3(CH_2CHO)_2$ require 4 $KMnO_4$

(*c*) $(CH_3O)_2C_6H_3CHO + H_2O \longrightarrow (CH_3O)_2C_6H_3COOH + 2H^+ + 2e$

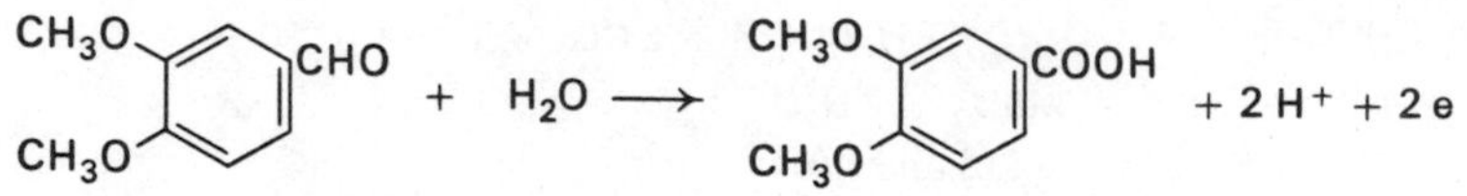

See (*a*) for ratio, aldehyde/$KMnO_4$ (= 3/2)

Exercise 10.5

(*a*) $H(H_3C)C{=}C(H)CH_3 + OsO_4 \longrightarrow$ H, OH, H_3C–C–C–HO, H, CH_3

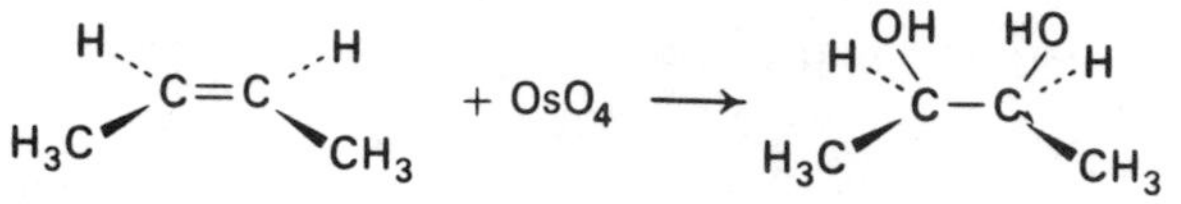

(*b*) $H(H_3C)C{=}C(C_2H_5)H + KMnO_4 \longrightarrow$ H, OH, H_3C–C–C–HO, C_2H_5, H + H, H_3C, OH–C–C–HO, C_2H_5, H

(*c*) $CH_3(CH_2)_7(H)C{=}C(H)(CH_2)_7COOH \xrightarrow{OsO_4}$ HO, OH, $CH_3(CH_2)_7$–C–C–$(CH_2)_7COOH$, H, H

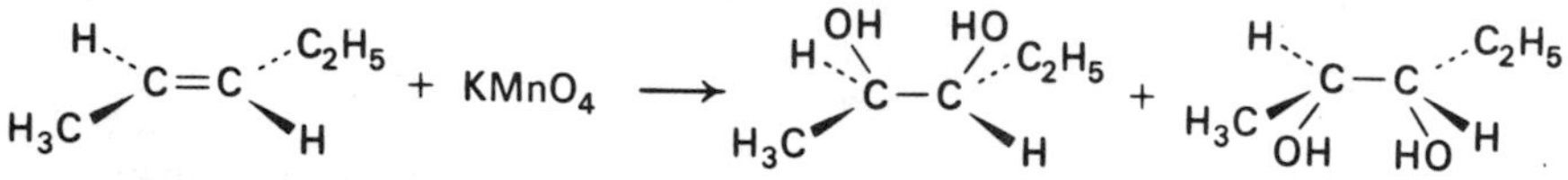

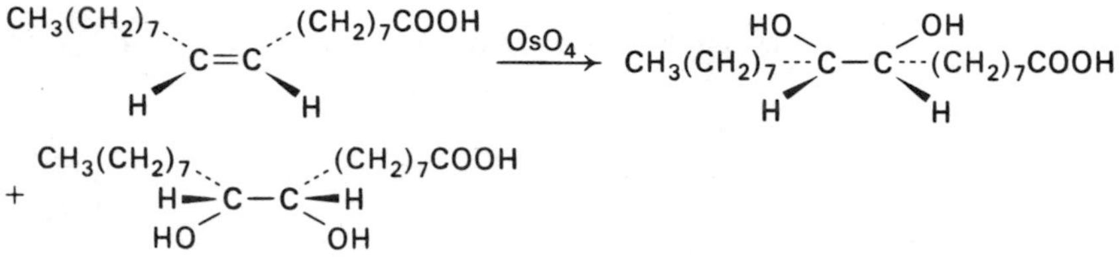

+ $CH_3(CH_2)_7$, $(CH_2)_7COOH$, H–C–C–H, HO, OH

(*d*) H, CH_3 (3-methylcyclohexene) + $OsO_4 \longrightarrow$ H, CH_3, H, OH, OH, H + H, CH_3, H, OH, OH, H

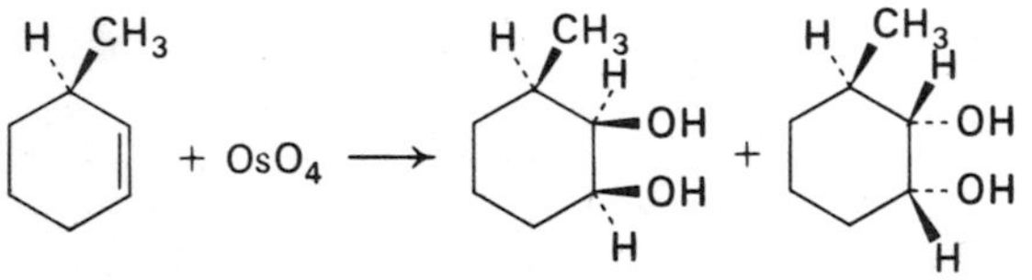

Exercise 10.6

R, O, H, C–C, H, R (*b*) (*a*) $\xrightarrow{\text{attack at (a)}}$ R, OH, H, C–C, H, HO, R (*i*)

$\xrightarrow{\text{attack at (b)}}$ R, OH, H–C–C–H, HO, R (*ii*)

attack by H_2O

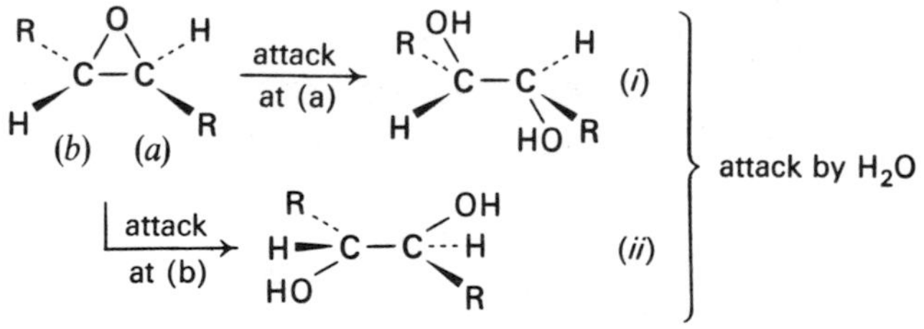

(*i*) and (*ii*) are identical: a *meso* glycol.

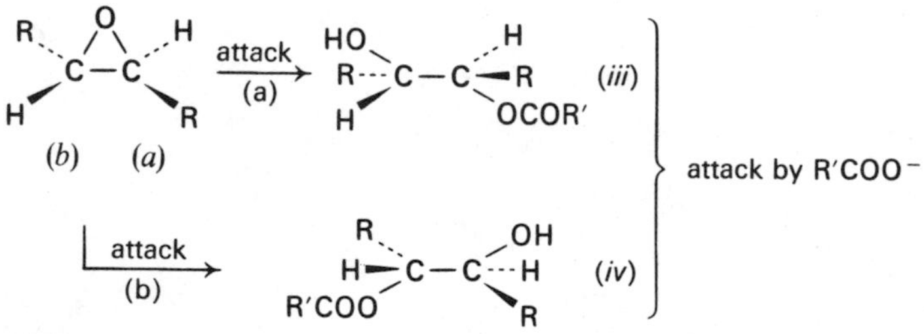

R, O, H, C–C, H, R (*b*) (*a*) $\xrightarrow{\text{attack (a)}}$ HO, H, R–C–C–R, H, OCOR′ (*iii*)

$\xrightarrow{\text{attack (b)}}$ R, OH, H–C–C–H, R′COO, R (*iv*)

attack by $R'COO^-$

(*iii*) and (*iv*) are enantiomers.

Exercise 10.7

(*a*) $(CH_3)_2C{=}CH_2$

(*b*) $CH_3CH_2\overset{\overset{\large CH_3}{|}}{C}{=}CHCH_3$

(*c*) [cyclohexane]=CHCH$_3$

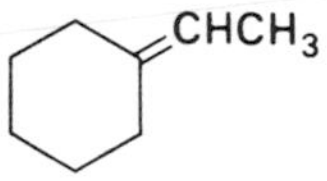

(*d*)

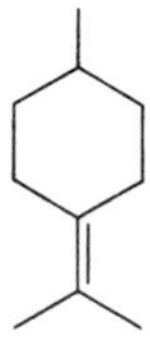

(*e*) C$_6$H$_5$, CH$_3$

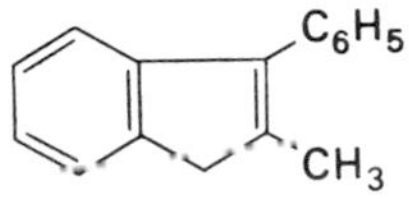

(*f*)

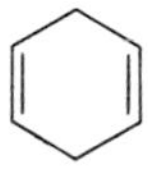

(*g*) CH$_3$, CH$_3$, CH$_3$

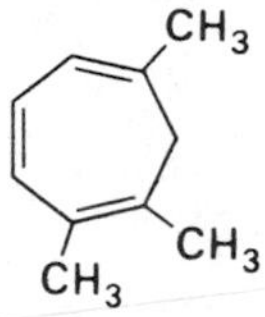

(*h*) H$_3$C, H$_3$C

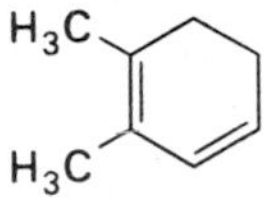

(*i*) O, CH$_2$

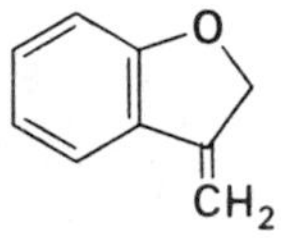

(*j*) =CHC$_6$H$_5$

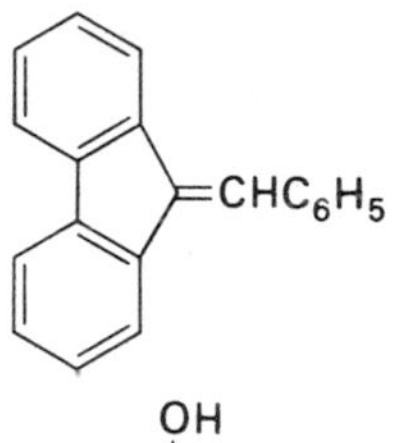

(*k*) 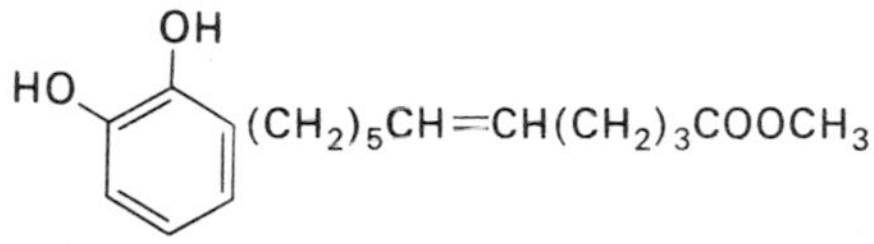 OH, HO, $(CH_2)_5CH{=}CH(CH_2)_3COOCH_3$

(*l*) $\overset{\overset{\large COOCH_3}{|}}{(CH_2)_7}CH{=}CH(CH_2)_7COOCH_3$

PART 11

Exercise 11.1

(*a*) $BrCH_2COOEt + EtO^- \longrightarrow EtOCH_2COOEt \xrightarrow{LiAlH_4} EtOCH_2CH_2OH$

(*Note*: The product cannot be prepared by first reducing the ester to $BrCH_2CH_2OH$, followed by reaction of the bromohydrin with NaOEt. Why?)

(*b*) [cyclohexanol] OH $\longrightarrow$ [cyclohexene] $\xrightarrow{CrO_3}$ COOH, COOH $\xrightarrow{LiAlH_4}$ $HOCH_2(CH_2)_4CH_2OH$

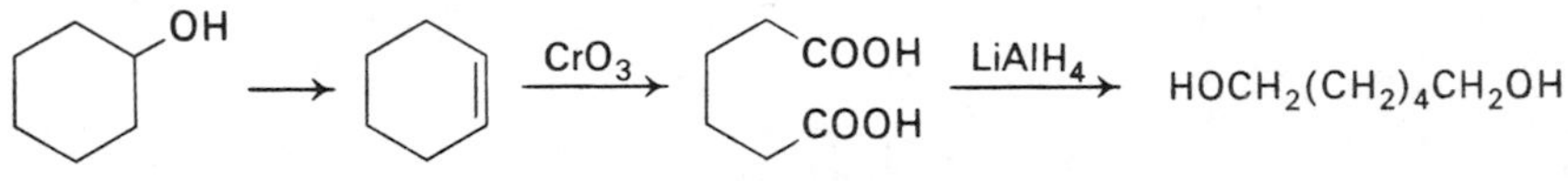

(*c*) O, COOEt from cyclohexanone, *via* the enamine, and ClCOOEt $\xrightarrow[CH_2{=}CHCH_2Br]{NaOEt}$

O, $CH_2CH{=}CH_2$, COOEt $\xrightarrow{\text{(1) saponify}}$ (2) decarboxylate (3) reduce → OH, $CH_2CH{=}CH_2$

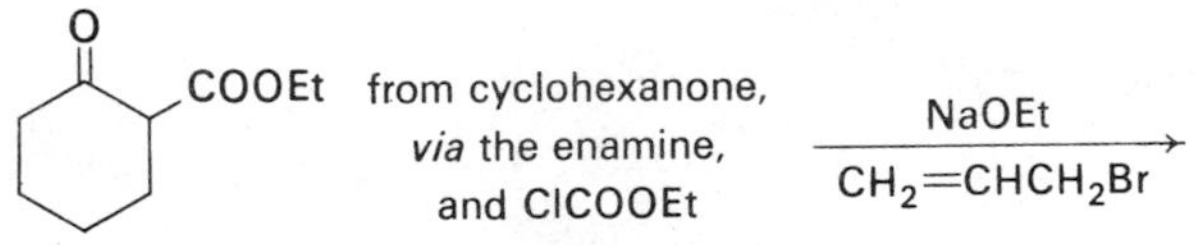

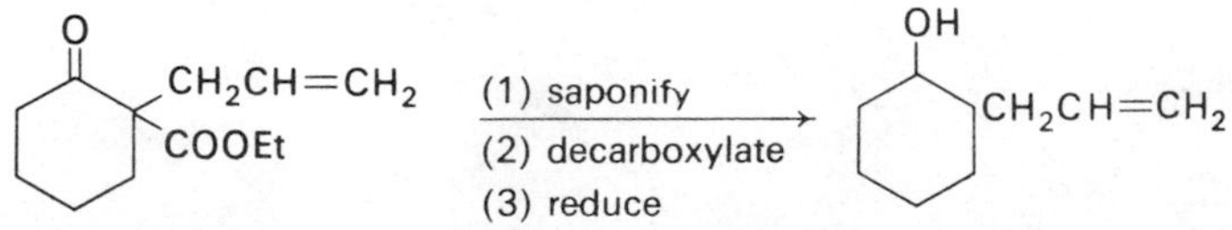

(*d*) $CH_3COOEt + C_6H_5MgBr \longrightarrow CH_3C(\phi)_2{-}OH \xrightarrow{-H_2O} CH_2{=}C(\phi)_2$

(*e*) $(CH_3O)_2C_6H_3CHO$ (veratraldehyde) $\xrightarrow[\text{reaction}]{\text{Perkin}} (CH_3O)_2C_6H_3CH{=}CHCOOH \xrightarrow[Pt]{H_2}$

$(CH_3O)_2C_6H_3CH_2CH_2COOH \xrightarrow[(2)\ AlCl_3]{(1)\ SOCl_2}$ dimethoxyindanone (CH_3O, CH_3O on the benzene ring; C=O in the five-membered ring)

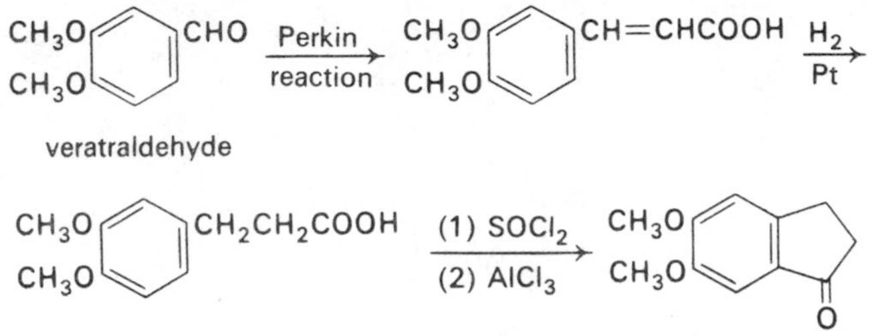

(*f*) $\phi CHO + BrCH(CH_3)COOEt \xrightarrow{Zn} \phi CH(OH)CH(CH_3)COOEt \xrightarrow[(2)\ H_2/Pt;\ (3)\ \text{saponify}]{(1)\ -H_2O} \phi CH_2CH(CH_3)COOH$

(*g*) $CH_2(COOEt)_2 \xrightarrow[CH_3I]{NaOEt} CH_3CH(COOEt)_2 \xrightarrow[NaOEt]{BrCH_2COCH_3}$

$CH_3COCH_2C(CH_3)(COOEt)_2 \xrightarrow[(2)\ -CO_2]{(1)\ \text{saponify}} CH_3COCH_2CH(CH_3)COOH$

(*h*) $BrCH_2CH_2CH_2CH_2Br$ ($H_2C(CH_2Br){-}H_2C(CH_2Br)$) $+ CH_2(COOEt)_2 \xrightarrow{NaOEt}$ cyclopentane-1,1-(COOEt)$_2$ $\xrightarrow{NaOH}$ cyclopentane-1,1-(COOH)$_2$ $\xrightarrow{-CO_2}$

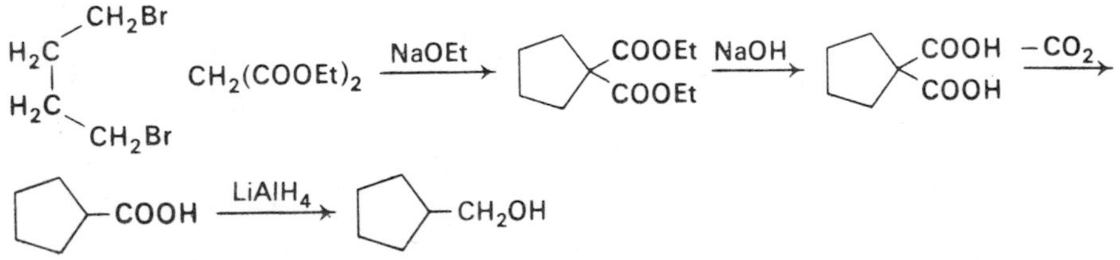

cyclopentyl–COOH $\xrightarrow{LiAlH_4}$ cyclopentyl–CH_2OH

(*i*) 2-oxocyclohexane–COOEt $\xrightarrow[NaOEt]{(CH_3)_2CHCH_2Br}$ 2-oxo-1-COOEt-1-$CH_2CH(CH_3)_2$-cyclohexane $\xrightarrow[(c)]{\text{see}}$

2-$CH_2CH(CH_3)_2$-cyclohexanol (OH) $\xrightarrow[(H^+)]{-H_2O}$ 1-$CH_2CH(CH_3)_2$-cyclohexene

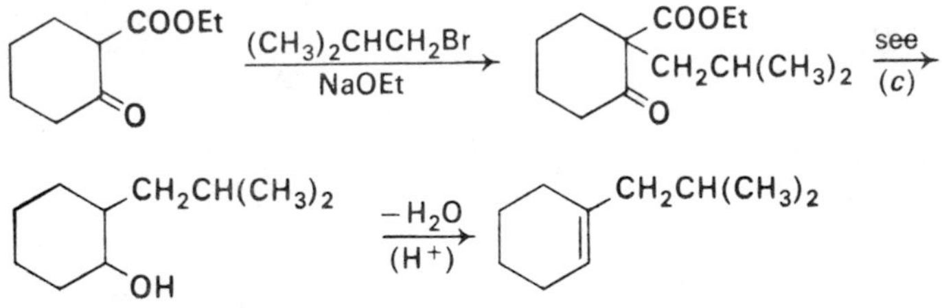

(*j*) $C_6H_5CH{=}CHCOC_6H_5 \xrightarrow{HCN} C_6H_5CH(CN)CH_2COC_6H_5 \xrightarrow[H^+]{H_2O} C_6H_5CH(COOH)CH_2COC_6H_5$

(*k*) $(CH_3O)_2C_6H_3CHO + CH_3MgI \longrightarrow (CH_3O)_2C_6H_3CH(OH)CH_3 \xrightarrow{CrO_3} (CH_3O)_2C_6H_3COCH_3$

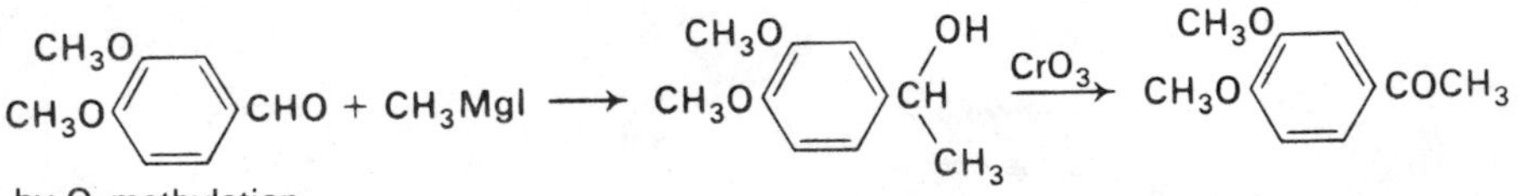

by O-methylation of vanillin

(*l*) From $CH_3CH_2CH_2COOEt + CH_3MgI \longrightarrow$ alcohol $\longrightarrow$ olefin $\longrightarrow$ alkane.

(*m*) ethyl pimelate $\xrightarrow{\text{Dieckmann}}$ 2-(COOEt)cyclohexanone $\xrightarrow{LiAlH_4}$ 2-(CH_2OH)cyclohexanol (OH, CH_2OH)

Exercise 11.2

The preparation of C_6H_5–C(OH)(CH_2OH)–$C_6H_4OCH_3$ (HO, CH_2OH, OCH_3) is not as straightforward as the preparation of the compound in the Example, in which both aryl groups are the same.

An alternate, simpler route is the following; it will be noted that acid-catalyzed rearrangement of the epoxide is equivalent to the corresponding reaction of the glycol:

CH_3O–C_6H_5 $\xrightarrow[AlCl_3]{CH_3COCl}$ CH_3O–C_6H_4–$COCH_3$ $\xrightarrow{C_6H_5MgBr}$

C_6H_5–C(HO)(CH_3)–$C_6H_4OCH_3$ $\xrightarrow{-H_2O}$ C_6H_5–C(=CH_2)–$C_6H_4OCH_3$ $\xrightarrow{C_6H_5CO_3H}$

C_6H_5–C(–O–CH_2)–$C_6H_4OCH_3$ $\xrightarrow{H^+}$ C_6H_5–CH(CHO)–$C_6H_4OCH_3$

Exercise 11.3

(*a*) 3-ethoxycyclohex-2-enone (O, OEt) $\xrightarrow[(2)\ H^+/H_2O]{(1)\ C_6H_5MgBr}$ 3-(C_6H_5)cyclohex-2-enone (O) $\xrightarrow{C_6H_5MgBr}$

3-(C_6H_5)-1-(C_6H_5)cyclohex-2-enol (OH, C_6H_5) $\xrightarrow[(2)\ \Delta,\ S(-H_2)]{(1)\ -H_2O}$ C_6H_5–C_6H_4–C_6H_5 (meta)

(*b*) 3-ethoxycyclohex-2-enone (O, OEt) $\xrightarrow{\text{same sequence as in } (a),}$ but with CH_3O–C_6H_4–MgBr in the first stage, Cl–C_6H_4–MgBr in the second. → Cl–C_6H_4–C_6H_4–C_6H_4–OCH_3 (meta-terphenyl with Cl and OCH_3)

(*c*) 3-ethoxycyclohex-2-enone $\xrightarrow[(2)\ H^+/H_2O]{(1)\ CH_3MgI}$ 3-methylcyclohex-2-enone $\xrightarrow{LiAlH_4}$ 3-methylcyclohex-2-enol $\xrightarrow{-H_2O}$ 1-methylcyclohexa-1,3-diene (CH_3)

Exercise 11.4

(*a*) $C_6H_5CH{=}CHCOOH \xrightarrow[Pt]{H_2} C_6H_5CH_2CH_2COOH \xrightarrow[(2)\ AlCl_3]{(1)\ SOCl_2}$ 1-indanone $\xrightarrow{\phi CO_3H}$

dihydrocoumarin $\xrightarrow[(2)\ HCl]{(1)\ NaOH}$ o-$HOC_6H_4CH_2CH_2COOH$

(*b*) $CH_3COCH_2COOEt + BrCH_2COOEt \xrightarrow{NaOEt} CH_3COCH(COOEt)CH_2COOEt \xrightarrow{saponify}$

$CH_3COCH(COOH)CH_2COOH \xrightarrow[-CO_2]{\Delta} CH_3COCH_2CH_2COOH$

(*c*) $C_6H_5COCH_3 + HCHO + CH_3NH_2 \longrightarrow (C_6H_5COCH_2CH_2)_2NCH_3$

(*d*) $(H_3C)_2CHCHO + HC{\equiv}CMgBr \longrightarrow (H_3C)_2CH{-}CH(OH)C{\equiv}CH \xrightarrow[(2)\ CH_3COCH_2CH_3]{(1)\ EtMgBr}$

$(H_3C)_2CH{-}CH(OH)C{\equiv}C{-}C(OH)(CH_2CH_3)CH_3 \xrightarrow[Pd]{1\ H_2} (H_3C)_2CH{-}CH(OH){-}CH\overset{c}{=}CH{-}C(OH)(CH_2CH_3)CH_3 \xrightarrow{-2\ H_2O}$

(*e*) 2-carbethoxycyclohexanone ($COOEt$) $+ C_6H_5CH_2Cl \xrightarrow{NaOEt}$ 2-carbethoxy-2-benzylcyclohexanone ($COOEt$, $CH_2C_6H_5$) $\xrightarrow[decarboxylate]{saponify}$ 2-benzylcyclohexanone ($CH_2C_6H_5$)

(*f*) $CH_3NH_2 + CH_2{=}CHCOOEt \longrightarrow CH_3N(CH_2CH_2COOEt)_2 \xrightarrow{NaOEt}$

1-methyl-3-carbethoxy-4-piperidone ($COOEt$, $N{-}CH_3$) $\xrightarrow[(2)\ -CO_2]{(1)\ saponify}$ 1-methyl-4-piperidone ($N{-}CH_3$) $\xrightarrow[(2)\ -H_2O]{(1)\ C_6H_5MgBr}$ 1-methyl-4-phenyl-1,2,3,6-tetrahydropyridine (C_6H_5, $N{-}CH_3$)

(*g*) $C_6H_5CH_2COOH \longrightarrow C_6H_5CH_2COOEt \xrightarrow{EtMgBr} C_6H_5CH_2-C(OH)(CH_2CH_3)-CH_2CH_3 \xrightarrow{-H_2O}$

$C_6H_5CH{=}C(CH_2CH_3)_2$

(*h*) $CH_2{=}CHCH_2Cl \longrightarrow CH_2{=}CHCH_2MgCl \xrightarrow{CO_2} CH_2{=}CHCH_2COOH \xrightarrow{LiAlH_4}$

$CH_2{=}CHCH_2CH_2OH \xrightarrow[\text{pyridine}]{(CH_3CO)_2O} CH_2{=}CHCH_2CH_2OCOCH_3$

(*i*) $C_6H_5CHO \longrightarrow C_6H_5CH_2OH \longrightarrow C_6H_5CH_2Cl \xrightarrow[NaOEt]{CH_2(COOEt)_2}$

$C_6H_5CH_2CH(COOEt)_2 \xrightarrow[\substack{(2)\ -CO_2 \\ (3)\ LiAlH_4}]{(1)\ \text{saponify}} C_6H_5CH_2CH_2CH_2OH \xrightarrow[\text{pyridine}]{CH_3SO_2Cl}$

$C_6H_5CH_2CH_2CH_2OSO_2CH_3 \xrightarrow[\text{as above}]{\substack{CH_2(COOEt)_2; \\ \text{then (1), (2), (3),}}} C_6H_5CH_2CH_2CH_2CH_2OH$

(*j*) $p\text{-}CH_3OC_6H_4CHO \xrightarrow{HCN} p\text{-}CH_3OC_6H_4CH(OH)CN \xrightarrow{LiAlH_4} p\text{-}CH_3OC_6H_4CH(OH)CH_2NH_2$

(*k*) $C_6H_5CH_3 \longrightarrow p\text{-}CH_3C_6H_4NO_2 \longrightarrow p\text{-}CH_3C_6H_4NH_2 \longrightarrow p\text{-}CH_3C_6H_4NHCOCH_3 \longrightarrow$

$4\text{-}CH_3\text{-}2\text{-}NO_2C_6H_3NHCOCH_3 \longrightarrow 4\text{-}CH_3\text{-}2\text{-}NH_2C_6H_3NHCOCH_3 \xrightarrow[(2)\ CuCN]{(1)\ HNO_2} 4\text{-}CH_3\text{-}2\text{-}CNC_6H_3NHCOCH_3 \xrightarrow{H_2O/H^+}$

$4\text{-}CH_3\text{-}2\text{-}CNC_6H_3NH_2 \xrightarrow[(2)\ H_3PO_2]{(1)\ HNO_2} m\text{-}CH_3C_6H_4CN \xrightarrow{EtMgBr} m\text{-}CH_3C_6H_4COCH_2CH_3$

(*l*) $C_6H_5CH_2CN \xrightarrow[NaH]{ClCH_2CH_2N(CH_3)CH_2CH_2Cl}$ 4-cyano-4-phenyl-1-methylpiperidine [ring: $C(C_6H_5)(CN)$, H_2C, CH_2, H_2C, CH_2, $N-CH_3$] $\xrightarrow{H_2O/H^+}$ 4-carboxy-4-phenyl-1-methylpiperidine [C_6H_5, HOOC on C-4; $N-CH_3$]

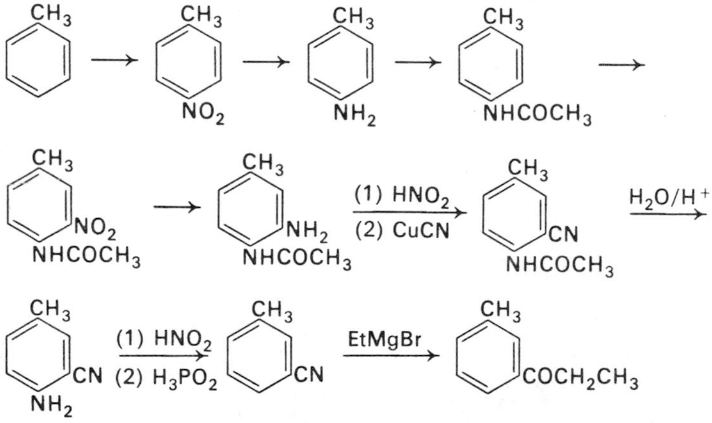

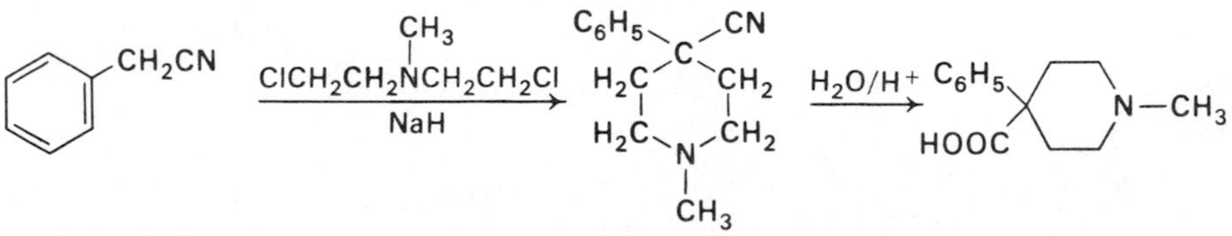

PART 12, B

16. 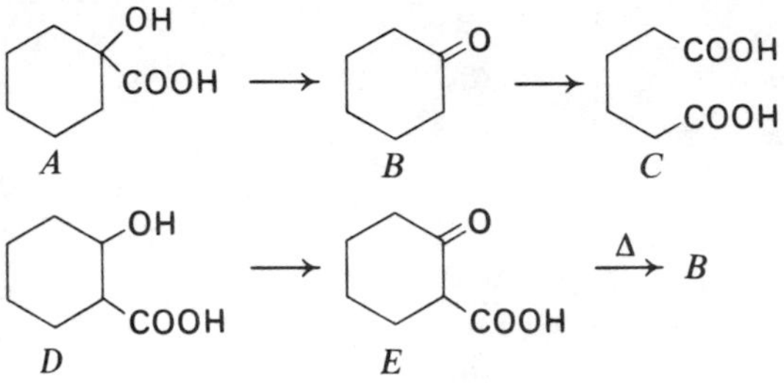 1-hydroxycyclohexanecarboxylic acid (A) $\longrightarrow$ cyclohexanone (B) $\longrightarrow$ $HOOC(CH_2)_4COOH$ (C)

2-hydroxycyclohexanecarboxylic acid (D) $\longrightarrow$ 2-oxocyclohexanecarboxylic acid (E) $\xrightarrow{\Delta}$ B

17. 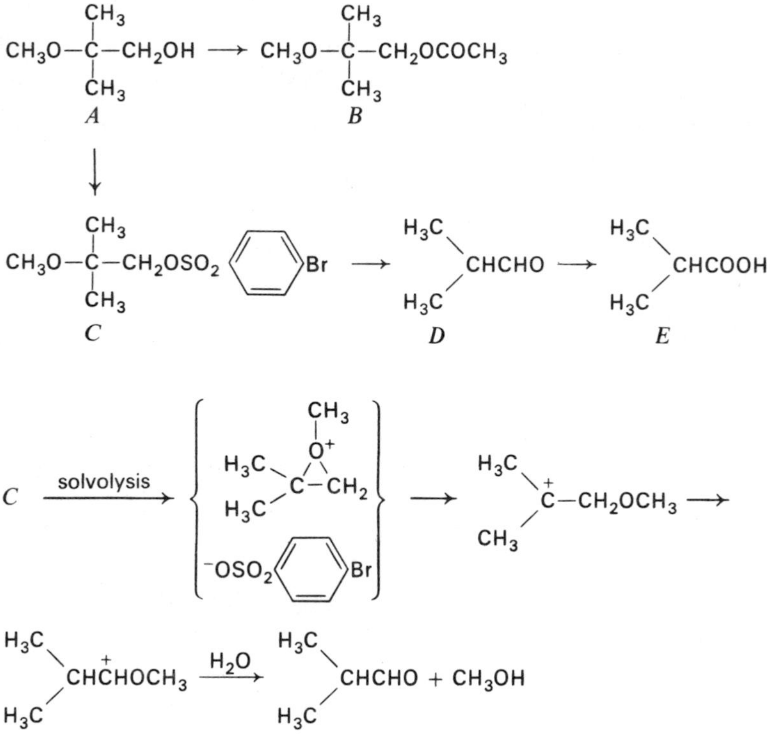 $CH_3O{-}C(CH_3)_2{-}CH_2OH$ (A) $\longrightarrow$ $CH_3O{-}C(CH_3)_2{-}CH_2OCOCH_3$ (B)

A $\downarrow$

$CH_3O{-}C(CH_3)_2{-}CH_2OSO_2C_6H_4Br$ (C) $\longrightarrow$ $(H_3C)_2CHCHO$ (D) $\longrightarrow$ $(H_3C)_2CHCOOH$ (E)

$C \xrightarrow{\text{solvolysis}} \left\{ (H_3C)_2C{-}CH_2 \text{ bridged by } \overset{+}{O}{-}CH_3 \quad {}^{-}OSO_2C_6H_4Br \right\} \longrightarrow (H_3C)(CH_3)\overset{+}{C}{-}CH_2OCH_3 \longrightarrow$

$(H_3C)_2CH\overset{+}{C}HOCH_3 \xrightarrow{H_2O} (H_3C)_2CHCHO + CH_3OH$

18. 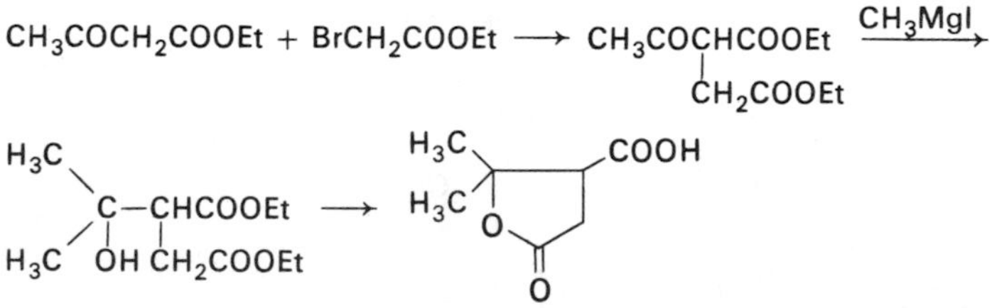 $CH_3COCH_2COOEt + BrCH_2COOEt \longrightarrow CH_3CO\underset{CH_2COOEt}{\underset{|}{C}}HCOOEt \xrightarrow{CH_3MgI}$

$(H_3C)_2\underset{OH}{\underset{|}{C}}{-}\underset{CH_2COOEt}{\underset{|}{C}}HCOOEt \longrightarrow$ 5,5-dimethyl-2-oxotetrahydrofuran-4-carboxylic acid (lactone ring: $(H_3C)_2C$–O–C(=O)–CH_2–CH(COOH))

19. (*a*) $CH_3CH_2CH_2COOCH_2CH_3$ $\qquad$ $CH_3CH_2COOCH(CH_3)_2$

$CH_3COOCH(CH_3)(CH_2CH_3)$ $\qquad$ $HCOOCH(CH_3)(CH_2CH_2CH_3)$

$(CH_3)_2CHCOOCH_2CH_3$

(*Note:* Iodoform is formed from the alcohol (CH_3CH_2OH or $RCH(OH)CH_3$) produced on saponification.)

(*b*) Final proof of structure by preparation of crystalline derivative of the acid. It will be seen that the acid components of all of the esters in (*a*) are different.

20.

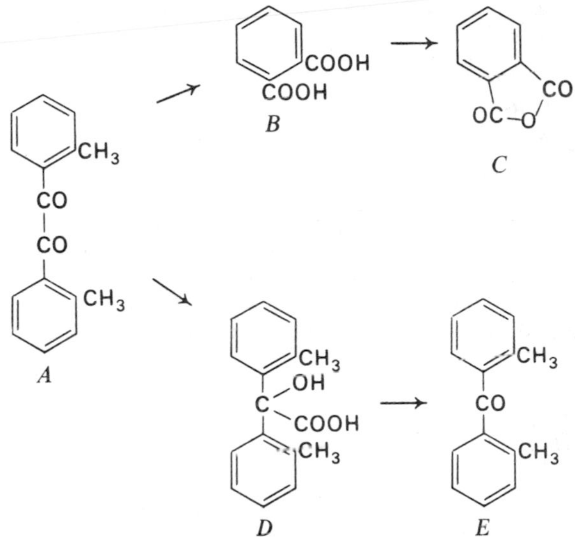

A: $CH_3C_6H_4$–CO–CO–$C_6H_4CH_3$ → *B*: $C_6H_4(COOH)_2$ (COOH, COOH) → *C*: $C_6H_4(CO)_2O$ (CO, OC–O)

A → *D*: $(CH_3C_6H_4)_2C(OH)COOH$ (CH_3, OH, C, COOH, CH_3) → *E*: $CH_3C_6H_4$–CO–$C_6H_4CH_3$ (CH_3, CO, CH_3)

Notice that *A* is described as a yellow compound. The presence of visible color is often a valuable clue to structure. In the present example the compound is an α-diketone, which are characteristically yellow in color.

21. $ICH_2CH_2COOEt + CH_2(CN)COOEt \longrightarrow (EtOOCCH_2CH_2)_2C(CN)COOEt$ (*A*) $\longrightarrow$

$(HOOCCH_2CH_2)_2CHCOOH \longrightarrow$ O=C_6H_9–COOH (*C*) $\longrightarrow$ HO, H_3C >C_6H_{10}–C(OH)(CH_3)$_2$ (*D*) $\longrightarrow$

E (or) $\longrightarrow$ *F*

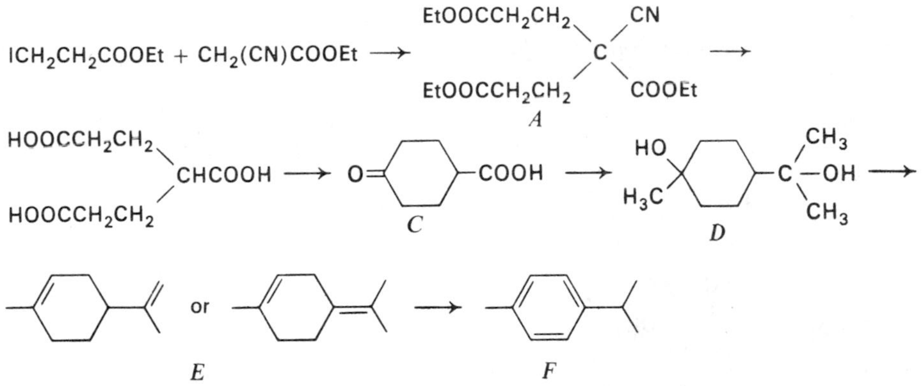

22. (+)-*A* → *C* + *D*

C ↙ ↘ *D*

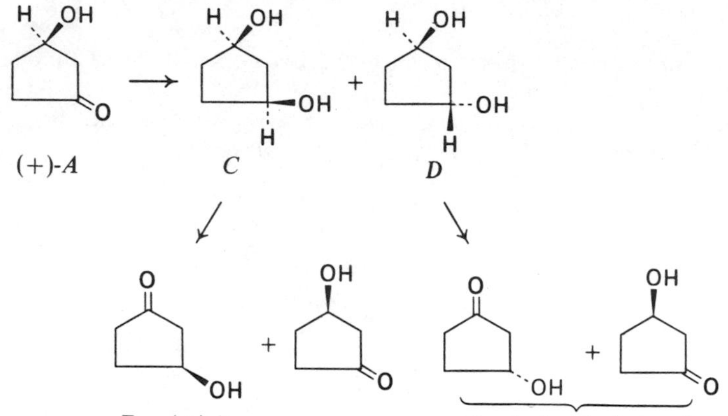

$E = (\pm)A$

identical = (+)-*A* = *F*

$G = HOOCCH_2CH_2COOH$

23. $BrCH_2CH_2Br \longrightarrow NCCH_2CH_2CN$ (*A*) $\longrightarrow EtOOCCH_2CH_2COOEt$ (*B*) $\longrightarrow$

$\{(Et)_2C(OH)CH_2CH_2C(OH)(Et)_2\} \xrightarrow{H^+}$ *C*

24. → *A* → $BrCH_2CH_2CH_2CH_2CH_2Br$ (*B*) → $Br(CH_2)_5OCOCH_3$ (*C*) →

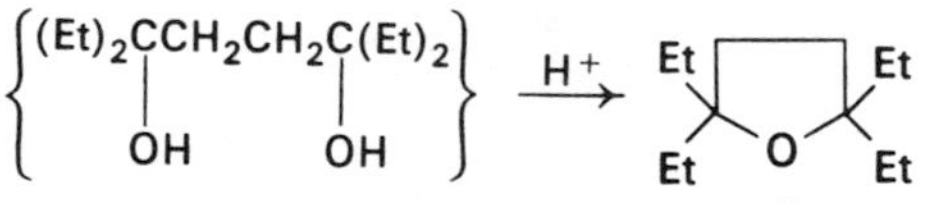

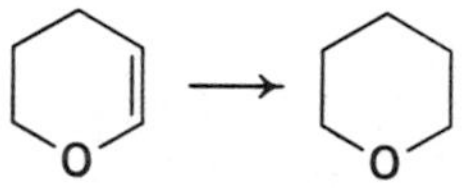

$Br(CH_2)_5OH$ (*D*) → $Br(CH_2)_5OSO_2CH_3$ (*E*) → *F* (N–CH_3)

25. $C_6H_5CH_2COCH_2CH_2COOEt$ (*A*) → $C_6H_5CH_2COCH_2CH_2COOH$ (*B*) →

C   *D* = C_6H_5COOH

↓ O_3

CH_2COOH—CH_2COOH (*G*) ← *F* + C_6H_5CHO (*E*)

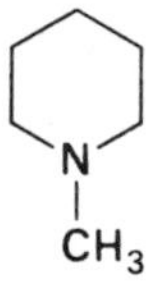

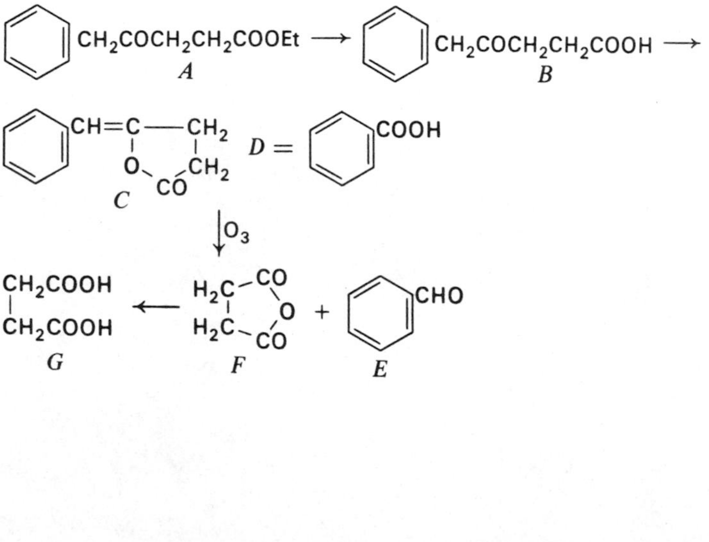

26. $C_6H_5C(CH_3){=}NOH$ → $C_6H_5C(CH_3){=}NOCOC_6H_5$

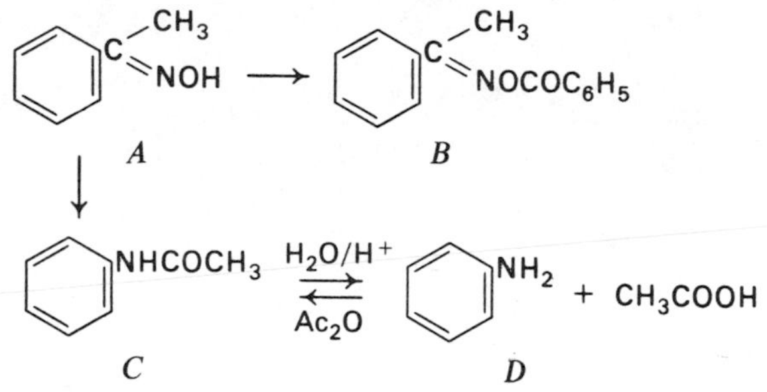

A B

↓

$C_6H_5NHCOCH_3$ $\underset{Ac_2O}{\overset{H_2O/H^+}{\rightleftarrows}}$ $C_6H_5NH_2$ + CH_3COOH

C D

27. CH_3O, CH_3O, N—CH_3, CH_3O → CH_3O, CH_3O, $\overset{+}{N}$ CH_3 CH_3 I^-, CH_3O →

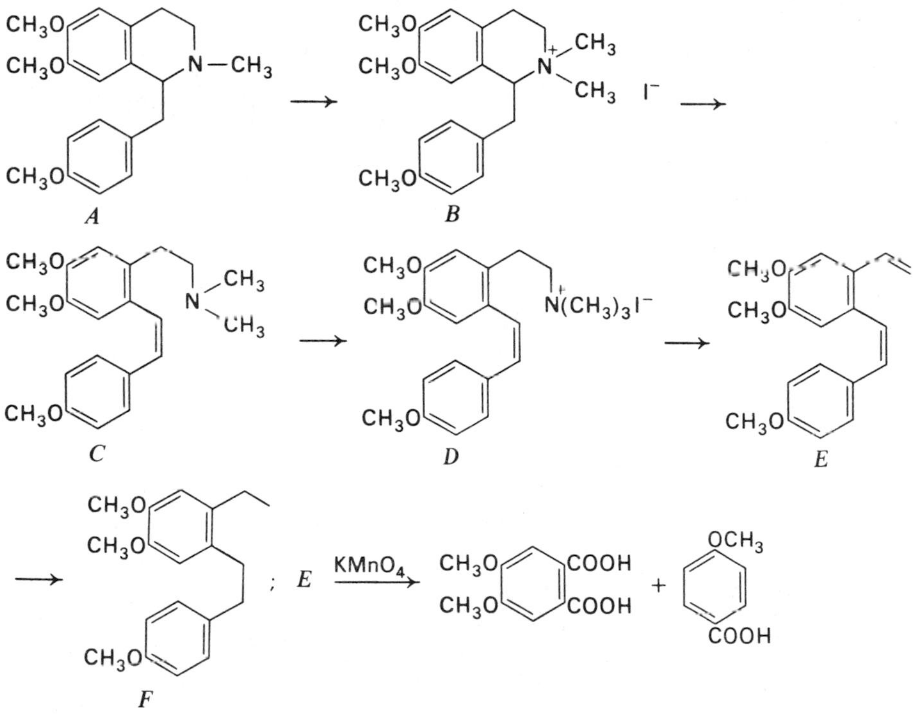

A B

CH_3O, CH_3O, N CH_3 CH_3, CH_3O → CH_3O, CH_3O, $\overset{+}{N}(CH_3)_3I^-$, CH_3O → CH_3O, CH_3O, CH_3O

C D E

→ CH_3O, CH_3O, CH_3O ; E $\xrightarrow{KMnO_4}$ CH_3O, CH_3O, COOH, COOH + OCH_3, COOH

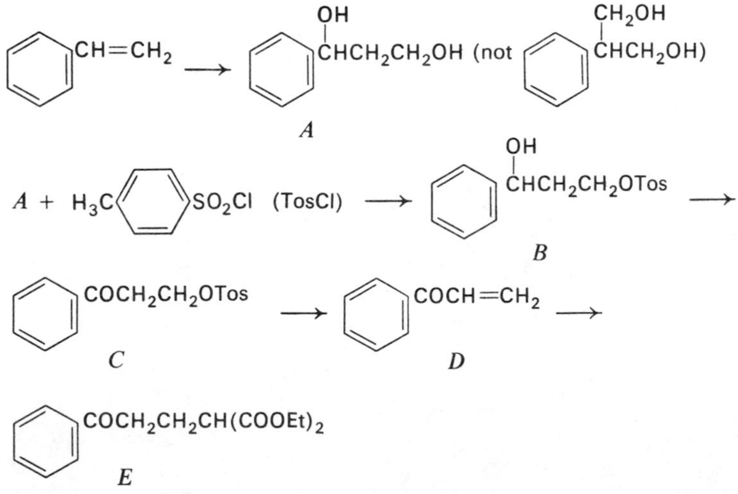

F

28. $C_6H_5CH{=}CH_2$ → $C_6H_5\underset{|}{\overset{OH}{C}}HCH_2CH_2OH$ (not $C_6H_5\overset{CH_2OH}{C}HCH_2OH$)

A

A + $H_3C\text{-}C_6H_4\text{-}SO_2Cl$ (TosCl) → $C_6H_5\overset{OH}{C}HCH_2CH_2OTos$ →

B

$C_6H_5COCH_2CH_2OTos$ → $C_6H_5COCH{=}CH_2$ →

C D

$C_6H_5COCH_2CH_2CH(COOEt)_2$

E

29. O + N H → N A $\xrightarrow{CH_3COCl}$ $\overset{+}{N}$ CH_3 C $OCOCH_3$ } Cl^- $\xrightarrow{H_2O}$

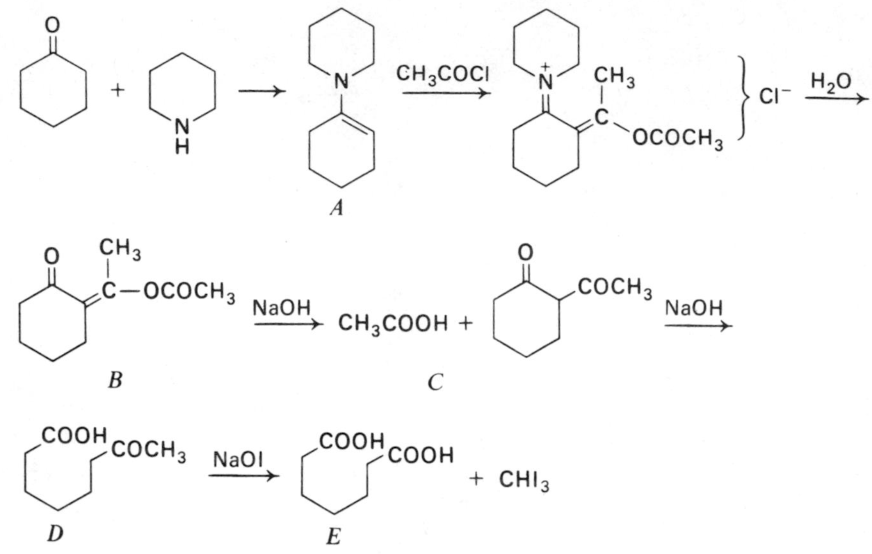

O CH_3 C—$OCOCH_3$ B $\xrightarrow{NaOH}$ CH_3COOH + O $COCH_3$ C $\xrightarrow{NaOH}$

COOH $COCH_3$ D $\xrightarrow{NaOI}$ COOH COOH E + CHI_3

30. O + $Me_2\overset{+}{N}H_2Cl^-$ + CH_2O → CH_2NMe_2 O A →

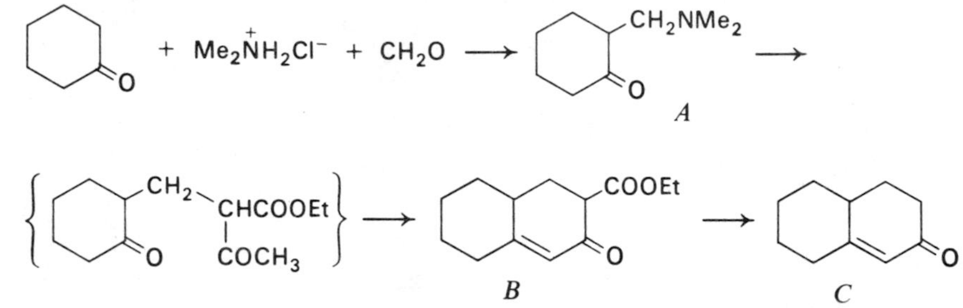

{ CH_2 CHCOOEt O $COCH_3$ } → COOEt O B → O C

31. O + Br_2 → O Br Br A $\xrightarrow{NaOH}$ { O Br } →

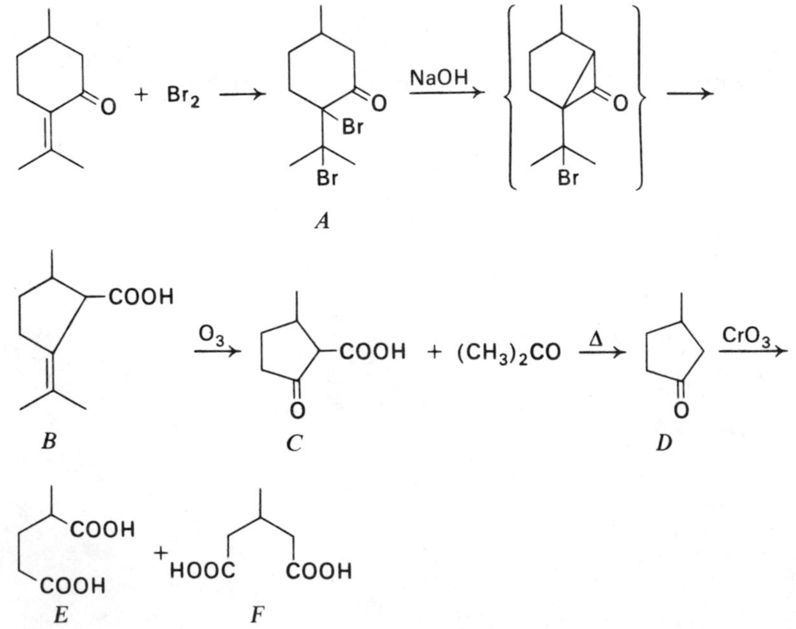

COOH B $\xrightarrow{O_3}$ COOH O C + $(CH_3)_2CO$ $\xrightarrow{\Delta}$ O D $\xrightarrow{CrO_3}$

COOH COOH E + HOOC COOH F

32. $CH_3OCH_2COOEt + C_6H_5MgBr \longrightarrow CH_3OCH_2C(C_6H_5)_2{-}OH$ (*A*) $\xrightarrow[H_2O]{H^+}$ $\left\{ CH_3OCH_2\overset{+}{C}(C_6H_5){-}C_6H_5 \longrightarrow CH_3O\overset{+}{C}HCH(C_6H_5)_2 \right\}$ $\xrightarrow{H_2O}$ $(C_6H_5)_2CHCHO$ (*B*) $\longrightarrow (C_6H_5)_2CHCOOH$ (*C*)

PART 13

1. $HC{\equiv}CH + CH_3CH_2COCH_3 \longrightarrow HC{\equiv}C{-}C(OH)(CH_3){-}CH_2CH_3 \xrightarrow{EtMgBr}$

$BrMgC{\equiv}C{-}C(OMgBr)(CH_3){-}CH_2CH_3 \xrightarrow{(CH_3)_2CHCHO} (H_3C)_2CH{-}CH(OH){-}C{\equiv}C{-}C(OH)(CH_3){-}CH_2CH_3 \xrightarrow[Pd]{H_2}$

$(H_3C)_2CH{-}CH(OH)CH{=}CH{-}C(OH)(CH_3){-}CH_2CH_3 \xrightarrow{-H_2O} (H_3C)_2C{=}CH{-}CH{=}CH{-}C(CH_3){=}CHCH_3$

2. $CH_3COCH_2COOEt + BrCH_2COOEt \xrightarrow{NaOEt} CH_3COCH(CH_2COOEt)COOEt$ (*A*) $\xrightarrow[NaOEt]{CH_3I}$

$CH_3COC(CH_3)(CH_2COOEt)COOEt$ (*B*) $\xrightarrow{CH_3MgI} (H_3C)_2C(OH){-}C(CH_3)(CH_2COOEt){-}COOEt$ (*C*) $\xrightarrow{NaOH}$

$(H_3C)_2C(OH){-}C(CH_3)(CH_2COOH){-}COOEt$ (*D*) $\xrightarrow{\text{further}} (H_3C)_2C(OH){-}C(CH_3)(CH_2COOH){-}COOH$ (*E*)

F: lactone — ring of $(H_3C)_2C$, $C(CH_3)(COOEt)$, CH_2, CO, O

3. 4-isopropylanisole (OCH_3) $\xrightarrow[NH_3]{Na}$ *A* (dihydroanisole, OCH_3) $\xrightarrow[H_2O]{H^+}$ *B* (4-isopropylcyclohex-2-enone, =O) $\xrightarrow[Pt]{H_2}$ *C* (3-isopropylcyclohexanone, =O) $\xrightarrow[EtO^-]{HCOOEt}$

D (=CHOH, =O) $\longrightarrow$ *E* (=CHOiBu, =O) $\xrightarrow{LiAlH_4}$ *F* (=CHOiBu, OH) $\xrightarrow[H^+]{H_2O}$ *G* (CHO, cyclohexene)

4. 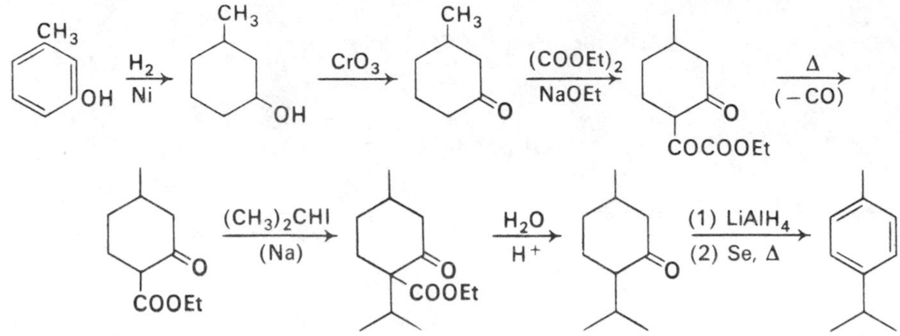 m-cresol (CH_3, OH on benzene) $\xrightarrow[Ni]{H_2}$ 3-methylcyclohexanol $\xrightarrow{CrO_3}$ 3-methylcyclohexanone $\xrightarrow[NaOEt]{(COOEt)_2}$ (COCOOEt derivative) $\xrightarrow[(-CO)]{\Delta}$ (COOEt derivative) $\xrightarrow[(Na)]{(CH_3)_2CHI}$ (isopropyl, COOEt derivative) $\xrightarrow[H^+]{H_2O}$ menthone $\xrightarrow[(2)\ Se,\ \Delta]{(1)\ LiAlH_4}$ p-cymene

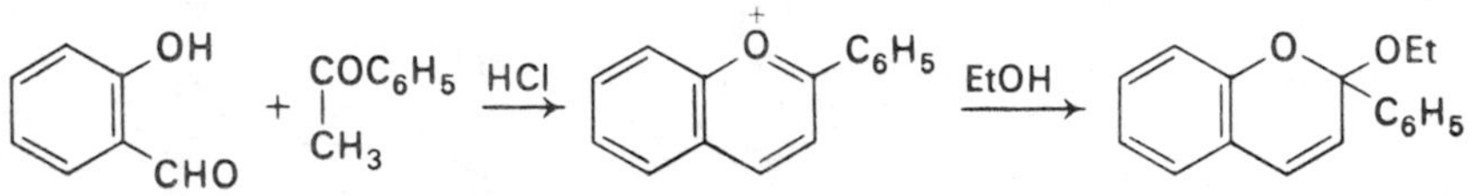

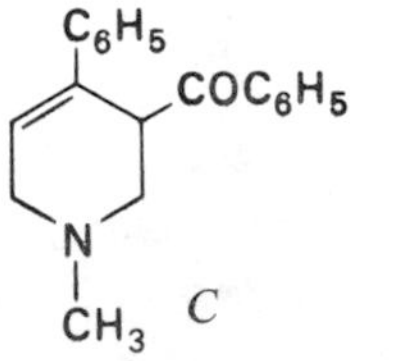

5. salicylaldehyde (OH, CHO) + COC_6H_5–CH_3 $\xrightarrow{HCl}$ flavylium ($\overset{+}{O}$, C_6H_5) *A* $\xrightarrow{EtOH}$ (O, OEt, C_6H_5) *B*

6. 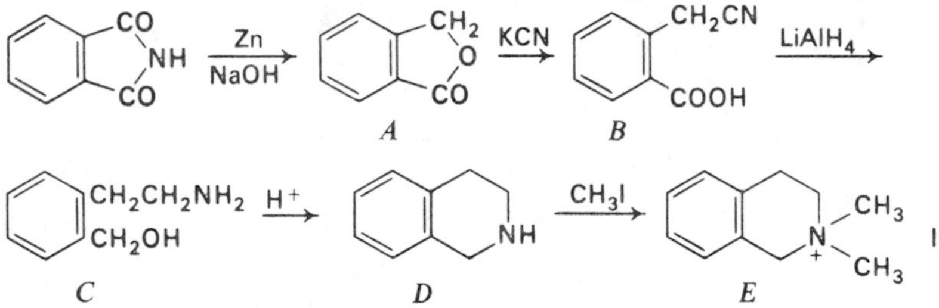 phthalimide (CO, NH, CO) $\xrightarrow[NaOH]{Zn}$ phthalide (CH_2, O, CO) *A* $\xrightarrow{KCN}$ (CH_2CN, COOH) *B* $\xrightarrow{LiAlH_4}$

($CH_2CH_2NH_2$, CH_2OH) *C* $\xrightarrow{H^+}$ (tetrahydroisoquinoline, NH) *D* $\xrightarrow{CH_3I}$ ($\overset{+}{N}$, CH_3, CH_3) I^- *E*

7. $C_6H_5COCH_3 + CH_2O + CH_3NH_2{\cdot}HCl \longrightarrow (C_6H_5COCH_2CH_2)_2NCH_3$ *A*

A + NaOH $\longrightarrow$ (piperidine: HO, C_6H_5, COC_6H_5, N–CH_3) *B*

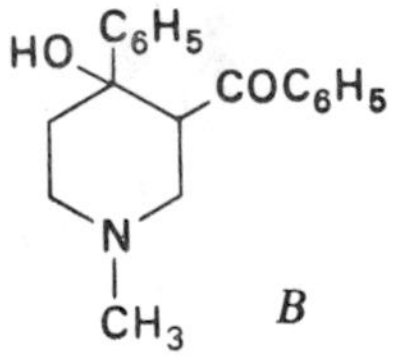

B + HCl $\longrightarrow$ (tetrahydropyridine: C_6H_5, COC_6H_5, N–CH_3) *C* instead of (tetrahydropyridine: C_6H_5, COC_6H_5, N–CH_3)

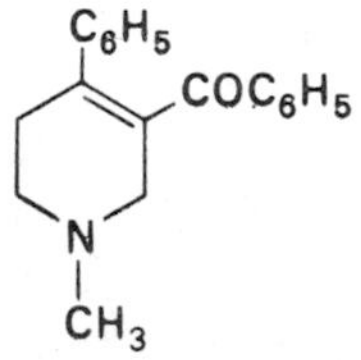

The conjugated unsaturated ketone expected from the dehydration of *B* is not formed, probably because of the interference between the phenyl and benzoyl groups on the *cis* double bond.

8. 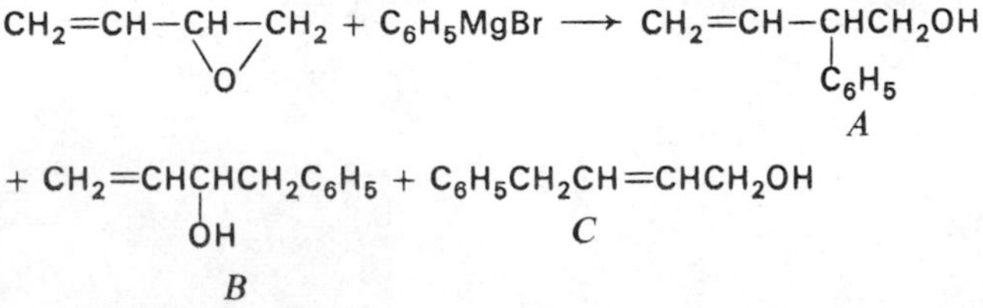 $CH_2{=}CH{-}CH{-}CH_2$ (epoxide, O) $+ C_6H_5MgBr \longrightarrow CH_2{=}CH{-}CH(C_6H_5)CH_2OH$ *A*

$+ CH_2{=}CHCH(OH)CH_2C_6H_5 + C_6H_5CH_2CH{=}CHCH_2OH$
B *C*

Chromic acid oxidation of *B* and *C* would give rise to the system C=C—CO, which would show high intensity absorption in the neighborhood of 220–230 nm.

9. 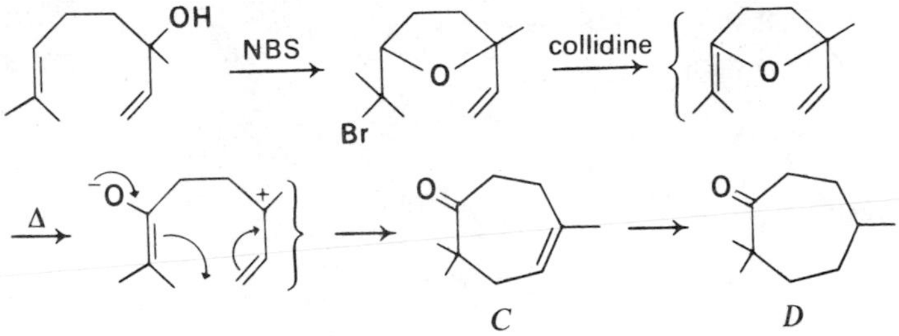 OH; NBS; Br; O; collidine; O

Δ; ^{-}O; +; O; O

C *D*

10. 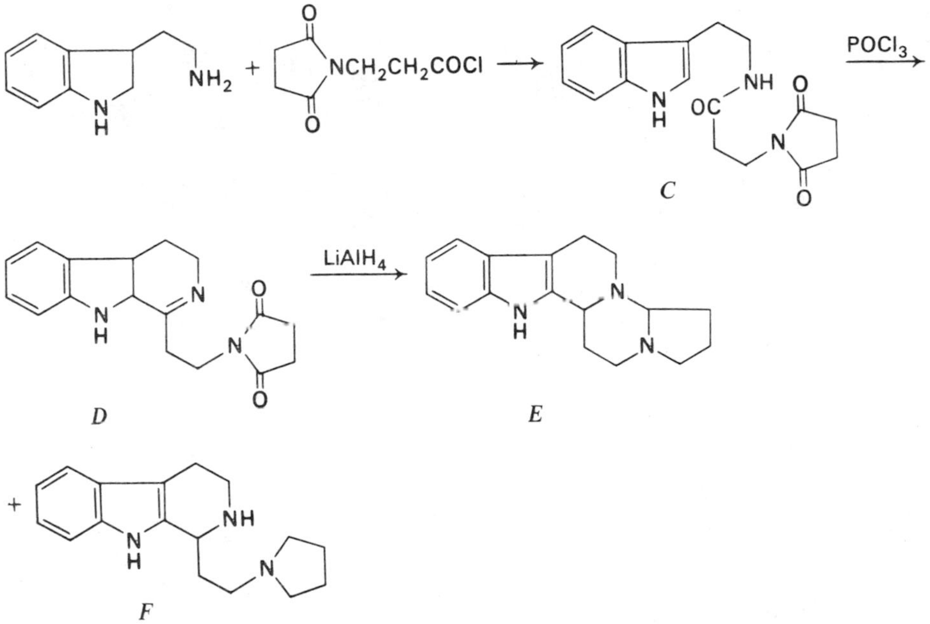 NH_2 + O; NCH_2CH_2COCl; O; N H; NH; OC; O; N; O; N H; $POCl_3$

C

N H; N; O; N; O; $LiAlH_4$; N H; N; N

D *E*

+ N H; NH; N

F

11. 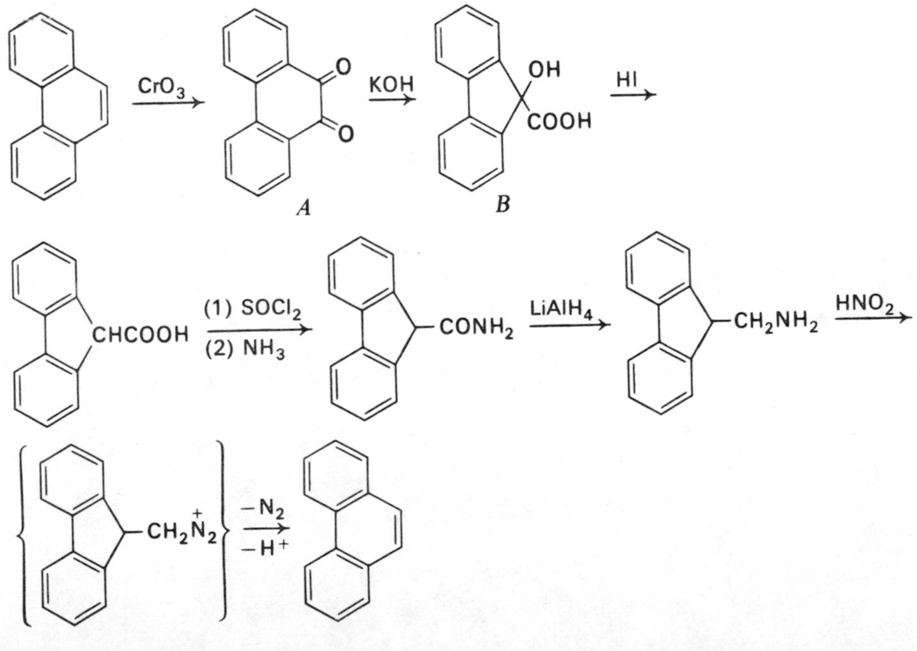 CrO_3; O; O; KOH; OH; COOH; HI

A *B*

CHCOOH; (1) $SOCl_2$ (2) NH_3; $CONH_2$; $LiAlH_4$; CH_2NH_2; HNO_2

$CH_2\overset{+}{N}_2$; $-N_2$ $-H^+$

12. 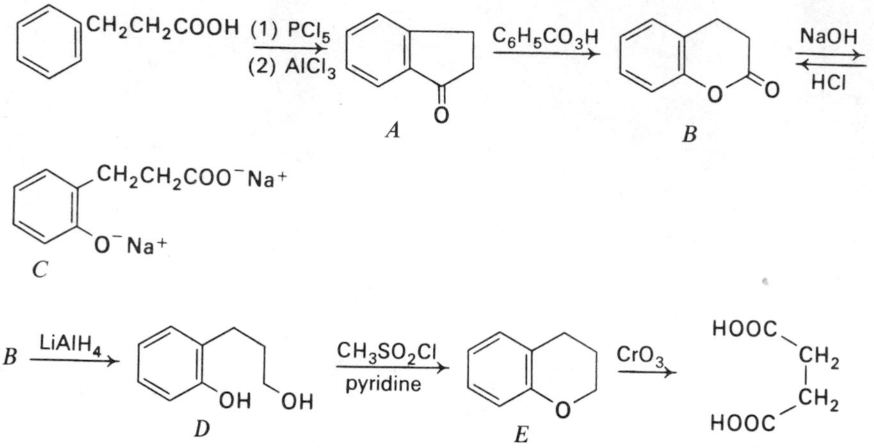 CH_2CH_2COOH (1) PCl_5 (2) $AlCl_3$ → *A* → $C_6H_5CO_3H$ → *B* ⇌ (NaOH / HCl)

$CH_2CH_2COO^-Na^+$, O^-Na^+
C

B → $LiAlH_4$ → OH OH *D* → CH_3SO_2Cl, pyridine → O *E* → CrO_3 → HOOC CH_2 CH_2 HOOC

13. 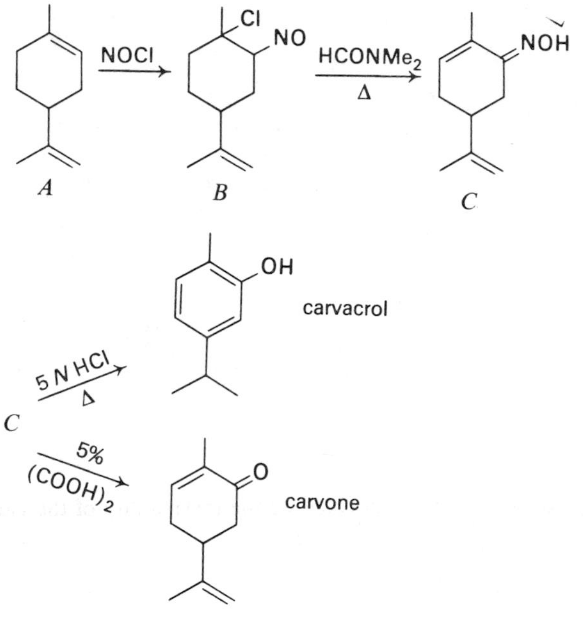 *A* → NOCl → Cl NO *B* → $HCONMe_2$, Δ → NOH *C*

OH carvacrol

C → 5 N HCl, Δ → carvacrol

C → 5% $(COOH)_2$ → O carvone

(*Note:* *B* is written as the C-nitroso compound, but probably has the tautomeric $\rangle C{=}NOH$ structure.)

14. $C_6H_5CHNH_2$ ($CH_2CH{=}CH_2$) $\xrightarrow[(H^+)]{H_2CO}$ $C_6H_5CH{-}\overset{+}{N}H{=}CH_2$ ($CH_2CH{=}CH_2$) ⟶

$C_6H_5CH{=}\overset{+}{N}H{-}CH_2CH{=}CH_2$ $\xrightarrow{H_2O}$ $C_6H_5CHO + H_2NCH_2CH_2CH{=}CH_2$ (*C*) $\xrightarrow[HCOOH]{H_2CO}$

$CH_2{=}CHCH_2CH_2NMe_2 + CO_2$
A

$\xrightarrow{H_2(Pt)}$ $CH_3CH_2CH_2CH_2NMe_2$
B

Synthesis of A:

$CH_3CH_2CH_2CH_2OH + H_2CO + Me_2NH \longrightarrow$

$$\underset{D}{CH_3CH_2CH_2CH_2{-}OCH_2NMe_2} \xrightarrow{CH_2{=}CHCH_2MgBr} \underset{A}{CH_2{=}CHCH_2CH_2NMe_2}$$

15. 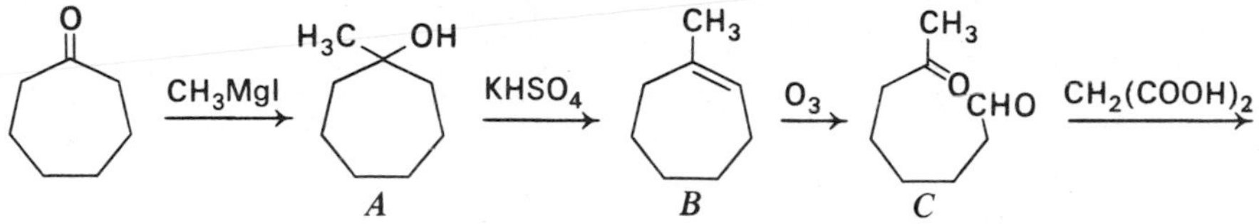 Cycloheptanone $\xrightarrow{CH_3MgI}$ 1-methylcycloheptanol (H$_3$C, OH) *A* $\xrightarrow{KHSO_4}$ 1-methylcycloheptene *B* $\xrightarrow{O_3}$ *C* (CH$_3$CO–(CH$_2$)$_5$–CHO) $\xrightarrow{CH_2(COOH)_2}$

$$\underset{D}{CH_3CO(CH_2)_5CH{=}CHCOOH} \longrightarrow \underset{E}{CH_3CO(CH_2)_7COOH} \xrightarrow{NaOI}$$

$$\underset{F}{HOOC(CH_2)_7COOH} + CHI_3.$$

16. 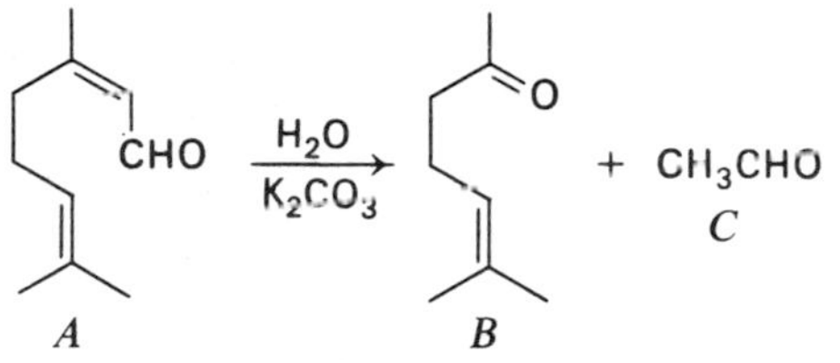 *A* (CHO) $\xrightarrow[K_2CO_3]{H_2O}$ *B* (O) + $\underset{C}{CH_3CHO}$

B + NaOI $\longrightarrow$ *D* (COOH) $\xrightarrow{O_3}$ $H_3C{-}CO{-}CH_3$

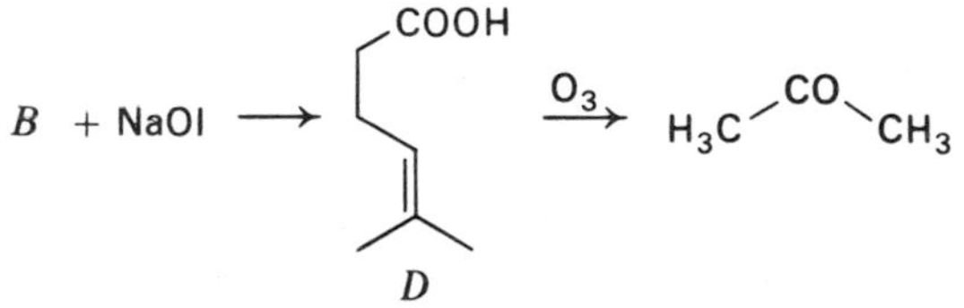

The base-catalyzed cleavage of *A* to give *B* and *C* shows the reversibility of the aldol condensation. Although *A* cannot in fact be prepared by the aldol condensation of *B* and *C* (side reactions intervene), *A* can be regarded as having been so derived.

17. (*a*) (tetrahydropyran-2-yl)CHO $\longrightarrow$ (tetrahydropyran-2-yl)COOH $\xrightarrow[-H^+]{-2e}$ $\{$(tetrahydropyran-2-yl)COO$^+\}$ $\longrightarrow$

(oxocarbenium ion, $\overset{+}{O}$) + CO_2 $\xrightarrow{H_2O}$ (tetrahydropyran-2-yl)OH $\longrightarrow$ (lactone, =O)

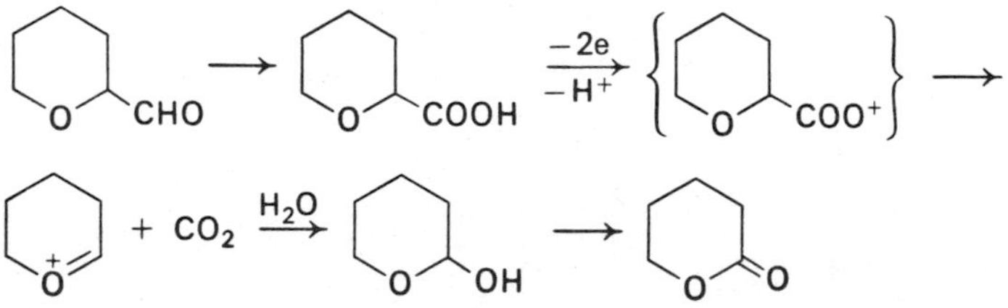

Steps shown separately may be concerted reactions.

(*b*) N-methylpyridinium ($\overset{+}{N}$–CH$_3$) $\underset{}{\overset{OH^-}{\rightleftharpoons}}$ 2-hydroxy-1,2-dihydro-1-methylpyridine (H, OH; N–CH$_3$) $\longrightarrow$ 1-methyl-2-pyridone (N–CH$_3$, =O)

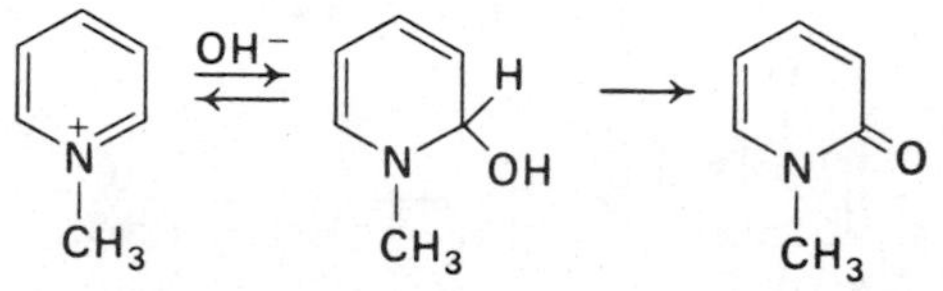

(c) [structures: decalin with OH → O⁺ → ring-opened oxonium → ketone with alkene]

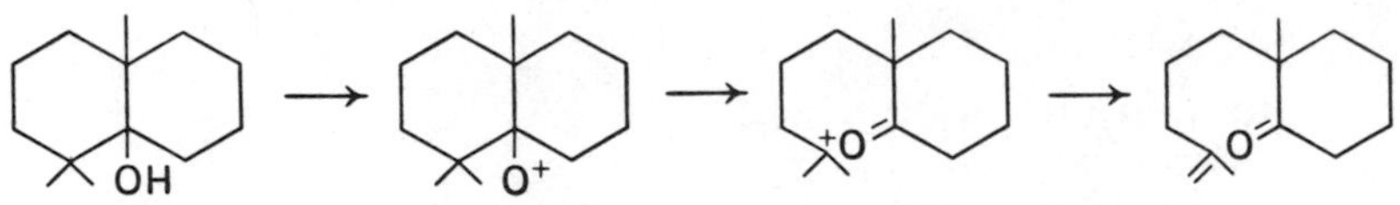

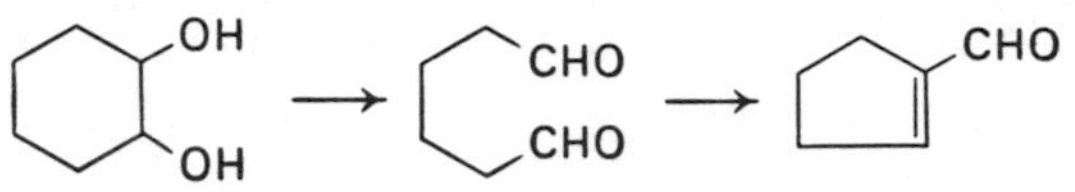

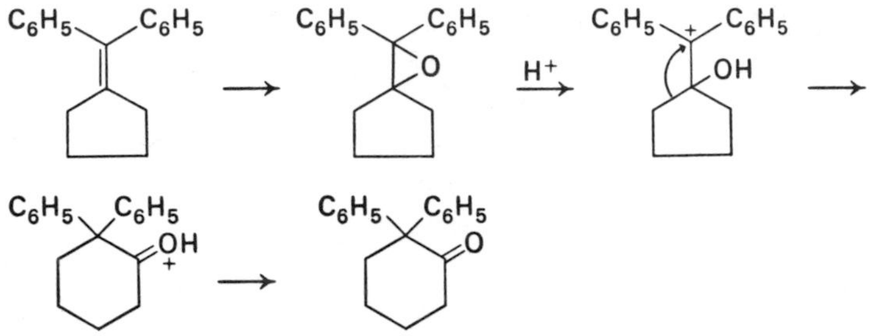

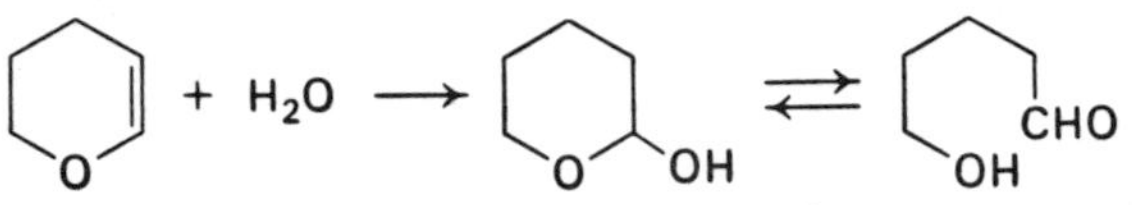

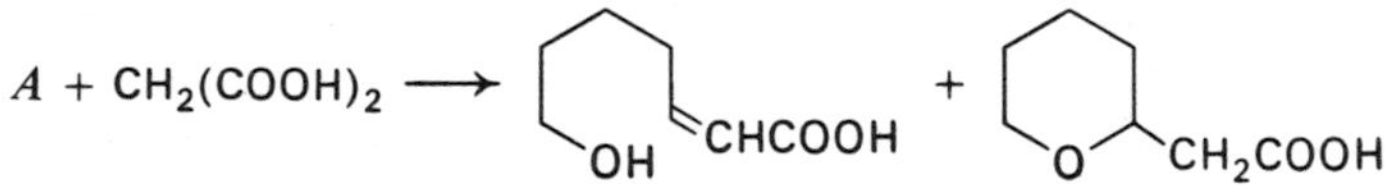

(d) [cyclohexane-1,2-diol] → $OHC(CH_2)_4CHO$ → [cyclopentene-1-carbaldehyde] CHO

(e) $(C_6H_5)_2C$=[cyclopentylidene] → [epoxide] $\xrightarrow{H^+}$ $C_6H_5\overset{+}{C}C_6H_5$–[1-hydroxycyclopentyl] →

[2,2-diphenylcyclohexanone, protonated: $\overset{+}{O}H$] → [2,2-diphenylcyclohexanone] O

18. [3,4-dihydro-2H-pyran] + H_2O → [tetrahydropyran-2-ol] OH ⇌ $HO(CH_2)_4CHO$
A

A + $CH_2(COOH)_2$ → $HO(CH_2)_4CH{=}CHCOOH$ (*B*) + [tetrahydropyran-2-yl]CH_2COOH (*C*)

B + PBr_3 or *C* + HBr → $Br(CH_2)_4CH{=}CHCOOH$
D

D + CH_3COCH_2COOEt → $CH_3COCH(COOEt)(CH_2)_4CH{=}CHCOOH$
E

E → $CH_3COCH(COOH)(CH_2)_4CH{=}CHCOOH$ (*F*) $\xrightarrow{\Delta}$ $CH_3CO(CH_2)_5CH{=}CHCOOH$ (*G*)

19.

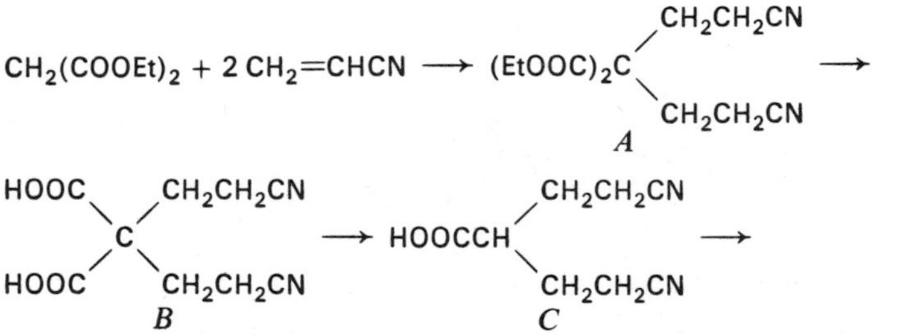

$CH_2(COOEt)_2 + 2\,CH_2{=}CHCN$ → $(EtOOC)_2C(CH_2CH_2CN)_2$ →
A

$(HOOC)_2C(CH_2CH_2CN)_2$ (*B*) → $HOOCCH(CH_2CH_2CN)_2$ (*C*) →

$CH_3OCOCH(CH_2CH_2COOCH_3)_2$ (*D*) → [2-oxocyclohexane with 3-$COOCH_3$ and 5-$COOCH_3$] (*E*) →

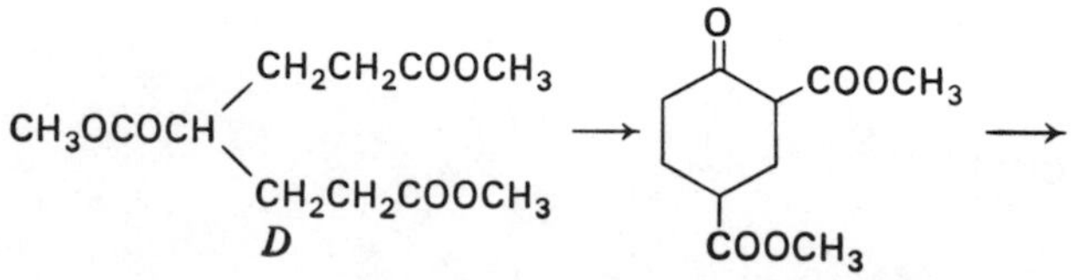

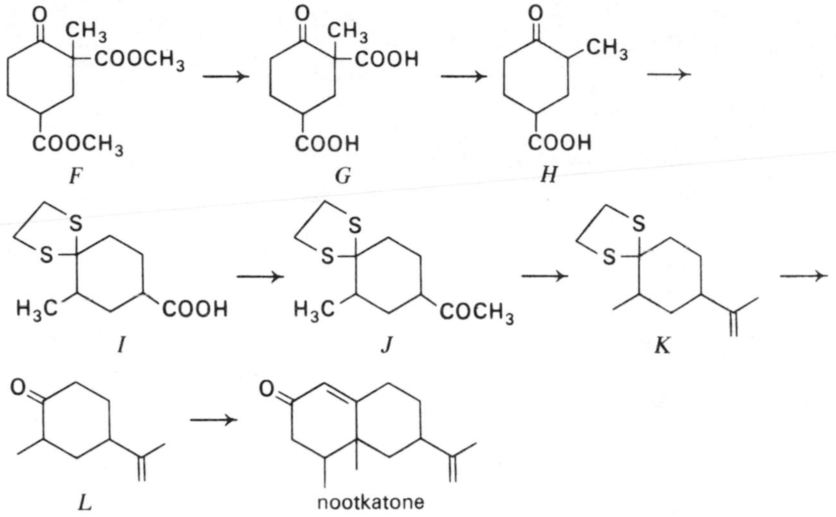

F → G → H → I → J → K → L → nootkatone

20.

$\xrightarrow{MeO^-}$ → $\xrightarrow{MeO^-}$ ≡

21.

A + CH_3COCH_2COOEt → ≡

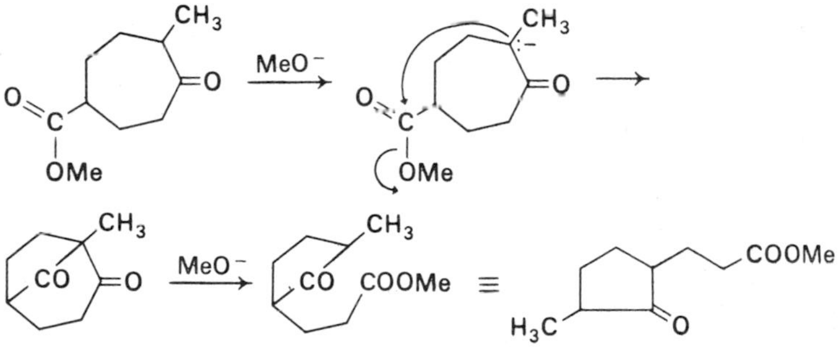

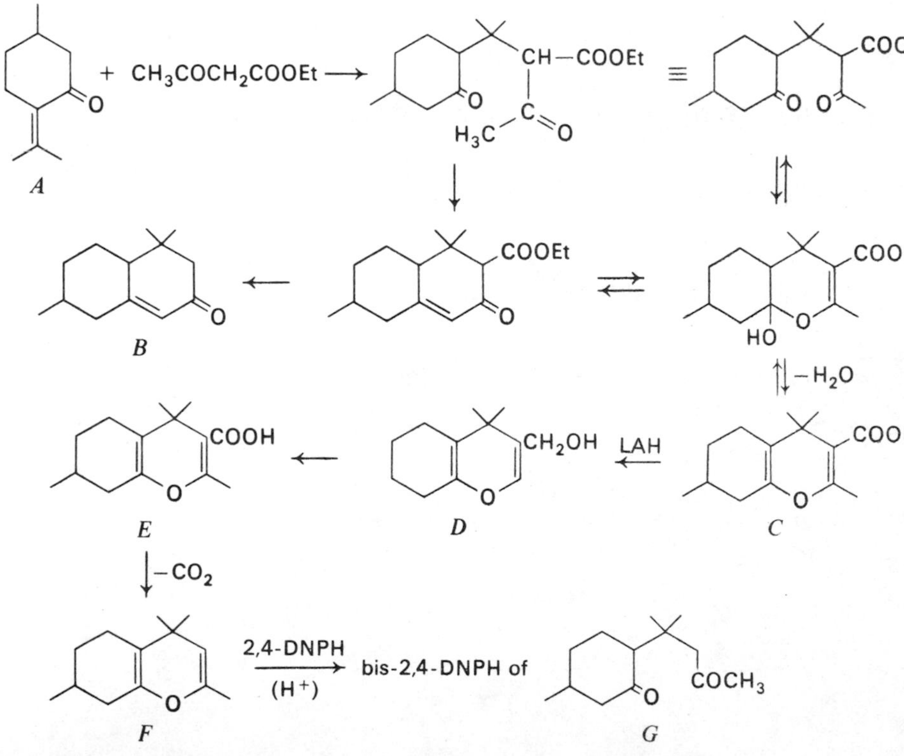

B ← ⇌

E ← D $\xleftarrow{LAH}$ C ($-H_2O$)

E $\xrightarrow{-CO_2}$ F $\xrightarrow[(H^+)]{2,4\text{-DNPH}}$ bis-2,4-DNPH of G

22. $$C_6H_5COCH_3 + Me_2\overset{+}{N}H_2Cl^- + CH_2O \longrightarrow \underset{A}{C_6H_5COCH_2CH_2NMe_2} \longrightarrow$$

$$\underset{B}{(C_6H_5)_2C(OH)CH_2CH_2NMe_2} \xrightarrow{Ac_2O} (C_6H_5)_2C(OAc){-}CH_2{-}CH_2{-}NMe_2 \longrightarrow$$

$$\underset{C}{(C_6H_5)_2C{=}CH_2} + \left\{CH_2{=}\overset{+}{N}Me_2 \xrightarrow{H_2O} CH_2O + Me_2NH\right\}$$

$$\longrightarrow \underset{D}{(C_6H_5)_2C{=}O} \longrightarrow$$

$$\underset{E}{(C_6H_5)_2C{=}NOH} \longrightarrow \underset{F}{C_6H_5NHCOC_6H_5} \longrightarrow \underset{G}{C_6H_5COOH} + \underset{H}{C_6H_5NH_2}$$

23. The reaction leading to *A* is a Darzens condensation, perhaps proceeding by way of an initial replacement of —Cl by $—NEt_3$:

(*a*) $CH_3COCH_2Cl + NEt_3 \longrightarrow CH_3COCH_2\overset{+}{N}Et_3Cl^-$

(*b*) $CH_3COCH_2\overset{+}{N}Et_3 + NEt_3 \rightleftarrows CH_3CO\underset{+NEt_3}{\underset{|}{C}}H\!:^- + H\overset{+}{N}Et_3$

(*c*) $CH_3CO\overset{-}{C}H\overset{+}{N}Et_3 + C_6H_5CHO \rightleftarrows C_6H_5\underset{O_-}{\underset{|}{C}}HCH(\overset{+}{N}Et_3)(COCH_3) \longrightarrow$

$C_6H_5CH{-}CHCOCH_3$ (epoxide, $-O-$ bridging) *A*

$$C_6H_5CH{-}CHCOCH_3\ (\text{epoxide}) \xrightarrow{AcO^-} C_6H_5CH{-}\overset{..-}{C}COCH_3\ (\text{epoxide}) \longrightarrow$$

$$C_6H_5CH{=}\underset{OH}{\underset{|}{C}}COCH_3 \rightleftarrows \underset{B}{C_6H_5CH_2COCOCH_3}$$

$$B \xrightarrow{o\text{-}C_6H_4(NH_2)_2} \underset{C}{\text{2-}CH_2C_6H_5\text{-3-}CH_3\text{-quinoxaline}}$$

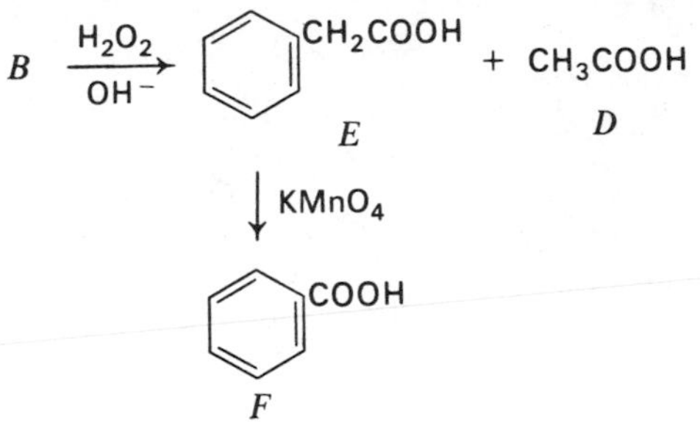

$$B \xrightarrow[OH^-]{H_2O_2} C_6H_5CH_2COOH\ (E) + CH_3COOH\ (D)$$

$$E \xrightarrow{KMnO_4} C_6H_5COOH\ (F)$$

24. o-$HOC_6H_4COCH_3 \longrightarrow o$-$C_6H_5COOC_6H_4COCH_3\ (A) \longrightarrow \{ o$-$(C_6H_5C(=O)O)C_6H_4COCH_2{:}^- \rightleftarrows o$-$(^-O)C_6H_4COCH_2COC_6H_5 \} \longrightarrow o$-$HOC_6H_4COCH_2COC_6H_5\ (B) \xrightarrow{HCl}$

$$\text{flavone } (C) \xrightarrow{KOH} B \longrightarrow o\text{-}HOC_6H_4COCH_3 + C_6H_5COOH$$

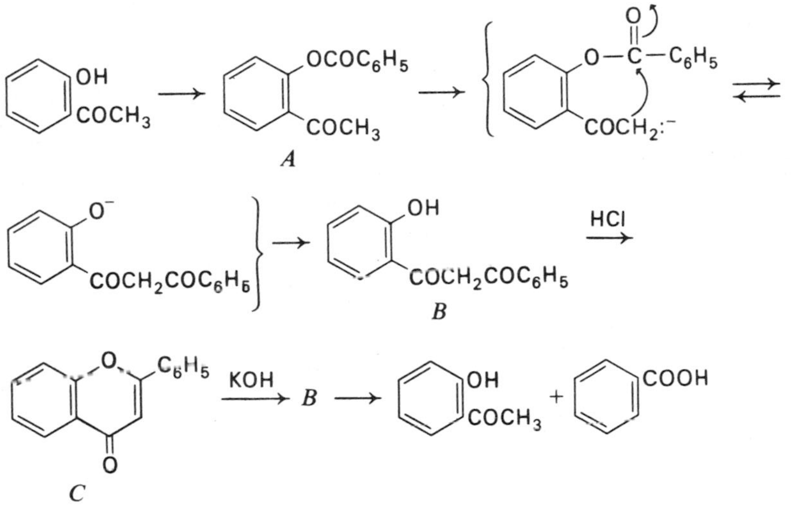

Infrared values: flavone (*C*) is a resonance hybrid in which structure *D* makes an important contribution; thus, the enhanced single-bond character of the C=O group gives rise to an unusually low stretching frequency.

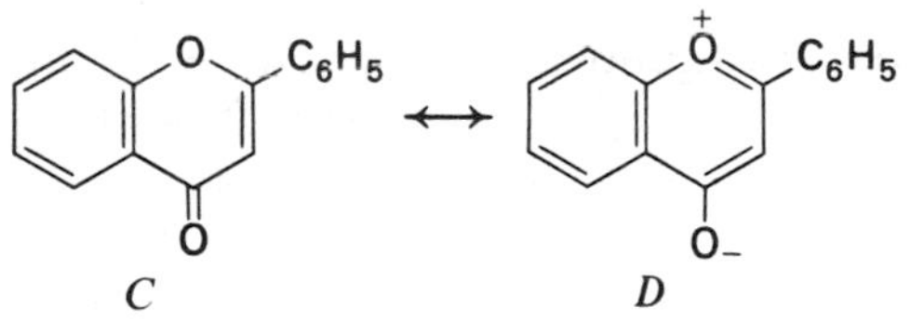

$$C \longleftrightarrow D$$

25. $HOC_6H_4OH + C_6H_5CH_2CN \xrightarrow[HCl]{ZnCl_2} HO\text{-}C_6H_3(OH)\text{-}C(=\overset{+}{N}H_2)CH_2C_6H_5\ Cl^-\ (A) \xrightarrow{H_2O}$

$$HO\text{-}C_6H_3(OH)COCH_2C_6H_5\ (B) \xrightarrow[K_2CO_3]{CH_3I} CH_3O\text{-}C_6H_3(OH)COCH_2C_6H_5\ (C) \xrightarrow[Na]{HCOOEt}$$

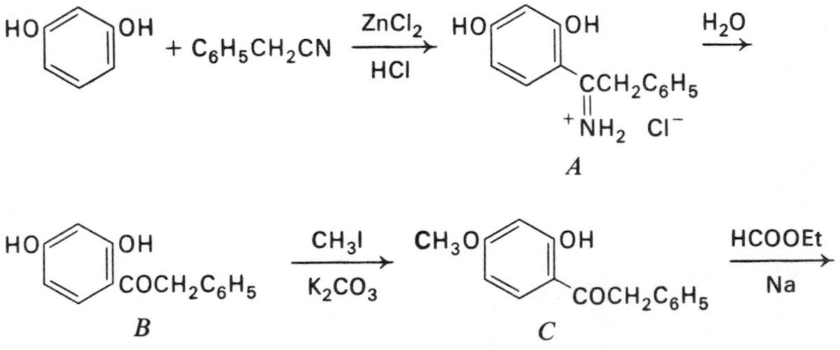

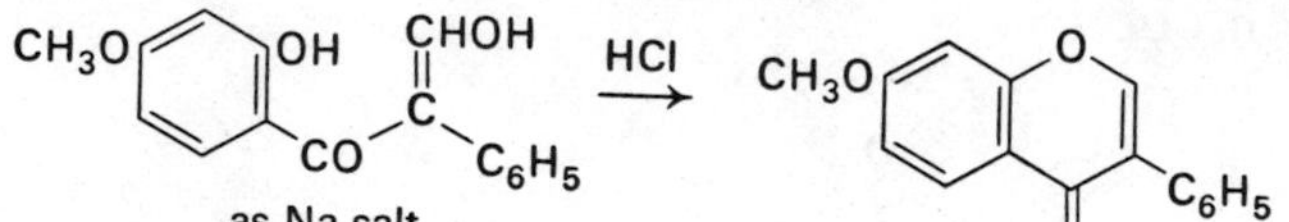

CH_3O OH CHOH C CO C_6H_5 $\xrightarrow{HCl}$ CH_3O O C_6H_5 O

as Na salt

D *E*

$E + KOH \longrightarrow C + HCOOH.$

Index